A D A
NESOTA
L. Superior
MICHIGAN
L. Huron
L. Michigan
L. Ontario
L. Erie
WISCONSIN
IOWA
ILLINOIS
INDIANA
OHIO
T A T E S
Missouri R.
MISSOURI
Ozark Mts.
KENTUCKY
Ohio R.
WEST VIRGINIA
VIRGINIA
MARYLAND
DELAWARE
NEW JERSEY
PENNSYLVANIA
NEW YORK
Adirondack Mts.
Catskill Mts
Appalachian Mountains
Blue Ridge
Mt Mitchell
CONNECTICUT
RHODE ISLAND
Long Island
MASSACHUSETTS
VERMONT
Green Mts
NEW HAMPSHIRE
MAINE
Mt Washington
White Mts
NORTH CAROLINA
TENNESSEE
SOUTH CAROLINA
GEORGIA
ALABAMA
MISSISSIPPI
Mississippi
ARKANSAS
Red R.
LOUISIANA
FLORIDA
Everglades
GULF OF MEXICO
ATLANTIC OCEAN

Wild Flowers of The United States

Volume One *Part Two of Two Parts*

Wild Flowers of

General Editor WILLIAM C. STEERE
Director, The New York Botanical Garden

COLLABORATORS

Rogers McVaugh
Robert B. Mohlenbrock
Gerald B. Ownbey
Reed C. Rollins
John W. Thomson
Robert E. Woodson

Publication of THE NEW YORK BOTANICAL GARDEN

Harold William Rickett

The New York Botanical Garden

The United States

Volume One *Part Two of Two Parts*

THE NORTHEASTERN STATES

FROM THE ATLANTIC TO MINNESOTA AND MISSOURI
AND FROM THE CANADIAN BORDER TO
VIRGINIA AND MISSOURI

McGRAW-HILL BOOK COMPANY · NEW YORK

WILD FLOWERS OF THE UNITED STATES:
VOLUME 1: THE NORTHEASTERN STATES

To DAVID and PEGGY ROCKEFELLER,
without whose enthusiasm and generosity
these books could not have been written,
illustrated or published

Printed in England by
W. S. COWELL LTD, BUTTER MARKET, IPSWICH

Library of Congress Catalog Card Number 66–17920
First Edition: Second Printing 52614

CONTENTS

GROUP VIII

GROUP IX

GROUP X

GROUP XI

GROUP XII

GROUP XIII

GROUP XIV

GROUP VIII

SEPALS two, four, or five, petals four or five; petals separate and radially symmetric. Stamens as many as the petals or twice as many. One style in a flower.
Exceptions: petals are lacking in some species of the evening-primrose family. Petals are joined in *Schrankia*. Petals are slightly bilateral in symmetry in *Cuphea* and *Cassia*. Stamens may be more numerous in *Portulaca* and *Talinum*. *Portulaca* may have more than one style.

I. *Petals five; leaves undivided (except in* Erodium), *but perhaps lobed or cleft.*

A. Sepals two; stamens five or more: purslane family.

B. Sepals five; stamens five or ten; style a long beak: geranium family.

C. Sepals five; stamens ten; leaves evergreen, or lacking and the plant without green color: pyrola family.

II. *Petals five; leaves pinnately divided.*

A. Flowers small, pink or white, in a dense head; stamens five or ten (or sometimes twelve); leaf-segments again divided pinnately into many small segments: mimosa family.

B. Flowers yellow, in small clusters; leaf-segments not again divided: cassia family.

III. *Petals variable, usually six or seven; stamens from six to fourteen:* loosestrife family.

IV. *Petals four (two in* Circaea; *occasionally five or six); leaves undivided and unlobed.*

A. Leaves paired, with three or five veins running lengthwise: stamens eight, discharging pollen through a pore in the end of their long head: melastome family.

B. Leaves with branching veins; stamens four or eight; pollen discharged by splitting of the stamen-heads: evening-primrose family.

C. Leaves with curved veins branching from the lower half of the midrib; small flowers in a cluster surrounded by four white (or pink) petal-like bracts: dogwood family.

THE MIMOSA FAMILY (MIMOSACEAE)

The mimosa family has very numerous species in warm countries, but few in North America – and those mostly trees and shrubs. The family, with the following one, really forms part of the vast bean family; for in the tropics the lines of separation fade out. But for present purposes it is easier to consider the three groups separately. Although their seed-pods have features in common, their flowers differ. The acacias and mimosas belong here; also the mesquite of the Southwest.

The herbaceous *Mimosaceae* of our range have five sepals, five separate or joined petals radially symmetric, and five or ten stamens or a few more. The leaves are pinnately divided and the segments again pinnately divided into very numerous small elliptic segments.

SENSITIVE-BRIERS (SCHRANKIA)

The sensitive-briers are sprawling prickly plants, with small rose-pink flowers in spherical heads. The petals are joined to make a five-toothed tube. The stamens are usually ten or twelve, sometimes more.

CAT'S-CLAW, S. NUTTALLII, has hooked prickles. The flower-heads are an inch or slightly less in diameter. The fruit is a long, narrow, prickly pod.

May to September: in sandy soil in fields and woods from Illinois to South Dakota and southward to North Carolina, Arkansas, and Texas. *Plate 61*.

S. MICROPHYLLA is very like *S. nuttallii*, smaller, with shorter leaf-segments which are less veiny. The heads of flowers are less than an inch in diameter. The pods are downy as well as prickly.

June to September: in sandy soil from Virginia to Kentucky and southward to Florida and Texas.

THE PRAIRIE-MIMOSAS (DESMANTHUS)

The prairie-mimosas are erect plants without thorns. The five minute, greenish petals are practically separate. There are five stamens, projecting from the perianth. The pods are flat.

PRICKLE-WEED, D. ILLINOENSIS, grows up to 8 feet tall. The pods are in dense heads, curved or twisted to form a sort of ball.

June to August: on prairies, river-banks, etc. from Ohio to Minnesota, North Dakota, and Colorado and southward to Florida, Texas, and New Mexico. The southwestern *D. leptolobus*, differing in having straight and very narrow pods, occurs in southern Missouri.

THE CASSIA FAMILY (CAESALPINIACEAE)

The temperate genera of *Caesalpiniaceae* are mostly trees and shrubs, such as redbud, honey locust, royal poinciana, and tamarind. But the very large and mainly tropical genus *Cassia* furnishes several common and easily distinguished species to our northeastern flora.

For the relation of this family to the bean family, see under the mimosa family.

THE SENNAS (CASSIA)

The species of *Cassia* have five sepals, scarcely joined; five petals, always yellow in our species, generally slightly unequal; and five or ten stamens. The stamens are commonly unequal, and some may form no pollen. The leaves are pinnately divided. The segments in some species fold together at night.

Senna is derived from an Arabic name for a tropical species of *Cassia;* it was used medicinally by the Arabs as early as the ninth century. Since then, other species have been used in the same way; the American kinds apparently share the medicinal properties of the tropical African and Indian plants.

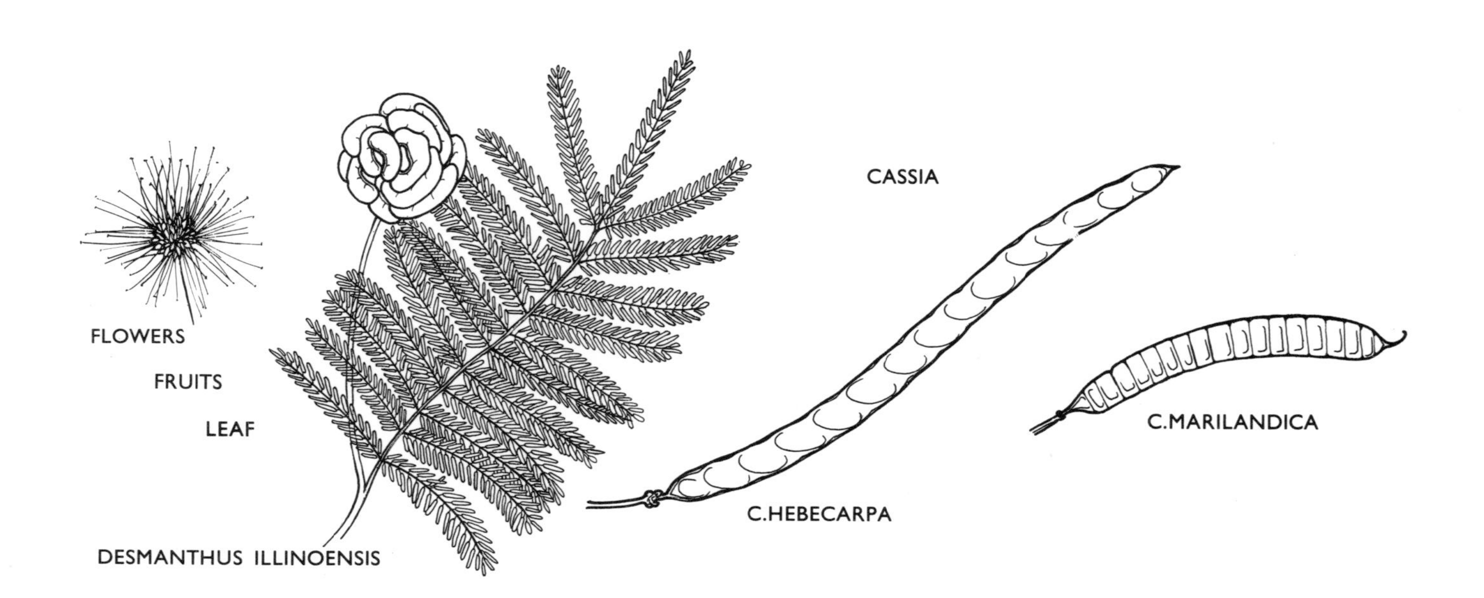

We have two distinct groups of species.

I. *Species whose leaves are divided into ten or fewer pairs of segments, each usually not less than an inch long.*

WILD SENNA, C. HEBECARPA, grows up to 4 feet tall, with leaves borne singly throughout the stem. The leaf-segments are from five to ten pairs, each about an inch long. The flowers are in short clusters on stalks from the axils of leaves. The pod is slightly curved, flat, divided into visible joints as long as they are broad.

July and August: on stream banks and in open woods from Massachusetts to Wisconsin and southward to North Carolina and Tennessee. *Plate 61.*

WILD SENNA, C. MARILANDICA, is quickly distinguished from *C. hebecarpa* by its fewer and larger leaf-segments: from three to six pairs, up to 3 inches long. The plants are about the same height. The pod of *C. marilandica* is curved, rather thick, divided into many joints much shorter than they are broad. *Plate 61.*

SICKLEPOD, C. TORA, is a tropical or semi-tropical species, extending through the southern United States and entering our southern borders. It grows about 3 feet tall, with leaves divided generally into three pairs of segments, the end one up to 3 inches long. The pod is strongly curved, very narrow.

July to September: moist woods and waste places from Pennsylvania to Kansas and southward to Florida and Texas.

The Old-World coffee senna or styptic-weed, *C. occidentalis*, with four to six pairs of leaf-segments each up to 3 inches long, is naturalized in the South and extends northward to Virginia, Illinois, Iowa, and Kansas. It flowers in August and September.

II. *Species whose leaves are divided generally into from ten to twenty pairs of segments less than an inch long.*

PARTRIDGE-PEA, C. FASCICULATA, is a common and pretty roadside plant, standing up to 3 feet tall but often less. The leaf-segments are from ten to fifteen pairs, up to $\frac{4}{5}$ inch long. The petals often have a purple spot at the base. Four of the stamens have yellow heads, the others purple; they are unequal in size.

July to September: in dry, especially sandy soil from Massachusetts to Minnesota and South Dakota and southward to Florida and Texas. *Plate 61.* The leaf-segments fold together at night, as in many of the "sensitive plants."

WILD SENSITIVE-PLANT, C. NICTITANS, is a smaller edition of *C. fasciculata*, with ten to twenty pairs of leaf-segments scarcely $\frac{1}{2}$ inch long, petals about $\frac{1}{4}$ inch long, and only five stamens.

July to September: in sandy soil from Massachusetts and Vermont to Kansas and southward to Florida and Texas.

THE PURSLANE FAMILY (PORTULACACEAE)

The purslanes and their kin are rather succulent plants. The flowers have two sepals, generally five petals, and five, ten, or more stamens.

THE SPRING-BEAUTIES (CLAYTONIA)

COMMON SPRING-BEAUTY, C. VIRGINICA, grows up to a foot tall, the stem rising from a subterranean corm. There are two narrow, rather thick leaves halfway up the stem. The flowers are in a false raceme which ends the stem; "false" because the single bract that may be present is *opposite* the lowest flower, not below it; and the flowers are in two rows *on one side* of the main stem. The five petals are pink with red veins. The stamens have pink heads. The small seed-pod is enclosed by the two sepals.

March to May: in fields, lawns, and open woods from Newfoundland to Minnesota and southward to Georgia, Alabama, and Texas. *Plate 68.* A favorite spring flower, decorating lawns, the borders of parkways, and other suburban precincts. There are forms with broad leaves and with minute petals.

C. CAROLINIANA is distinguished by much broader leaf-blades on distinct stalks.

March to July: open woods, etc. from Newfoundland to Saskatchewan and southward to North Carolina, Tennessee, and Minnesota. *Plate 68.*

The genus *Montia*, resembling *Claytonia* but with from three to five petals and stamens, and more leaves, has two species that touch our northern borders, in Maine and Minnesota.

PURSLANE (PORTULACA)

The purslanes have much-branched stems, often lying on the ground, bearing small succulent leaves. The flowers have two sepals, about five petals, and usually numerous stamens. The so-called moss-rose of gardens belongs to this genus.

PURSLANE, P. OLERACEA, known as pussley by the southern countryfolk, is a plant that lies on the ground, often in our gardens. The leaves, borne paired or singly, are thick, succulent, broadest towards the tip. The flowers are small, in the axils of leaves, with yellow petals.

June to November: a native of Europe or Asia, now abundant throughout this country. This common weed needs no introduction to gardeners. It is almost impossible to kill by ordinary cultivation, pieces of its succulent stems withstanding the sun on the surface of the soil and quickly rooting. Its redeeming feature is that it may be cooked for "greens."

FAMEFLOWERS (TALINUM)

The plants called fameflower have very short, branched stems covered with numerous very narrow leaves. From the tips of these branches arise leafless flowering stems 8 or 12 inches tall, bearing a branched cluster of pink or white flowers. The flowers have two sepals, five petals, and from five to forty-five stamens.

T. RUGOSPERMUM has from ten to twenty-five stamens. The sepals fall as the flower opens, which is in the afternoon. The style is terminated by three narrow stigmas.

June to August: in sand and on sandstone from Indiana to Minnesota and Iowa. *Plate 67*.

T. TERETIFOLIUM is very like the preceding species, differing chiefly in having a single round stigma, and seeds with smooth surface.

June to September: on dry rocks and sand from Pennsylvania to West Virginia and southward to Georgia and Alabama. *Plate 67*.

I have not discovered the origin of the English name. The species are also called rock-pinks.

Two other species, *T. parviflorum* and *T. calycinum* (*Plate 67*) are found in Missouri and Minnesota and thence westward, the first with only a few stamens and small, pale petals, the second with thirty or more stamens and larger, redder petals.

THE GERANIUM FAMILY (GERANIACEAE)

The plant that most people know as geranium is actually *Pelargonium;* it is in the geranium family. The true geranium is the genus *Geranium*, with many species all over the world.

The family is characterized by numerical precision: our species have five sepals, five petals, ten stamens in two circles of five (one circle sterile in some), and a pistil composed of five divisions. The leaves of our species are palmately lobed or cleft or pinnately divided. The pistil forms a curious fruit; the ovary separates at maturity into five parts, each containing a seed, and each attached to a part of the long style, which curls upward. This long style forms a beak or bill on the fruit which gives its name to the family. *Geranium* is derived from the Greek word for the long-billed bird we call a crane; and the English name for several species in this family is crane's-bill.

THE WILD GERANIUMS OR CRANE'S-BILLS (GERANIUM)

Our species of the genus *Geranium* have leaves palmately lobed, cleft, or divided. The five parts of the ovary at maturity are lifted up by the curving of the corresponding parts of the style, which remain attached at the tip to the central, undivided part.

WILD GERANIUM, G. MACULATUM, is a common flower of spring and early summer. The stem grows from a rhizome to a height of 2 feet. There are several long-stalked leaves at the base of the stem, and a pair of short-stalked leaves on the stem. All the leaf-

PLATE 67

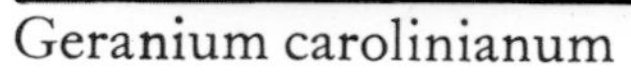

Geranium carolinianum *Allen*

Geranium maculatum *Gottscho*

Talinum calycinum *Johnson*

Talinum teretifolium *Elbert*

Geranium robertianum *Gottscho*

Talinum rugospermum *Johnson*

Chaerophyllum teinturieri *Johnson*

blades are palmately cleft into five or seven lobes. The flowers have rose-purple petals from ½ to 1 inch long.

April to June: in woods and meadows from Maine to Manitoba and southward to Georgia, Tennessee, Arkansas, and South Dakota. *Plate 67.*

HERB-ROBERT, G. ROBERTIANUM, has a stem up to 2 feet tall, usually hairy. The leaves are not truly palmately divided, since the end segment is stalked; the segments, moreover, are pinnately cleft. The flowers, about ½ inch across, are usually in pairs, with pinkish petals.

May to October: introduced from Europe and now growing wild in rocky woods and banks from Newfoundland to Manitoba and southward to Maryland, West Virginia, Indiana, and Nebraska. The plant was dedicated to Robin Goodfellow, *Knecht Ruprecht*, the Brownie of English folk-lore, and, perhaps to Robin Hood; also to the little Robin redbreast, which, like the plant, was constantly about the house (it brought bad luck if he flew *in* the house). Some of the local names in England are death-come-quickly, dragon's-blood, granny's-needles, Robin redshank [the stems are often red], and stinking-Bob [the plant has a rank odor].

G. CAROLINIANUM has its flowers in a dense cluster. It is a small bushy plant, about 2 feet tall, with leaves palmately cleft into from five to nine narrow lobes, each lobe pinnately lobed or cleft. The petals are pink, less than ½ inch long.

May to August: in dry soil in woods and waste places from Maine to British Columbia and southward to Florida and California. *Plate 67.*

G. BICKNELLII resembles *G. carolinianum* in stature and leaf shape, but the flowers are generally in pairs on long stalks. The stem and its branches are hairy, with stalked glands among the pointed hairs. The petals are rose-pink, about ⅓ or ⅖ inch long.

May to September: in open woodlands and fields from Newfoundland to Alaska and southward to Connecticut, Pennsylvania, Indiana, Iowa, Utah, and California. *Plate 68.*

G. COLUMBINUM differs from *G. bicknellii* only in having stems and branches smoothish, with minute hairs lying flat on the surface.

May to September: in fields and waste places from New York to Iowa and southward to South Carolina and Tennessee; and in South Dakota; a weed from the Old World.

G. PUSILLUM is much branched, the branches up to 2 feet long. The leaves are deeply cleft into from five to nine lobes which are themselves lobed at the end. The petals are red-violet and only about ⅕ inch long. There are five stamens instead of the usual ten.

June to October: an Old-World species, now established as a weed in fields and waste places from Massachusetts to British Columbia and southward to North Carolina, Arkansas, and Oregon.

Besides these, other Old-World species turn up in scattered places in our range: *G. dissectum*, similar to *G. bicknellii; G. sibiricum*, with weak, hairy stems up to 3 feet long, and lilac or white petals, ¼ inch long, marked with violet lines; *G. sphaerospermum*, with flowers crowded together and rose petals about ½ inch long; *G. pyrenaicum*, with deeply notched petals more than ½ inch long; dove's-foot crane's-bill, *G. molle*, much like the preceding species but softly (*molle*) downy, and petals about ¼ inch long; bloody crane's-bill, *G. sanguineum*, often cultivated in rock gardens and sometimes found in the wild, with deep rose flowers an inch or more across; and *G. pratense*, the common roadside or meadow crane's-bill of Europe, with woolly flower-stalks and deep blue-purple flowers.

STORK'S-BILL (ERODIUM)

Of this genus of the Mediterranean countries we have one species now common throughout the United States; and another species established in a few places. The botanical name is derived from the Greek for stork, and, like the name *Geranium*, refers to the long beak of the fruit. In *Erodium*, however, the beaks of the five parts into which the fruit splits separate entirely from the central column and become twisted like a corkscrew. The leaves are pinnately divided, cleft, or lobed. Only five stamens bear pollen.

ALFILARIA or FILAREE, E. CICUTARIUM, is the common species. The leaves usually overwinter as a rosette on the ground. They are pinnately divided and the segments pinnately cleft. The flowering stems rise from the midst of these leaves, at first often only a few inches tall. The flowers are rose-red or purple, less than ½ inch across.

March to November, occasionally all winter: in fields and waste places from Quebec to Michigan and southward and southwestward to Virginia, Tennessee, Texas, and Mexico. *Plate 68.*

Musk stork's-bill, *E. moschatum*, is much larger, with leaf-segments toothed but not cleft; it occurs in a few places from New England to Delaware, and on the Pacific Coast. Several other species have been reported at seaports, but are probably not established.

PLATE 68

Erodium cicutarium *D. Richards*

Claytonia caroliniana *Johnson*

Claytonia virginica *Elbert*

Decodon verticillatus *Elbert*

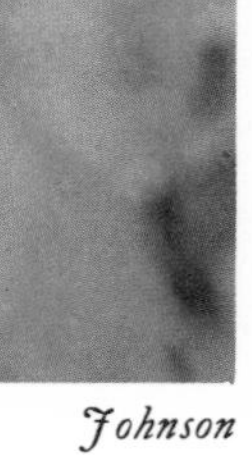

Lythrum alatum *Johnson*

Geranium bicknellii *Williamson*

Cuphea petiolata *Rhein*

Lythrum salicaria *Gottscho*

THE LOOSESTRIFE FAMILY (LYTHRACEAE)

The loosestrife family includes a number of species which at first sight have little in common. The plants of our range usually have narrow leaves with short stalks or none, mostly in pairs. The flowers are in the axils of leaves, extending up into a spike in some species. The petals and sepals vary in number from four to seven; the stamens are as many or twice as many (with some irregularity). The color is generally some shade of crimson or purple. Perianth and stamens grow on a tubular or cup-shaped flower-base (receptacle), the stamens on its inner surface. For the name loosestrife see under *Lysimachia* in the primrose family.

THE PURPLE LOOSESTRIFES (LYTHRUM)

The flowers of the genus *Lythrum* have from four to seven sepals and petals, the stamens of the same number or twice as many. The chief interest of some of these flowers is that, in one species, there may be two or three types of flowers. These are explained under the species below.

PURPLE LOOSESTRIFE, L. SALICARIA, is a tall and beautiful weed of wet places, an immigrant from the Old World now very much at home in the eastern states. It may grow up to 6 feet tall. The leaves, which may be in pairs or in threes and fours, are lanceolate, without stalks. The flowers are clustered in the upper axils, forming a spike at the top where the leaves are so small as to be termed bracts. The flowers are of three types. There are two circles of stamens, each equaling the petals in number, and one pistil. The two circles of stamens and the stigma of the pistil are at three different levels, and different combinations exist on different plants between tall stamens and short pistil, short stamens and tall pistil, and so on. This arrangement favors cross-pollination rather than self-pollination; that is, the pollen of one plant is likely to be deposited on the stigma of another plant. This induces variability and hence evolution.

June to September: in marshes and wet meadows from Newfoundland and Quebec to Minnesota and southward to Virginia, Ohio, and Missouri. *Plate 68.*

L. ALATUM is a native species, up to 4 feet tall, with flowers borne singly in the axils of leaves. The leaves vary from lanceolate to ovate. The upper leaves are single instead of paired. There are two types of flowers (not three as in *L. salicaria*), with long stamens and short style, or vice versa.

June to September: in swamps and meadows from New York and Ontario to British Columbia and southward to Georgia, Louisiana, and Texas; also as a stray in New England. *Plate 68.*

L. HYSSOPIFOLIA is narrow-leaved, about 2 feet tall. The flowers are single or in pairs, in the axils of leaves – even low on the stem. The petals are pale purple.

June to September: in marshes and other wet places, mostly near the coast, from Maine to New Jersey and Pennsylvania; inland in Ohio; also on the West Coast, in South America, and in the Old World.

L. LINEARE grows up to 4 feet tall, with very narrow ("linear") leaves in pairs. The white or pale purple flowers are in the axils of the upper leaves. They are of two types as in *L. alatum.*

July to September: in salt marshes from New Jersey to Florida and Texas.

DECODON

There is but one species of *Decodon.*

WATER-WILLOW, D. VERTICILLATUS, is, of course, not a willow, but has willow-like leaves and grows in water. The stems are generally bent and up to 8 feet long. The leaves are paired or in circles of three or four, with short stalks and lanceolate blades. The flowers are clustered in the axils of the upper leaves. As in *Lythrum salicaria*, they are of three types according to the lengths of the two circles of five stamens each and the length of the style. The petals are narrow, purplish-pink. The base of the flower is bell-shaped.

July to September: in swamps and shallow water from Maine to Minnesota and southward to Florida and Louisiana. *Plate 68.* The arching stems may root at the tips, forming new plants.

CUPHEA

Cuphea is a large genus in tropical America. Only one species reaches the northeastern region.

CLAMMY or BLUE WAXWEED, C. PETIOLATA, is sometimes 2 feet tall but generally less. It is covered all over with clammy or sticky hairs. The leaves are paired, with lanceolate blades on fairly long stalks. The flowers are single or in pairs in the axils of leaves, stalked. The base of the flower is a tube, quite long, with a curious saclike expansion on the upper side so that the stem seems to be attached at the side not at the end. The sepals are six triangular teeth, and there are six reddish-purple petals unequal in size, all on the rim of the tube.

July to October: in dry soil in open woodlands from New Hampshire to Illinois and Iowa, and southward to Georgia, Louisiana, and Kansas. *Plate 68*.

There are several other genera of *Lythraceae* in our range, *Didiplis*, *Rotala*, and *Ammania*, with a total of five species. They all have very inconspicuous flowers with very small pink, purple, or green petals. All grow in swamps and other wet places.

THE MELASTOME FAMILY (MELASTOMATACEAE)

This vast tropical family is represented in our range by three species of one genus, with two more just touching our borders. Elsewhere it includes trees, shrubs, and vines.

MEADOW-BEAUTIES (RHEXIA)

The parts of the flower are in fours. Four sepals and four petals are seated on the rim of the tubular base of the flower. There are eight stamens, the pollen-bearing heads of which are long, narrow, and curved, and appear to be set on their stalks at right angles. The pistil becomes a small pod (capsule) enclosed in the enlarged flower-base, which changes to a vase-like form, broad at the bottom and tapering to a narrower neck; Thoreau compared these to "perfect little cream-pitchers." The vivid pink flowers are seen in profusion in some moist sandy meadows. The leaves, in pairs, have small stalks or none, and are toothed on the margins. They generally have three main veins.

R. MARIANA has a round, hairy stem up to 2 feet tall, growing from slender horizontal runners. The leaf-blades are elliptic, on short stalks.

June to September: in moist sandy soil on the coastal plain from Massachusetts to Florida; and inland from Virginia and Georgia to Missouri and Texas.

R. VIRGINICA is distinguished by tubers on the roots and the absence of runners. The stem, up to 2 feet tall, smooth or nearly so, has four conspicuous thin lengthwise ridges or "wings." The leaves are more or less ovate, without stalks.

July to September: in wet sandy and peaty places from Nova Scotia to Ontario and southward to Florida and Louisiana. *Plate 69*.

R. ARISTOSA, a much less common species than the preceding two, has a smooth stem about 2 feet tall, four-angled but not "winged." The leaves are narrow, lanceolate, with no stalks. The petals taper sharply to narrow points (*aristae*).

July to September: in sandy bogs and wet pinelands from New Jersey to Georgia.

R. interior is known only in wet land and prairies from southwestern Missouri to Kansas, Arkansas, and Oklahoma. It has a square, hairy stem, and lanceolate leaves without stalks. *R. ventricosa*, a southern species similar to *R. mariana* but taller and hairy on the leaves, and *R. petiolata*, also southern, with bristle-edged leaves, grow in southeastern Virginia.

THE EVENING-PRIMROSE FAMILY (ONAGRACEAE)

The evening-primroses and their relatives form a large, widely distributed family. In the United States they are mostly herbaceous plants. The flowers generally have four petals and eight stamens, but there are many exceptions. The ovary is inferior; it forms (in our species) a dry fruit, which commonly splits into four parts.

Fuchsia is a widely cultivated member of this family. *Clarkia* (including *Godetia*) is often grown in gardens; so are the species of *Oenothera* called sundrops.

In our range there are few genera of wild flowers of this family, but some of these have many species.

I. *One genus has only two sepals and two petals (the petals forked):* Circaea.

II. *The remaining genera mostly have four sepals and petals; some species have five; a few lack petals.*

A. Two genera have a long hollow tube extending from the summit of the inferior ovary and bearing the sepals and petals at its upper end (this may be mistaken for the flower-stalk until the swollen part that contains the ovary is noticed at its base): *Oenothera* (many seeds in the capsule; petals yellow, white, or pink); *Gaura* (from one to four seeds in a dry fruit that does not open; petals small, white turning pink).

B. Two genera have no hollow tube or a short one above the ovary: *Ludwigia* (petals four or five or none, yellow or greenish); *Epilobium* (petals four, white, pink, or red-purple).

CIRCAEA

The enchanter's nightshades are unique among the wild flowers of the northeastern United States in having only two sepals, two petals, and two stamens. (*Maianthemum* in the lily family has two sepals, two petals, and four stamens.) The petals may be cleft so as to give the illusion of four. The fruit is characteristic: a small shell formed from the inferior ovary, covered with bristles, not opening when mature; it is bent downwards when mature.

These are insignificant plants with small white or pink-tinged flowers. Dr Fernald characterizes them as "modest – except for clinging fruits." They are named for the enchantress Circe of ancient legend, and various magical properties have been attributed to the European variety of the first species below.

ENCHANTER'S NIGHTSHADE, C. LUTETIANA, is the common species, growing from 8 inches to 3 feet tall with leaves from 1 to 6 inches long. The leaves are slightly wavy-edged but scarcely toothed, and taper to a point. The flowers are numerous in the racemes and rather widely spaced.

June to August: in woods from Nova Scotia to Ontario and North Dakota and southward to Georgia, Tennessee, and Oklahoma. *Plate 69*. *Lutetiana* means "of Paris"; this is a variety of the magical herb of medieval Parisian botanists.

C. ALPINA is a smaller plant, not over a foot tall with leaves not much more than 2 inches long. The leaves are quite coarsely toothed and have an abrupt rather than a tapering point. The flowers are less numerous and more closely clustered.

June to September: in moist woods from Labrador to Alaska and southward to New York, in the mountains to Georgia, to Illinois, South Dakota, New Mexico, and Washington. *Plate 69*. *C. intermedia* or *C. canadensis* is intermediate between the above two species and rarely forms fruit; for which reasons it is thought to be a hybrid.

EVENING-PRIMROSES AND SUNDROPS (OENOTHERA)

The evening-primroses and sundrops form a large genus native in America, particularly numerous in the West and Southwest. They have been extensively studied by botanists, who have shown that they exist in numerous local races which differ only slightly and sometimes start new races through hybridization. This has created much confusion in the past, many groups having been named as species that are impossible to distinguish in such a large and complex assemblage of plants. The reader must therefore expect to find plants that do not closely fit the descriptions given, and to include very different-appearing plants under one name.

All have eight stamens and all but one have four narrow stigmas, crosslike, at the tip of the style. All our species have a tube – very long in the commoner species – above the inferior ovary and carrying the sepals and petals at its end. The sepals fold back along this tube as the flower opens. The fruit is a pod, the shape of which is useful in identification; fruits and flowers are often present together, the fruits below the flowers. Since the tube, the sepals, and the petals are all shed from the fruit, this somewhat resembles an unopened bud, and care must be taken to make a distinction. The leaves are borne singly and are mostly narrow.

PLATE 69

Oenothera parviflora *Johnson*

Oenothera missouriensis *Lee*

Oenothera biennis *Gottscho*

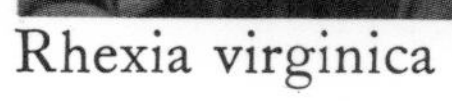
Rhexia virginica *Miller*

Circaea lutetiana *Scribner*

Oenothera missouriensis *Johnson*

Circaea alpina *Johnson*

They are not related to true primroses. Many of them open in the evening, their pale yellow, fragrant flowers attracting night-flying insects; they wither the following morning. Others – the sundrops – are day-bloomers.

I. *Species with yellow petals* (*see also* O. triloba *in group II*).

A. Of these, five species have a pod (capsule) which has no sharp angles or projecting flanges (wings); the leaves are not lobed or cleft (they may have small teeth; compare C).

1. In three species the small projections on the ends of the sepals in the flower-bud are parallel and close together, at least at their base (see the drawings and compare 2).

COMMON EVENING-PRIMROSE, O. BIENNIS, has petals not more than an inch long. It ranges in height from 3 to 6 feet. Commonly the stem is red and the leaves and flowers very crowded, but in this and other characteristics the plants vary greatly. The flowers are in a raceme, on the stem of which there are usually some gland-tipped hairs.

June to October: in dry open places, on roadsides, etc. from Newfoundland to Alberta and southward to North Carolina and Texas. *Plate 69*.

O. STRIGOSA differs from *O. biennis* chiefly in being covered with gray hairs (either standing out straight or lying flat on the surface). It is mainly mid-western.

July to October: in fields and waste places from Michigan to British Columbia and southward to the Ohio valley, Arkansas, New Mexico, and California; occasionally found farther to the east.

O. ERYTHROSEPALA is distinguished by much larger petals (from 1 to 2 inches long). It ranges from 2 to 4 feet in height. The leaves are strongly crinkled, and often have a red midrib. The whole plant is hairy or downy, the hairs, as seen under a lens, often with enlarged red bases. The flower-buds are angled and hairy. The sepals are usually red (*erythro-*).

July to October: scattered in various open situations from Quebec to British Columbia and southward to New Jersey and California; mainly in the eastern and western coastal states, but occasional also in the central states; perhaps of garden origin and escaped from cultivation, in the Old World as well as America.

2. In two species of group A the projections at the ends of the sepals arise just *behind* the tip of the sepal instead of prolonging that tip. Consequently they are not close together at their base, and generally spread apart in the bud (see the drawings).

O. PARVIFLORA has petals less than an inch long. It varies from a few inches to 5 feet tall. The stem is often red, like that of *O. biennis*. The stem and leaves may be smooth or beset with hairs lying flat on the surface; the flower-buds may be hairy.

July to October: in sandy soil, often along streams, and waste places from Newfoundland to On-

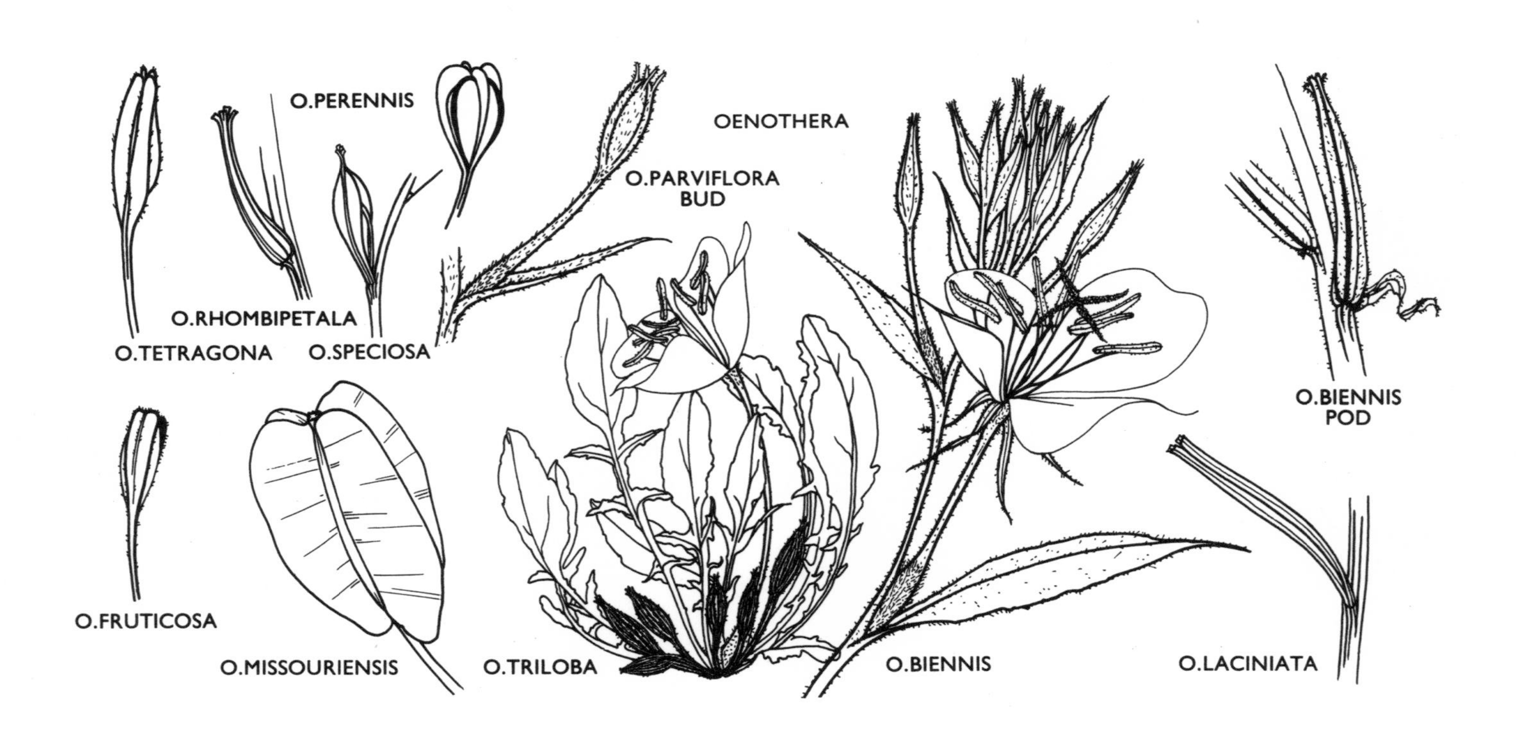

tario (perhaps farther west) and southward to Virginia and Illinois. *Plate 69*. Certain plants have narrow petals which form a cross; these have been named *O. cruciata*, but they seem to be only a variant of *O. parviflora*.

O. ARGILLICOLA is a much less widespread species but a handsome one, with petals nearly 2 inches long. The stems, from 2 to 5 feet long, often sprawl. The leaves are narrow.

July to October: in barren rocky places in Pennsylvania, Virginia, and West Virginia.

B. Six species with yellow petals have pods (capsules) strongly ribbed at the four angles or provided with thin projecting wings; the leaves are not noticeably toothed or lobed.

1. One of these may be at once distinguished by its very large petals, from 2 to 3 inches long; and by the very broad wings on its fruit (up to an inch wide).

MISSOURI-PRIMROSE or GLADE-LILY, O. MISSOURIENSIS, is a low plant, the stem varying from an inch or two to 2 feet tall. It is usually whitened with hairs lying flat on the surface. The buds are spotted with red. The sepals are almost as conspicuous as the petals; they may be nearly 2 inches long, and their tips cohere even after the flower opens. The tube which bears the sepals and petals, at the base of which is the inferior ovary, may be 6 inches long, easily mistaken for the stem of the flower.

May to July: in dry open places from Illinois to Colorado and Texas. *Plate 69*. This species is often cultivated. Before Missouri became a state, the name was often applied to a very large and vaguely defined territory west of the Mississippi.

2. The remaining five species of group I, with petals less than 2 inches long, are the sundrops. The flowers open in the morning. Some of these find a place in herbaceous borders.

O. FRUTICOSA may be erect and 3 feet tall or taller, or branched and spreading, forming bushes less than a foot high. The leaves are narrow and lanceolate. The plant is usually hairy with short white hairs lying flat or longer hairs extending out. The petals are from ½ to 1 inch long. The pod is widest near the summit, tapering downward to a sort of stalk, with four sharp angles.

May to August: in various situations, moist and dry, from New England to Michigan and Missouri and southward to Florida and Oklahoma. *Plate 70*. This and the following species are the most frequent of the cultivated sundrops. They are both variable.

O. TETRAGONA is similar to *O. fruticosa*, differing in having gland-tipped hairs on the tube of the flower, and in the shape of the pod which is mainly oblong and four-angled (*tetragona*), and tapers not gradually but sharply to a stalk. The plants range from 8 inches to more than 3 feet tall. The petals are from ½ to more than 1 inch long.

June to August: in fields, meadows, and open woodland from Nova Scotia and New England to Michigan and Iowa and southward to Florida and Louisiana. A very variable species, forms of which are frequent in cultivation.

O. PERENNIS is distinguished by the drooping tip of the inflorescence before all the buds are open. The plant is from 8 inches to 2 feet tall. The leaves are elliptic, or wider towards the tip than in the basal part. The petals are about ⅓ inch long. The pod tapers downward and has four narrow wings.

May to August: in fields, meadows, and open woodland from Newfoundland to Manitoba and southward to Virginia, the mountains of Georgia, Indiana, and Missouri. *Plate 70*.

O. PILOSELLA is conspicuously hairy. The stem grows from 1 to 3 feet tall, bearing a few flowers at its tip and the tips of branches. The leaves are elliptic or lanceolate. The petals are from ½ to 1 inch long. The pod tapers downward, is four-winged, and hairy when young.

May to July: in open woodland, meadows, and prairies from southern Ontario to Michigan and Iowa and southward to West Virginia, Ohio, Illinois, and Arkansas; also occasionally further eastward. *Plate 70*.

C. Three species with yellow petals have a pod without sharp angles or wings, usually curved; at least some of the leaves are generally sharply lobed, cleft, or toothed.

O. LACINIATA is usually a low plant, the branches tending to spread sideways; but it may stand erect, up to 2 feet or more tall. The leaves distinguish it from others of this group; they are mostly pinnately lobed and cleft, some perhaps having merely a wavy or even a plain outline. The flowers grow from the upper axils. The petals are from ¼ to ¾ inch long; they become reddish as they age. The pod is cylindric but usually curved.

May to October: in fields and waste places from Maine to South Dakota and southward to Florida and Texas; also in Mexico and South America. *Plate 70*.

O. RHOMBIPETALA grows from 1 to 3 feet tall, with few or no branches, the stem often silky with hairs lying flat on the surface. The leaves vary from oblong to ovate; the basal leaves may be toothed, wavy-edged, or plain. The numerous flowers are in a spike. The petals are about ½ inch long. The pod is usually curved.

June to October: in sandy places from New York to Minnesota and South Dakota and southward to Florida and Texas. *Plate 70*.

O. HUMIFUSA is bushy, with branches lying on the ground or spreading upward at an angle, up to 2 feet long. The plant is gray with fine hairs lying flat. The leaves are numerous, sometimes with a few teeth; the basal leaves are pinnately lobed or cleft. The flowers grow from the axils of leaves. The petals are about ⅖ inch long. The pod is narrow and generally curved.

On beaches and dunes along the coast from New Jersey to Florida and Louisiana.

The western O. *serrulata*, with narrow, finely toothed leaves, petals ½ inch long or less, and narrow fruit with four blunt angles, is found from Wisconsin to Missouri (*Plate 178*). The southwestern *O. linifolia*, not more than 18 inches tall, with threadlike leaves, and petals only about ⅕ inch long, occurs in southern Illinois and Missouri.

II. *Species with white or pink petals* (*but see* O. triloba).

WHITE EVENING-PRIMROSE, O. SPECIOSA, is a small plant, not more than 2 feet tall, with few but large and handsome flowers in the upper axils. The petals are white or pale pink, varying from less than an inch to nearly 2 inches long. The buds droop, rising to an erect position as they open. The pod is narrow, tapering downward, and strongly four-ribbed, the ribs being separated by deep grooves.

May to July: in prairies and along roadsides from Missouri and Kansas to Mexico; escaped from cultivation in the eastern states from Pennsylvania to Illinois and southward to Florida. *Plate 70*.

O. TRILOBA is practically without an erect stem, all the leaves arising in a cluster near the ground. They are deeply pinnately cleft. The flowers rise from among the leaves to a height of 6 to 8 inches. The petals are white or pink or pale yellow, up to an inch long. The pod is quite distinctive, the four wings pointed.

April to June: in dry woods and prairies from West Virginia to Indiana and Kansas and southward to Alabama and Texas.

GAURA

Gaura is an unpretentious genus with no English name. The plants are rather straggling with long branches bearing rather small leaves and ending in racemes or spikes of small white or pink flowers. The tube of the receptacle is very long, like a flower-stalk; but the ovary and later the pod is at the base of this tube and has scarcely any stalk. There are eight stamens. The narrow petals are not quite radial in their symmetry. The fruit is a "pod that does not open" – really a sort of nut.

G. BIENNIS is hairy and downy; it grows up to 10 feet tall. The petals are about ⅕ inch long.

June to October: in meadows and prairies from Quebec to Minnesota and Nebraska and southward to North Carolina, Tennessee, Louisiana, and Texas. *Plate 71*.

G. PARVIFLORA reaches 6 feet in height, the stem softly hairy and the leaves downy. The petals are pink, only about $\frac{1}{12}$ inch long (*parvi-*).

May to October: in dry prairies and fields and on roadsides from Indiana to Washington and southward to Texas and Mexico.

The western *G. coccinea*, with white or red petals ¼ inch long, and four-sided fruits which are pear-shaped in outline, reaches eastward to Minnesota and Missouri. The southern *G. filipes*, with very slender branches, very narrow leaves, petals ⅕ inch long, and four-angled fruit on a short stalk, extends northward to Kentucky and southern Indiana.

THE PRIMROSE-WILLOWS (LUDWIGIA)

The plants of *Ludwigia* grow in water or wet places. Most have leaves borne singly, without teeth. The mostly yellow flowers are single in their axils or at the tip of the stem. A number of species lack petals. Others have four or five petals. The number of stamens varies from four to ten – as many as the petals or twice as many.

We may classify the rather numerous species first by the number of stamens. The first two groups below have generally been treated as a separate genus, *Jussiaea;* but according to the most expert opinion there is no real basis for making two genera here.

I. *Plants with eight stamens.*

PRIMROSE-WILLOW, L. DECURRENS, is of course neither a primrose nor a willow. It grows erect to a height of 4 feet or more. The word *decurrens* refers to

PLATE 70

Oenothera perennis *Scribner*

Oenothera rhombipetala *Horne*

Oenothera pilosella *Stees*

Oenothera speciosa *Becker*

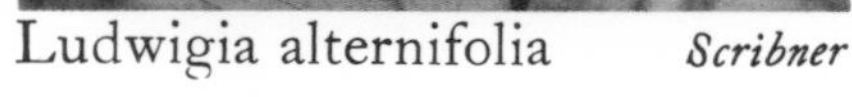

Ludwigia alternifolia *Scribner*

Oenothera laciniata *Murray*

Oenothera fruticosa *Gottscho*

Ludwigia peploides *Johnson*

the projecting "wings" which "run down" from the base of each leaf on the stem, making it four-angled. There are typically four yellow petals.

May to October: in wet places from Virginia to Missouri and southward to Florida and Texas.

II. *Plants with ten stamens.*

PRIMROSE-WILLOW, L. PEPLOIDES, is a wide-ranging, largely tropical species of which one variety occurs in our range. This has creeping or floating stems with leaves borne singly; the leaves are oblong or elliptic. The petals are yellow, notched, about $\frac{1}{2}$ inch long.

June to October: in ditches, pools, swamps, etc., our variety from Indiana to Kansas and southward to Alabama and Texas; also as a stray from New Jersey and Pennsylvania to North Carolina. *Plate 70.*

The southern *L. leptocarpa* is found in Missouri. It has hairy stems and yellow petals up to $\frac{2}{5}$ inch long.

III. *Plants with four stamens. The false loosestrifes (true loosestrifes are in the primrose family).*

A. Four species have yellow petals.

L. ALTERNIFOLIA grows erect from 1 to 4 feet tall; it is smooth or almost so. The petals are no longer than the sepals ($\frac{1}{8}$ inch). The fruit is almost a cube, with four sharp angles.

June to August: in swamps from Massachusetts to Ontario and Michigan and southward to Florida and Texas. *Plate 70.*

L. HIRTELLA has erect hairy stems from 1 to 3 feet tall. The leaves also are hairy; they are without stalks. The petals are about $\frac{1}{2}$ inch long. The fruit is square, sharply angled.

June to September: in swamps and bogs in pinelands on the coastal plain from New Jersey to Florida and Texas; inland in Kentucky and Arkansas.

L. LINEARIS is from 1 to 3 feet tall, minutely downy or smooth. The leaves are very narrow ($\frac{1}{5}$ inch wide or less). The petals are narrow, not exceeding the sepals in length. The fruit is cylindric or bell-shaped.

July to September: in wet pinelands, ditches, swamps, etc. from New Jersey to Florida and Texas, mostly on the coastal plain; inland in Tennessee and Arkansas.

L. BREVIPES forms mats of creeping or floating stems bearing paired leaves which are widest towards their ends. The petals are about equal to the sepals in length. The fruit is nearly cylindric.

June to September: in shallow water and wet sand on the coastal plain from New Jersey to Florida. *L. lacustris*, found in Rhode Island and Connecticut, seems to be only a small form of *L. brevipes*.

B. Four species with four stamens have very minute petals or none at all.

L. PALUSTRIS has a floating or creeping stem bearing paired, stalked leaves with lanceolate or elliptic blades. The flowers are minute. The fruit is nearly cylindric and marked with four lengthwise green stripes.

June to September: in shallow water and mud practically throughout North America; also in the Old World.

L. POLYCARPA is an erect, smooth plant from 6 inches to 3 feet tall, with lanceolate leaves borne singly. The petals, when there are any, are minute and greenish. The fruit is four-sided but not sharply angled.

July to September: in wet places in Maine and Massachusetts and from Ontario to Minnesota and southward to Tennessee and Kansas.

L. SPHAEROCARPA has smooth stems up to 3 feet tall with narrow lanceolate leaves borne singly. Minute petals may be present. The fruit (*-carpa*) is nearly spherical (*sphaero-*) and very finely downy.

July to September: in wet places from Massachusetts to Florida and Texas.

L. GLANDULOSA is a bushy plant from 6 inches to 5 feet tall. The name is derived from four glands which take the place of petals. The fruit is cylindric.

June to September: in wet soil and shallow water from Virginia to Kansas and southward to Florida and Texas.

WILLOW-HERBS AND FIREWEEDS (EPILOBIUM)

This is a large genus of plants most characteristic of cooler latitudes. The flowers vary greatly in size but not in pattern. They have eight stamens. The four petals are pink, purplish, or white, notched at the end in all but the first of the species described below. They are seated on a long receptacle within which is the inferior ovary – the whole simulating a flower-stalk. The fruit is a slender pod. The seeds bear each a tuft of hairs. To quote the herbalist Gerard, the pod "is full of downy matter, which flieth away with the wind . . .".

PLATE 71

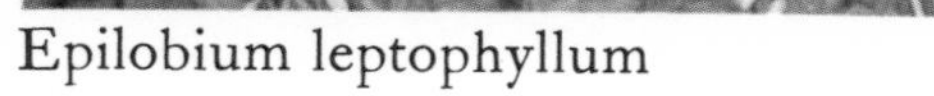
Epilobium leptophyllum *Elbert*

Epilobium angustifolium *Johnson*

Gaura biennis *McDowell*

Epilobium hirsutum *Smith*

Epilobium hornemannii *Rhein*

Epilobium adenocaulon *Scribner*

Epilobium palustre *Gottscho*

Epilobium coloratum *Gottscho*

These species flower all summer from June or July to September.

I. *In two common species the style bears a cross at its upper end, the four arms being four stigmas.*

FIREWEED, E. ANGUSTIFOLIUM, is a familiar tall plant (from 4 inches to 6 feet or more) with spires of rose-magenta flowers. Unopened buds at the tip point downward, but rise as they expand. The stamens emerge first, the style with its four stigmas projecting after pollen is shed. The petals are about ½ inch long.

In clearings and thickets and along roads in the woods, in burned-over land, etc. from Newfoundland to Alaska and southward to Maryland, the mountains of North Carolina, Ohio, Illinois, Iowa, South Dakota, Arizona, and California. *Plate 71*. The name fireweed signifies its appearance after fires. In England it became abundant with the industrial revolution, along railroads and new roads; there it is generally called rosebay-willow-herb ("willow" from the shape of the leaves). After World War II it appeared in profusion in the bombed and burned parts of London. The American Indians called it wickup or wickopy; but this name was also given to leatherwood and other plants which were perhaps used in making the shelters called wickiup or wigwam.

HAIRY WILLOW-HERB, E. HIRSUTUM, is a much-branched, bushy plant from 2 to 7 feet tall. The stem and leaves are hairy, and toothed along the edges. The flowers are in several racemes. They are smaller than those of fireweed, the petals not exceeding ⅗ inch; they are notched at the end, purplish-red.

This is an immigrant from the Old World, now established in this country in waste places, on moist roadsides, etc. from Quebec to Ontario and Michigan and southward to New York, Ohio, and Illinois. *Plate 71*. In England it is called codlins-and-cream. Codlins are cooking apples; the cream would be the white stigmas. The plants are generally larger there.

II. *In the rest of our species the style is terminated by a single round stigma.*

A. Three of these species have leaves with numerous fine sharp teeth along the margins.

E. COLORATUM is a very bushy plant up to 3 feet tall, very finely downy all over. The leaves are narrowly lanceolate. The flowers are tiny, the petals only about ⅕ inch long; they are pink or white. The receptacle below the petals is long, and forms a slim pod up to 2 inches long; there are so many of these that the plant seems full of branches that have no leaves. It is important in identifying this species to notice that the hairs on the seeds are white. The *coloratum* of the name is from the red color of the mature stem.

In wet grounds from Quebec to Minnesota and South Dakota and southward to Georgia, Alabama, Arkansas, and Kansas. *Plate 71*.

E. ADENOCAULON resembles the preceding species in general aspect and in the size of its flowers. The leaves are broader, tending to be ovate. And the hairs on the seeds are cinnamon-colored.

In wet ground from Newfoundland to Alaska and southward to Maryland, Illinois, and Nebraska. *Plate 71*.

E. CILIATUM is a small plant, from a few inches to a foot tall, and quite smooth. The leaves are ovate or elliptic. The petals are from ⅛ to ¼ inch long, pink.

In damp and springy places from Newfoundland to British Columbia and southward to New England, Pennsylvania, Wisconsin, New Mexico, and California.

B. Three other species with a round stigma have leaves practically without teeth. They form threadlike runners at the base of the stem.

E. STRICTUM has a stem covered with a grayish down and from 1 to 2 feet tall; it is not much branched. The leaves are narrowly lanceolate. The petals are from ⅕ to ⅖ inch long, pink.

In meadows and bogs from Quebec to Minnesota and southward to Virginia, Ohio, and Illinois.

E. LEPTOPHYLLUM grows from 8 inches to 4 feet tall. Its stem and leaves are hoary with dense, short, curved hairs. The leaves are very narrow – not more than ⅛ inch wide. The flowers are pink or white, the petals from ⅛ to ¼ inch long.

In marshes and meadows from Quebec to Alberta and southward to Virginia, Ohio, Illinois, Kansas, and Utah. *Plate 71*.

E. PALUSTRE has very narrow leaves on a stem from 4 to 16 inches tall, which bears curved hairs. There is often only one flower, at the tip of the stem; but there may be several. The petals, from ⅛ to ⅕ inch long, vary greatly in color from violet through pink to white.

In bogs and marshes from Greenland to Alaska and southward to Long Island, Wisconsin, South Dakota, Colorado, and Oregon. *Plate 71*.

Besides all these, there are several far northern species some of which enter the mountains of our northernmost states. *E. anagallidifolium* is a dwarf only a few inches tall, with from one to three small pink or purplish flowers. The tongue-twisting name refers to *Anagallis*, the scarlet pimpernel, which has similar leaves. *E. lactiflorum* has from one to six white flowers on a plant not more than 8 inches tall. *E. hornemannii* (*Plate 71*) has from three to six larger flowers, the lilac-pink or white petals up to ⅖ inch long. The plant has only from three to six pairs of ovate or elliptic leaves.

THE DOGWOOD FAMILY (CORNACEAE)

The family is represented with us only by the genus *Cornus*.

THE DOGWOODS (CORNUS)

The dogwoods are mostly trees and shrubs and so – regrettably – omitted from this book. One species, however, though it grows perennially from a woody base, passes as herbaceous.

BUNCHBERRY or DWARF CORNEL, C. CANADENSIS, grows from 4 to 12 inches tall, the erect stem arising from a forking woody rhizome. There are one or two pairs of small leaves or scales on the lower part of the stem, and a pair of larger leaves at the summit, in the axils of which arise branches each consisting only of a pair of leaves – the whole forming an apparent circle of usually six leaves. The leaves bear veins which branch from the midvein near its base and run parallel to the margins. Above the leaves grows a single head of small yellowish flowers surrounded by four white (or pink), petal-like bracts – the whole easily mistaken for a single flower. The actual flowers have four minute sepals, four petals, four stamens, and an inferior ovary which develops into a red stone-fruit – the "bunchberry." Much variation occurs in form and color of flowers and fruit.

June to October: in moist acid woods, in bogs, and on upland slopes from Greenland to Alaska and southward to Maryland, Ohio, northern Indiana and Illinois, South Dakota, New Mexico, and California; also in Asia. *Plate 74.*

THE SHINLEAF FAMILY (PYROLACEAE)

The shinleaf family (see under *Pyrola* for an explanation of the rather absurd name) is a group of rather small plants, some evergreen, some without green color at any time, all generally woodland plants. The leaves are undivided. There are generally five sepals, five petals, ten stamens, and one pistil with five chambers in the ovary. A distinctive feature is that the pollen-bearing heads of the stamens do not split open lengthwise as in most families but discharge the pollen through tubes in the end. Many of these characteristics are shared by the heath family (*Ericaceae*); the two families are often merged.

The two groups of genera in this family are easily distinguished.

I. *Plants with green leaves.*

Chimaphila has leaves on the stem, flowers hanging at the ends of stalks at or near the summit.

Moneses has leaves only at or near the base, and a single flower at the summit.

Pyrola has leaves at and near the base, flowers along the end part of the stem (in a raceme). (Leaves may be lacking in some species.)

II. *Plants without green color (see also under* Pyrola *in I.); leaves are represented by scales on the stem. These plants live, as fungi do, on organic matter in the soil or on the living roots of other plants.*

Monotropa may be white, red, or tawny, with one or a few flowers on the bent-over tip of the stem.

Monotropsis is purplish-brown, with several rose-pink flowers on the bent-over end of the stem.

Pterospora is purplish-brown, and clammy, with numerous white or red flowers in a raceme.

PIPSISSEWA (CHIMAPHILA)

The visible plants of *Chimaphila* grow from underground stems (rhizomes). The erect branches bear leaves which last all winter. The flowers are pink or white. The style of the pistil is very short and bears a round stigma.

SPOTTED PIPSISSEWA, C. MACULATA, has leaves striped rather than spotted with white; they are lanceolate and sharp-pointed, with teeth at rather wide intervals on the margin. The flower is nearly an inch across, white and fragrant; several hang face down at the ends of their stalks, which spread from the tip of the stem.

June to August: in woods from southern New Hampshire and southern Ontario to Michigan and Illinois and southward to Georgia and Alabama. *Plate 72*. This is also known as spotted-wintergreen; but this is confusing, since the plants whose leaves have the odor and taste of wintergreen are *Gaultheria procumbens* in the heath family. (Compare also flowering-wintergreen, *Polygala* paucifolia.)

PRINCE'S-PINE or PIPSISSEWA, C. UMBELLATA, has leaves without white marks, tending to be broader than those of *C. maculata* and wider in their outer half; they are sharply toothed all around.

June to August: in woods from Quebec to Alaska and southward to Georgia, Ohio, northern Illinois, Utah, and California. *Plate 72*. The flowers are white or pale pink, about ¾ inch across. The leaves of both these species are refreshing when chewed in the woods. Those of *C. umbellata* are used in making root beer.

MONESES

We have only one species of *Moneses*.

ONE-FLOWERED-WINTERGREEN, M. UNIFLORA, looks like a small *Chimaphila* with only one flower. It stands about 5 inches tall. The leaves are round at the end and very finely toothed. The flower hangs from the curved tip of the stem. It is white or rose, and fragrant; much like a flower of pipsissewa but with a stigma bearing five narrow lobes.

June to August: in mossy woods from Greenland to Alaska and southward to Pennsylvania, Michigan, Minnesota, New Mexico, and Oregon. *Plate 72*.

SHINLEAF (PYROLA)

The name is said to be derived from an early use of the leaves in making plasters for injured shins! (Why these leaves were considered especially efficacious I do not know.) The leaves, generally rather dark green, and lasting through winter, are all at or near the base of the stem. The flowers are in the long cluster called a raceme.

The species are easily confused, the differences between them being minor.

P. ELLIPTICA is probably our commonest shinleaf. The leaf-blades are dull green, broadly elliptic, on short stalks. The stem rises from 6 to 12 inches above the leaves. The petals are more than four times as long as the small teeth of the calyx.

June to August: in woods from Newfoundland to British Columbia and southward to Delaware, West Virginia, Indiana, northern Illinois, South Dakota, and New Mexico. *Plate 72*. This species is sometimes called wild lily-of-the-valley, a name also applied to *Maianthemum* in the lily family. The resemblance is not close.

P. ROTUNDIFOLIA is quite similar to *P. elliptica*. It is named for its round leaf-blades, but they may be elliptic, and the distinction between this and the preceding species is not always easy. The petals are not more than three times the length of the calyx teeth, often only twice.

June to August: in woods, thickets, and bogs from Greenland to Minnesota and southward to North Carolina, Kentucky, and Wisconsin; also in the Old World. *Plate 72*.

P. ASARIFOLIA has shining leaf-blades as broad as long or broader and indented at the base (not much like those of *Asarum*, wild ginger, for which the species is named). The petals vary from crimson to pale pink.

June to August: in woods and thickets from Newfoundland to Alaska and southward to New York, Indiana, Wisconsin, South Dakota, Colorado, and Oregon. *Plate 73*. A variety has dull leaves not indented at the base and is hence difficult to distinguish from preceding species.

ONE-SIDED PYROLA, P. SECUNDA, is easy to identify by the characteristic that gives it its name: the flowers are all directed to one side of the stem that bears them (the raceme is "secund"). The leaf-blades are shining, elliptic or ovate, toothed or scalloped, and on short stalks. The stem rises to a height of about 8 inches. The petals are greenish-yellow.

June to August: in woods from Greenland and Labrador to Alaska and southward to Virginia, Ohio, northern Illinois, South Dakota, New Mexico, and California; also in the Old World. *Plate 72*. Some plants have duller, rounder leaves, and shorter stems with cream-white flowers.

P. VIRENS has thick, broad leaf-blades with round ends, on long stalks. The stem is from 3 to 12 inches tall, with greenish flowers. The style is directed

PLATE 72

Moneses uniflora *Horne*

Pyrola elliptica *Elbert*

Pyrola rotundifolia 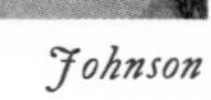 *Johnson*

Pyrola secunda *Scribner*

Chimaphila umbellata *Ryker*

Chimaphila maculata *Gottscho*

downward with the tip curved upward. In some plants leaves are lacking; the plant then lives on organic matter in the soil (like a fungus) or is parasitic on the roots of other plants.

June to August: in coniferous woods and thickets from Labrador to Alaska and southward to Delaware, Maryland, West Virginia, Indiana, Wisconsin, South Dakota, Arizona, and Oregon.

INDIAN-PIPE AND PINESAP (MONOTROPA)

These are white, tawny, or red plants with thick, fleshy stems, their leaves represented only by scales. The number of sepals, petals, and stamens varies. The flowers hang with their open end down, but as the fruit develops the stem straightens and the open end of the flower now points up.

INDIAN-PIPE or CORPSE-PLANT, M. UNIFLORA, is generally white, turning black as the fruit ripens. A pink form occurs which appears a little later. The stem is from 2 to 12 inches tall, and bears one (*uni-*) flower.

June to September: in the litter of woodlands from Newfoundland to Alaska and southward to Florida, California, and Central America; also in Asia. *Plate 73.*

PINESAP or FALSE BEECH-DROPS, M. HYPOPITHYS, is tawny, yellow, or red. The stem stands from 4 to 16 inches tall and bears several flowers at its tip.

June to October: in woodlands from Newfoundland to British Columbia and southward to Florida and Mexico; also in the Old World. *Plate 73.* This was formerly considered a distinct genus, named *Hypopithys.*

SWEET-PINESAP (MONOTROPSIS)

We have one species of *Monotropsis.*

SWEET-PINESAP or PYGMY-PIPES, M. ODORATA, is a purplish-brown plant somewhat resembling a species of *Monotropa.* The five petals are joined. There are several fragrant, rose-colored flowers at the bent summit of the stem.

February to May: in sandy woods, chiefly under pines, from Maryland to Kentucky and southward to Georgia and Alabama. *Plate 73.*

PTEROSPORA

There is only one species of *Pterospora.*

PINE-DROPS, P. ANDROMEDEA, is a purplish-brown, clammy plant up to 3 feet tall. The flowers vary from red to white; the joined petals form a small vase. The inflorescence is a raceme, with conspicuous bracts below the flowers.

June to August: in dry soils chiefly under pines from Prince Edward Island to British Columbia and southward to Pennsylvania, Michigan, and Wisconsin, and along the Rocky Mountains to Mexico. *Plate 73.* This is a "root-parasite"; that is, it obtains food directly from the living roots of other species. Fernald says it is "probably largely extinct" in our range.

PLATE 73

Monotropa hypopithys *Horne*

Pterospora andromedea *Johnson*

Monotropsis odorata *Horne*

Monotropa hypopithys *Rickett*

Pyrola asarifolia *Rhein*

Monotropa uniflora *Rickett*

GROUP IX

SEPALS three or five, petals three or five; all or some of the petals separate; petals bi-laterally symmetric; stamens five or ten. Exceptions: *Amorpha* has only one petal; in *Petalostemum* four of the petals are scarcely recognizable as such.

I. *Petals five (at least two joined); stamens ten:* bean family.

II. *Petals three; one sepal colored, sac-like; stamens five:* touch-me-not family.

III. *Petals five, all separate; stamens five:* violet family.

THE BEAN FAMILY (FABACEAE)

The bean family is one of the great families of plants: great in numbers, great in distribution over the earth, great in importance to man. It is estimated to contain 10,000 species. These grow all over the earth from the boreal zones to the tropics. Among them are beans and peas, soybeans and cow-pea, clovers and alfalfa, sweet-peas and lupines, peanuts and chick-peas, laburnum and wisteria and broom, indigo and various insecticides. There are trees, shrubs, vines, and herbs in the list.

This great assemblage of plant forms is brought into one family largely by the structure of the flowers. Typically there is a calyx of five sepals joined to form a cup or tube; and a corolla of five petals. One petal – the standard – stands erect; two at the sides are known as wings; and two below are joined along one edge to form a boat-shaped part called the keel. The stamens and pistil are within the keel. There are generally ten stamens, all joined by their stalks, or nine joined and one free, or all free. The single pistil is a miniature pea pod, developing into a fruit containing a single row of seeds (or only one seed) and in many species splitting lengthwise into two halves (in others not opening at all).

The type of flower outlined above is called papilionaceous, meaning "butterfly-like." It takes many forms: the standard in many species folded over the other petals; the wings in some joined with the keel. Two of our genera have unusual flowers with only one real petal, the standard. The inclusion of stamens and pistil in the keel acts as an aid to pollination. When an insect alights on the keel this is bent down and the stamens and pistils spring up, the stamens scattering pollen over the intruder and the pistil perhaps receiving pollen carried from a flower previously visited.

The leaves of all our species but one are divided, pinnately or palmately. It is important for identification to know just how much constitutes one leaf. It should be remembered that a branch bearing leaves has buds at the tip and in the axils, while a leaf-stalk bearing many small blades or segments carries no buds. A pair of stipules at the base of the leaf-stalk generally aids in determining the extent of a single leaf.

Besides the species here described, several that have been introduced from the Old World and have escaped from cultivation may be found in our range. They are mentioned below but not described or illustrated.

Guide to Common Genera of Fabaceae

I. *A number of genera include plants that climb on other plants or other supports, or trail on the ground.* (*Compare II; however, caution must be used in making this distinction, for some of these climbers and trailers, when not fully developed, may stand erect.*)

A. Of the climbers, some have tendrils, hairlike parts of their leaves which coil around supports: *Vicia* (from eight to twenty leaf-segments, pinnately arranged, and small stipules; flowers from purple to white); *Lathyrus* (from two to twelve leaf-segments, pinnately arranged, and mostly larger stipules; flowers blue, purple, red, or cream-colored).

B. The remaining climbers and trailers have stems that twine around their supports – or around themselves; the flowers of the first three have five nearly equal calyx-lobes, those of the next three have four; all but the last have leaves of three segments: *Phaseolus* (several or many small purple flowers, not more than ½ inch long, in a loose raceme; keel coiled; pods flat); *Clitoria* (one or two large blue flowers, up to 2 inches long, on a short stalk; pods flat); *Centrosema* (one to four violet flowers about an inch long in a short cluster on a long stalk; pods coiling after opening); *Amphicarpaea* (several or many small purplish or white flowers, about ½ inch long, in a raceme; pods twisted after opening); *Strophostyles* (several small purple flowers, less than ½ inch long, in a close cluster on a long stalk; pods twisted after opening); *Galactia* (several small purplish flowers, ⅖–⅗ inch long, in a long, narrow raceme; pods twisted after opening); *Apios* (from five to seven leaf-segments, pinnately arranged; flowers purplish-brown).

II. *The remaining genera are composed of plants that stand erect without support. These may be separated into four groups by their leaves.*

A. Genus with undivided leaves: *Crotalaria* (the stipules of our one widespread species form an arrowhead).

B. Genera with palmately divided leaves:* *Lupinus* (blue flowers in a tall raceme; from seven to eleven leaf-segments); *Baptisia* (white, yellow, or blue flowers in several racemes; three leaf-segments); *Thermopsis* (yellow flowers in racemes; three leaf-segments); *Trifolium* (most species, with red, pink, white, or yellow flowers in close heads; three leaf-segments); *Psoralea* (most species; blue flowers in racemes or spikes; three or five leaf-segments).

C. Genera with leaves pinnately divided into three segments (the end segment stalked, the side segments with shorter stalks or none): *Psoralea* (two species with blue flowers); *Trifolium* (two species with small yellow flowers in close heads on long stalks); *Melilotus* (small white or yellow flowers in long racemes); *Medicago* (blue or yellow flowers in small clusters on long stalk; pods one-seeded, spirally coiled); *Desmodium* (blue, lavender, or white flowers in racemes; pods of several joints which separate, stick to clothing); *Lespedeza* (purple or yellowish flowers in close clusters in axils of leaves); *Stylosanthes* (small orange flowers in short spikes); *Rhynchosia* (small yellow flowers in racemes; pods flat); *Petalostemum* (some plants of one species; see D).

D. Genera with leaves pinnately divided into five or more segments.

1. Two of these have only one apparent petal: *Petalostemum* (flowers in short round or cylindric spikes or heads); *Amorpha* (flowers in long, tapering racemes).

2. The remaining genera with many leaf-segments have the usual five petals: *Astragalus* (flowers purple or white in racemes or heads from axils of leaves; pods generally plump); *Oxytropis* (like *Astragalus* but the keel-petals ending in a sharp beak); *Dalea* (white or pink flowers with very hairy calyx in a dense spike; pod one-seeded); *Tephrosia* (yellow-and-pink or purple or white flowers in racemes; stems and pods hairy); *Glycyrhiza* (calyx with upper and lower lips; flowers yellow; pods prickly); *Lotus* (yellow or red-and-yellow flowers in an umbel); *Coronilla* (pink flowers in an umbel); *Anthyllis* (red-and-yellow flowers in a head).

*All leaf-segments without stalks or with equal stalks. *Thermopsis* has large stipules at the base of the leaf-stalk; compare *Lotus* (under D), which has a pair of leaf-segments in this position.

THE VETCHES (VICIA)

The vetches are slender climbing or straggling plants, familiar along many roadsides. Some species have long been cultivated for forage. *V. faba* is the broad-bean or horse-bean; its old name, *Faba*, gives the name to the family. Some are widely used as "green manure," to restore fertility to worn-out soil (bacteria that inhabit their roots can use the nitrogen of the air to form nitrates, which remain in the plants). The leaves are pinnately divided, one or more segments at the end being replaced by tendrils. The stipules are small. The flowers are purple, blue, or white, generally in racemes which grow from the axils. The technical distinction between this genus and *Lathyrus* is the tuft of hairs ("beard") on the style; it is at the end in *Vicia*, along the upper side in *Lathyrus*.

Several of the species that grow wild in our range are natives of the Old World, now well established as American plants. Some of them are known in England as tares, a name that found its way into the parables of the New Testament.

COW VETCH or CANADA-PEA, V. CRACCA, is common on roadside banks. The stem may be smooth or clothed with small hairs lying flat on the surface; sometimes these are silky. There are from eight to twelve pairs of leaf-segments (besides the tendrils). The flowers are numerous and densely clustered, pointing towards the base of the raceme, not more than ½ inch long; they are blue or purple (or sometimes white).

May to August: on roadsides and in fields and thickets from Newfoundland to British Columbia and southward to Virginia and Illinois. *Plate 74.* From Europe.

HAIRY or WINTER VETCH, V. VILLOSA, earns its name by being clothed with soft hairs (it is "villous"). The leaves have from five to ten pairs of segments. The flowers, in dense racemes as in *V. cracca*, are violet-and-white; they are somewhat larger, up to ⅘ inch long. Careful examination discloses another distinctive characteristic: the flower stalk is attached *under* the bulge of the calyx, not at the end.

May to October: in fields and on roadsides practically throughout the United States. *Plate 74.* From the Old World.

AMERICAN VETCH, V. AMERICANA, has a smooth stem. There are from four to nine pairs of very narrow leaf-segments, rather longer than those of the preceding two species. The small stipules are often toothed. The flowers are comparatively few in a raceme (rarely more than nine), but rather large (about an inch long), and closely clustered; they are bluish-purple. The whole raceme does not project beyond the adjacent leaves.

May to July: in meadows and moist woods from Quebec to Alaska and southward to Virginia, Ohio, Arkansas, New Mexico, and California. A native of North America.

WOOD VETCH, V. CAROLINIANA, has from six to twelve pairs of leaf-segments. The flowers are in rather loose racemes. They are mostly white, perhaps with some blue on the keel; they are small, not more than ½ inch long.

April to June: in woods and thickets from New York and southern Ontario to Minnesota and southward to Florida and Texas. *Plate 74.* A native of North America.

COMMON VETCH, V. ANGUSTIFOLIA, is smooth, with from two to five pairs of leaf-segments. The flowers are quite small, scarcely more than ½ inch long.

March to October: in waste ground and on roadsides practically throughout our range. *Plate 74.* This is close to *V. sativa*, long cultivated for forage.

Besides these, the southwestern *V. micrantha*, with small purplish flowers and very numerous leaf-segments, and *V. ludoviciana*, with not more than eight purplish flowers and few leaf-segments, are found in Missouri. *V. hirsuta* and *V. tetrasperma* are Old-World species with flowers up to ⅓ inch long. *V. dasycarpa* resembles *V. villosa* but has few hairs, which lie flat. And there are several other Old-World species, with one or more flowers in the axils, some with very few leaf-segments.

THE VETCHLINGS (LATHYRUS)

The vetchlings resemble the vetches in being climbing or trailing plants with pinnately divided leaves and tendrils. The leaf-segments and flowers are in general larger. The flowers are comparatively few in each raceme. Technically the genus is distinguished from *Vicia* by having hairs (a "beard") along the upper side of the style rather than at the tip. Identification of some species is aided by their large stipules.

The numerous races of sweet-peas were developed in relatively recent times from *L. odoratus*, a wild flower of the Mediterranean lands. Another European, the perennial sweet-pea, *L. latifolius*, is common in our gardens and sometimes found growing wild. It has flat, thin-edged ("winged") stems and leaf-stalks, and only one pair of leaf-segments (not counting the tendrils). The flowers are usually purplish-pink.

I. *In two of our species the stipules are long, leaflike, indented at the base where they are attached so that a lobe projects on either side; they are arrow-shaped or heart-shaped.*

BEACH-PEA, L. MARITIMUS, is a stout plant with branching stems up to 5 feet long. There are from four to twelve leaf-segments, not always paired, measuring between 1 and 2 inches long and about half as wide. The flowers are in long-stalked clusters of from three to ten; they are bluish or purple, about an inch long.

PLATE 74

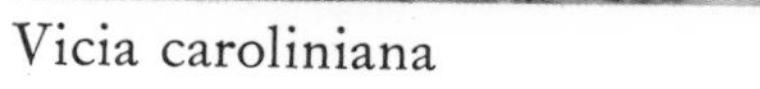

Vicia caroliniana *Elbert*

Vicia cracca *Johnson*

Vicia angustifolia *Elbert*

Vicia villosa *Johnson*

Cornus canadensis *Johnson*

Cornus canadensis *Johnson*

June to August: around the earth in northern Europe, Asia, and North America, extending southward along the Atlantic coast to New Jersey, along the Pacific Coast to California, and in the interior to the beaches of the Great Lakes and some other lakes. *Plate 75*. As might be expected from this vast range, the beach-pea is extremely variable in dimensions, hairiness, and size and shape of leaf-segments.

YELLOW VETCHLING, L. PRATENSIS, has a straggling, branching stem 2 feet long or longer. There are only two leaf-segments, generally more than an inch long, lanceolate and sharp-pointed, and one or more tendrils. The long-stalked racemes bear from four to ten bright, closely clustered yellow flowers. This is called lady's-slipper in some parts of England – a far cry from our flowers of that name; also, in some places, granny's-slipper-sloppers!

II. *Other species have stipules seemingly attached at the middle of one side, usually tapering to sharp points above and below; they may be broad or narrow.*

L. PALUSTRIS, which seems to have acquired no common name, has climbing stems up to 4 feet long, sometimes thin-edged or "winged." The leaves are divided into from four to six or sometimes ten usually narrow segments up to 3 inches long. The stipules are narrow. There are from two to nine flowers at the end of a long stalk, each from ½ to 1 inch long, with reddish-purple flowers.

June to September: in meadows and marshes around the earth in northern Europe, Asia, and North America, extending southward in America to New England, Ohio, Missouri, Colorado, and California. A very variable species, its races differing in the presence or absence of "wings" on the stem, in the size and shape and number of leaf-segments, in hairiness, and in other details.

L. VENOSUS is stouter than *L. palustris*, with a squarish stem up to 6 feet long. The leaves have from eight to twelve ovate, lanceolate, or elliptic segments from 1 to 2 inches long, or longer. The stipules are small and narrow. The racemes are dense, with from five to thirty purplish flowers.

May to July: in moist woods and thickets and along streams from Quebec to Saskatchewan and southward to Georgia, Tennessee, Louisiana, and Texas. *Plate 75*.

L. OCHROLEUCUS (the name means "yellow-white") has a stem up to 3 feet long. There are from four to ten elliptic or ovate, blunt leaf-segments, each from 1 to 2 inches long or longer. The stipules are broad, the two together making a sort of butterfly; they may be toothed at the side. The pale yellow flowers are in clusters of from five to ten, about ½ inch long; the stalks of the clusters are comparatively short.

May to July: in woods and on rocky slopes from Quebec to British Columbia and southward to Pennsylvania, Illinois, Iowa, South Dakota, Idaho, and Oregon. *Plate 75*.

TUBEROUS VETCHLING, L. TUBEROSUS, is named for the numerous small tubers on its underground stem (rhizome). It is a smooth plant, up to 3 feet tall. The very slender leaf-stalks carry only two segments, which are narrowly elliptic. The flowers are fragrant, pink or violet, about ½ inch long.

June to August: naturalized in fields and on roadsides from Vermont to Wisconsin and southward to Massachusetts and Ohio. *Plate 75*.

PHASEOLUS

This is the genus that includes all our cultivated true beans, which are mostly natives of the more tropical parts of the Americas and of Asia. We have one wild species.

WILD BEAN, P. POLYSTACHIOS, is a twining or trailing plant. The leaves are divided pinnately into three (the end segment having a long stalk, those at the side almost none, so that the arrangement is not palmate); the segments are ovate, with what look like minute stipules near the base of each. The flowers are pale purple, about ½ inch long, strung out loosely along a long stem. The keel in each (with the style inside) is coiled at the end. The two halves of the pod coil spirally after they separate.

July to September: in woods and thickets from Quebec to Manitoba and Montana and southward to Florida and Texas.

CLITORIA

This is mainly a tropical genus. One species is cultivated in the South. One species is wild in our range.

BUTTERFLY-PEA, C. MARIANA, is a low plant, generally twining but sometimes standing erect. The three leaf-segments are pinnately arranged, ovate, from

PLATE 75

Lathyrus maritimus *Allen*

Lathyrus maritimus *Johnson*

Clitoria mariana *Justice*

Lathyrus ochroleucus *Elbert*

Lathyrus tuberosus *Horne*

Amphicarpaea bracteata *Rickett*

Lathyrus venosus *Johnson*

1 to 2 inches long, with small, stipule-like appendages; the true stipules at the base of the leaf-stalk are small and narrow. The flower is handsome, pale blue, with a large, erect standard nearly 2 inches long; it may be single or there may be two or three on the stalk.

June to August: New York to southern Indiana and Iowa and southward to Florida and Arizona. *Plate 75.* Later flowers may pollinate without opening.

CENTROSEMA

Centrosema is a genus of tropical America; one species of our southern states extends northward into our range.

SPURRED-BUTTERFLY-PEA, C. VIRGINIANUM, has a twining or trailing stem, bearing leaves divided pinnately into three ovate or lanceolate segments with rather blunt tips. The pretty violet flowers, about an inch long, are single or up to four in a cluster on a stalk shorter than the adjacent leaf. The name refers to a very small hollow projection or "spur" (*centrum*) on the back of the standard near the base; this distinguishes *Centrosema* from *Clitoria*. The two halves of the pod, which is flat and up to 6 inches long, coil spirally after they open.

July and August: in sandy woods and open places from New Jersey, Kentucky, and Arkansas southward to Florida and Texas.

AMPHICARPAEA

We have one species of *Amphicarpaea*, the others being Asian.

HOG-PEANUT, A. BRACTEATA, is a straggling plant with twining stems up to a yard long, more or less covered with hairs (often reddish) which point downwards towards the base of the plant. The leaves have three pinnately arranged, ovate segments, the stalk and midrib hairy like the stem. The stipules are small, the appendages beneath the segments minute. There are usually a number of pale purple flowers (from two to fifteen) in the raceme, quite close together, each ½ inch long or a little more. The calyx has only four teeth. There are also flowers without petals or with rudimentary petals on creeping stems at the base of the plant. The flowers with petals form flat pods with several seeds; the two halves coil after the pod opens. The flowers without petals form round, rather succulent pods which do not open, each containing one seed, and more or less subterranean. These are the "peanuts" presumably relished by hogs; they were also eaten by the Indians, and are reported to be not unlike beans when properly prepared.

August and September: in woods and thickets from Quebec to Manitoba and Montana and southward to Florida and Texas. *Plate 75.* The plants differ in hairiness, the midwestern and western plants being more likely to have hairs standing out on the surface.

WILD-BEANS (STROPHOSTYLES)

These wild-beans (not the genus of cultivated beans; see *Phaseolus*) have trailing or twining, more or less hairy stems and leaves pinnately divided into three segments. The purplish-pink flowers are closely clustered in short racemes on long stalks. They often turn green as they age. The keel of each flower is strongly curved upwards (but not coiled like that of *Phaseolus*), forming a sort of beak. The calyx has only four teeth, the lowest the longest. The pod is not flat; its two halves twist after they separate. Our three species flower from June or July to October.

S. HELVOLA may have some leaf-segments three-lobed. The stem is sparsely hairy or practically smooth. The standard is about ½ inch broad and long.

In old fields and thickets from Quebec to Minnesota and South Dakota and southward to Florida and Texas; our most widespread species. *Plate 76.*

S. UMBELLATA has rather narrow ovate or lanceolate leaf-segments, hardly ever lobed. The flowers are similar to those of *S. helvola.*

In sandy open woods and fields from Long Island to Florida and Texas on the coastal plain and northward to Indiana, Missouri, and Oklahoma. *Plate 76.*

S. LEIOSPERMA is the hairiest of these three, the stems and leaves gray with the long soft hairs. The leaf-segments are narrow. The flowers are small, the standard only about ⅓ inch or less broad and long.

In dry soil from Ohio to Minnesota and Colorado and southward to Alabama and Texas.

PLATE 76

Baptisia leucophaea *Lee*

Lupinus perennis *Johnson*

Apios americana *Johnson*

Galactia regularis *Uttal*

Baptisia tinctoria *Rickett*

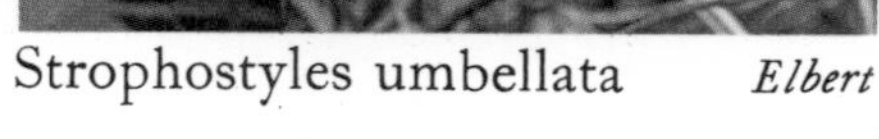
Strophostyles umbellata *Elbert*

Strophostyles helvola *D. Richards*

Baptisia leucantha *Johnson*

MILK-PEAS (GALACTIA)

Patrick Browne, an Englishman who named this mainly American genus, was wrong in thinking that the plants had milky juice: *Galactia* is derived from the Greek for "milk." The English name is just as wrong, but no one seems to have substituted a better one. These are low, twining or trailing plants with leaves divided pinnately into three segments and small purplish flowers. The calyx has only four teeth. The pod is flat, the two halves twisting spirally after they separate.

G. VOLUBILIS has hairy stems, much intertwined (*volubilis*). The flowers are from $\frac{1}{3}$ to $\frac{1}{2}$ inch long, very loosely arranged in a long raceme. The pod may be 2 inches long.

July to August: dry woods and open places from Long Island to Indiana and Kansas and southward to Florida and Texas.

G. REGULARIS usually lies on the ground, the nearly smooth stem scarcely twining. The flowers are $\frac{1}{2}$–$\frac{4}{5}$ inch long, in small, fairly dense racemes.

June to August: in sandy soil from southern New York and eastern Pennsylvania to Florida and thence to Louisiana; perhaps also in Tennessee, Missouri, and Kansas. *Plate 76.*

APIOS

Two species of this genus, one very rare, are reported from North America. Several others grow in Asia.

GROUND-NUT, A. AMERICANA, is conspicuous for its purplish-brown, fragrant flowers, $\frac{1}{2}$ inch long or longer, in a dense raceme. The keel is sickle-shaped and coiled. The four upper teeth of the calyx are very short. The leaves are pinnately divided into five or seven segments. These are climbing, twining plants. The English name refers to the underground stem (rhizome) which is thickened at intervals to form a series of small tubers.

July to September: in moist woods and thickets from Quebec to Minnesota and Colorado and southward to Florida and Texas. *Plate 76.* The tubers are edible, and even palatable when properly cooked, and were often gathered for food by the Indians. It is reported that the pilgrims had to rely on them during their first year in New England.

A. PRICEANA, in Kentucky and Tennessee, is rare and little known. It has greenish-white flowers with rose-purple tips, nearly an inch long. There is said to be but one tuber, several inches thick.

THE RATTLEBOXES (CROTALARIA)

The rattleboxes are at once distinguished from all other genera in the family by having undivided leaves. The flowers are yellow, in racemes. We have only one species in most of our range, but several others from the South (*C. purshii*, *C. spectabilis*) just enter southeastern Virginia and one (*C. angulata*) occurs also in southeastern Missouri.

C. SAGITTALIS is easily recognized by its stipules, which together form an arrowhead pointing downwards (*sagitta* is Latin for "arrow"). The plant is bushy, up to 16 inches tall, and hairy. The leaves are mostly lanceolate and blunt. The calyx is longer than the yellow petals. The English name is due to the large dry pod within which the seeds are loose.

June to September: in dry soil from Massachusetts and Ontario to Minnesota and South Dakota and southward to Florida and Texas.

LUPINES (LUPINUS)

Lupines form a large and difficult genus in the West and Southwest, but only one species is at all widespread in the Northeast. Two western species, *L. polyphyllus* and *L. nootkatensis*, are found in northern New England. The garden lupines are mostly hybrids. The leaves are mostly palmately divided into several narrow segments. The flowers are in a tall raceme which terminates the stem.

WILD LUPINE, L. PERENNIS, has a stem from 8 inches to 2 feet tall. There are from seven to eleven leaf-segments, about 2 inches long, all attached to the tip of the leaf-stalk. The flowers are generally purplish-blue. There are forms with white and with pink flowers, and with hairy stems.

April to July: in open woods and fields from Maine to Minnesota and southward to Florida and Louisiana. *Plate 76*.

FALSE INDIGOES AND RELATED SPECIES (BAPTISIA)

The species of *Baptisia* are much-branched, bushy plants, with white, cream-colored, yellow, or violet flowers in often handsome racemes. The erect stems grow from horizontal stems underground (rhizomes). The leaves are divided palmately into three segments (but in some the stipules are so large as to appear like two more segments). The pods are thick, rather woody, beaked, and stand out from the calyx on a stalk. The leaves of most of our species turn black when they wither or are dried. Some of the species hybridize.

I. *Species with yellow flowers.*

FALSE INDIGO, WILD-INDIGO, or RATTLEWEED, B. TINCTORIA, has the smallest leaves of these species, the leaf-segments rarely reaching an inch in length on a minute stalk; they make up for it in number. The flower-clusters also are very numerous. The flowers are about ½ inch long, and the pods scarcely longer.

May to September: in dry soil from Maine to Minnesota and southward to Florida and Louisiana. *Plate 76*. Children sometimes use the dry pods as rattles. As the botanical name suggests, the plants have been used for dyeing; in colonial times they were cultivated for that purpose. But the dye obtained is less satisfactory than true indigo, which comes from Asian and tropical American species of another genus in this family.

II. *Species with white or cream-colored flowers.*

PRAIRIE FALSE INDIGO, B. LEUCANTHA, makes a bush from 3 to 5 feet tall. The leaf-segments are about 2 inches long, the stipules small and sharp-pointed. The flowers are white, almost an inch long, in one or a few racemes (sometimes 2 feet long). The pods droop; they are from 1 to 2 inches long and black.

May to July: in prairies and open woodland from Ohio to Minnesota and Nebraska and southward to Mississippi and Texas. *Plate 76*. This species is known to cause poisoning of cattle in spring.

B. LEUCOPHAEA is smaller than *B. leucantha*, not more than 3 feet tall. The leaves have very short stalks, and segments (up to 3 inches long) which are wider towards their tips. The stipules are quite large, more than an inch long. The flowers are cream-colored, about an inch long, in racemes that tend to slant downward. The pod is up to 2 inches long, with a conspicuous beak.

May to June: in prairies and open woodland from Michigan to Minnesota and Nebraska and southward to Kentucky, Louisiana, and Texas. *Plate 76*.

III. *Species with indigo-blue flowers.*

BLUE FALSE INDIGO, B. AUSTRALIS, may grow 5 feet tall or taller, spreading into a bush. The leaf-segments may reach 3 inches in length and are widest near their tips. The flowers are an inch or more long, in erect racemes.

May and June: in woods and moist open places from Pennsylvania to Indiana and southward to Georgia and Tennessee; a variety with larger flowers and pods but smaller leaf-segments occurs from Missouri to Nebraska and southward to Arkansas and Texas. *Plate 77*.

Besides these species, two plants of the southern coastal plain extend northward into Virginia: *B. alba* with white flowers and brown pods (otherwise much like *B. leucantha*); and *B. cinerea*, with yellow flowers an inch long. The southwestern *B. sphaerocarpa*, also with yellow flowers an inch or more long, reaches Missouri.

The genus *Thermopsis* scarcely differs from *Baptisia;* the chief distinguishing characteristic is the flatness of the pod. One southern species, the bush-pea, *T. mollis*, with a raceme of yellow flowers, occurs in the mountains of Virginia.

THE CLOVERS (TRIFOLIUM)

At least some kinds of clover are known to most persons. Some are regularly mixed with grass in suburban lawns (but in my youth in England I spent many hours – by command – pulling out clover from the lawn); some are cultivated in farmers' fields; many are roadside weeds. Their leaves are divided, generally palmately, into three segments more or less toothed on their edges. The flowers are in heads or spikes. The uppermost petal, the standard, does not stand erect like that of a sweet-pea, but is folded over the wings and keel. The petals of many species remain attached even as they wither. The pods are very short, mostly hidden by the calyx and whatever remains of the corolla.

I. *Species with white, pink, red, or purple flowers.*

A. Among these we can group together several species whose flowers have distinct (though short) stalks and turn downward as they wither.

1. The first two species have stems that creep on the ground.

WHITE CLOVER, T. REPENS, is the common clover of lawns. Its main stem lies flat on the ground, sending up long-stalked, erect leaves, and, in the axils of some of these, branches each tipped by a head of flowers. The leaf-segments are broadly elliptic and toothed. The flowers, less than ½ inch long, turn from white to pinkish-brown as they turn downward.

All summer: in lawns and waste land throughout the United States and in Canada; from the Old World. *Plate 77.*

BUFFALO CLOVER, T. STOLONIFERUM, has stems that lie on the ground with their tips growing upward to bear leaves and flowers, besides forming long runners. The leaf-segments are almost wedge-shaped and toothed. The flowers are about ½ inch long, white tinged with purple.

May to August: in prairies and open woodlands from Ohio to South Dakota and southward to West Virginia, Tennessee, Missouri, and Kansas; not common, although widespread.

2. The next three species have erect stems.

BUFFALO CLOVER, T. REFLEXUM, has usually hairy stems up to nearly 2 feet tall. The leaf-segments are generally elliptic, toothed all around. The flowers are less than ½ inch long, red, pink, or white.

May to August: in sandy woods and fields, and prairies, from Virginia to South Dakota and southward to Florida and Texas. A native American species.

ALSIKE, T. HYBRIDUM (it is not really a hybrid), has smooth stems up to almost 3 feet tall. The leaf-segments are mostly elliptic, often more than 2 inches long. The flowering stems rise above the leaves. The flowers are pink or pink-and-white, soon turning brown.

May to October: on roadsides and in fields almost throughout North America. *Plate 77.* This is commonly planted in fields for forage.

T. VIRGINICUM has a dense tuft of hairy stems only a few inches long. The leaf-segments are narrow. The flowers are white, about ½ inch long, or less.

May and June: on rocky slopes in the mountains of Maryland, southern Pennsylvania, Virginia, and West Virginia.

B. Other species have white, pink, red, or purple flowers with very short stalks, which do not turn downwards as they age.

1. The first two of these species have heads not much, if at all, longer than they are wide.

RED CLOVER, T. PRATENSE, is perhaps our most abundant clover. It grows up to nearly 3 feet tall. The lower leaves have long stalks and segments often more than an inch long marked with a light V. The whole plant may be hairy. The flowers are pink, more than ½ inch long, in a usually large head.

May to September: extensively planted for hay and forage and escaped on roadsides and in fields and waste places throughout North America; originally from Europe. *Plate 77.* The flowers are fragrant. I have known fields of red clover in Wisconsin to sweeten the air with a perfume like that of Concord grapes.

ZIGZAG CLOVER, T. MEDIUM, somewhat resembles red clover but has narrower stipules and leaf-segments. It will scarcely reach 2 feet in height.

June to August: an European species sometimes planted and found growing wild in fields and roadsides and open woodland from Quebec and New Brunswick to Massachusetts.

2. Two other species have cylindric or ovate heads, longer than broad.

RABBIT-FOOT CLOVER, T. ARVENSE, is a softly hairy plant generally about a foot tall. The leaf-segments are narrow, and toothed only near their tips. The heads are cylindric, up to nearly 2 inches long. The sepals are beset with long hairs which conceal the small petals, the general effect being furry.

May to October: on roadsides and in old fields and waste places from Quebec to the Pacific and southward to Florida. *Plate 77.*

CRIMSON CLOVER, T. INCARNATUM, is readily distinguished by the bright color of its flowers. The plant is softly downy. The flower-heads are ovate in outline or cylindric. The flowers are about ½ inch long, bright red.

May to July: a cultivated European species occasionally found on roadsides and in waste ground and old fields. *Plate 77.*

II. *Species with yellow flowers; the hop clovers.*

Here are three Old-World species with small yellow flowers and leaf-segments less than an inch long. They may be confused with Medicago lupulina, *a common weed, which they closely resemble. In these clovers the upper teeth of the calyx are much shorter than the lower; in the* Medicago *they are equal. In the* Medicago *the pod is coiled; in the clovers straight.*

PLATE 77

Trifolium repens *Scribner*

Baptisia australis *Johnson*

Trifolium incarnatum *Phelps*

Trifolium hybridum *Gottscho*

Trifolium pratense *Gottscho*

Trifolium arvense *Gottscho*

YELLOW CLOVER or YELLOW TREFOIL, T. AUREUM, grows mostly erect, up to 18 inches tall. The leaf-segments are all practically without stalks. The flowers are about ¼ inch long. As they age they turn brown and are marked with fine lines.

May to September: naturalized in fields and waste places from Newfoundland to British Columbia and southward to South Carolina and Arkansas. *Plate 78*. The botanical names of this and the following two species have been much confused. The correct name of *T. aureum* may be *T. agrarium;* that of *T. campestre* may be *T. procumbens*.

LESSER YELLOW TREFOIL, T. DUBIUM, is a trailing plant, much branched. The leaf-segments are about ½ inch long, or less, the end one with a little longer stalk than those at the side. The flowers are only about $\frac{1}{10}$ inch long or a little longer.

April to October: in waste ground and old fields from Nova Scotia to Wisconsin and southward to Florida and Texas; on the Pacific Coast from British Columbia to California.

HOP TREFOIL, T. CAMPESTRE, is similar to *T. dubium*, but the end leaf-segment has a stalk much longer than those of the segments at the side.

May to September: on roadsides and in old fields and waste places from Quebec to North Dakota and southward to Georgia, Mississippi, Arkansas, and Kansas; and on the Pacific Coast. *Plate 78*. See the note under *T. aureum*.

Other Old-World species that have made themselves at home in fewer places in America are *T. hirtum*, with downy stem and purplish flowers; *T. striatum*, with pink flowers, *T. fragiferum*, strawberry clover, like *T. repens*, but with rose petals; and *T. resupinatum*, with purplish flowers turned upside-down ("resupinate")!

SCURF-PEAS (PSORALEA)

The scurf-peas are so called from the small scale-like glands which cover most parts of the plants (the botanical name refers to this same "scurf"). They are rather delicate-appearing plants with few branches, three or five leaf-segments, and long clusters (racemes or spikes) of generally blue-purple flowers.

I. *In the first two species the leaves are pinnately divided into three segments; i.e. the end leaflet is some distance along the midrib from those at the side, which have very short stalks. The plants are smooth or nearly so. The flowers are only about ¼ inch long.*

SAMPSON'S SNAKEROOT, P. PSORALIOIDES, grows from 1 to 3 feet tall. The leaf-segments may be 2 or 3 inches long but less than an inch wide. The most widespread variety is almost devoid of scurfy glands. The flower-clusters (spikes) are held aloft on long stems, rising well above the leaves. The flowers are bluish-purple.

May to July: in prairies and open woodland from Virginia to Kansas and southward to Georgia. The botanical name means "*Psoralea* like *Psoralea*"; this absurdity results from originally classifying the species in another genus but recognizing its similarity to a *Psoralea;* and we are not allowed to change the word when we place the species in *Psoralea!*

P. ONOBRYCHIS is practically free not only from hairs but also from the scurfy glands characteristic of the scurf-peas. The stem ranges from 3 to 5 feet tall; the leaf-segments are up to 4 inches long and 2 inches wide or wider. The flower-clusters do not rise much above the leaves, if at all. The flowers are blue.

June and July: in prairies and woodland from Ohio to Iowa and southward to western Virginia, Kentucky, and Missouri. *Plate 177*.

II. *In the remaining species the leaves are palmately divided, all the five or seven segments with equal stalks or none, attached at the tip of the leaf-stalk.*

PRAIRIE-TURNIP, P. ESCULENTA, has long been known for its starchy root-tuber. This was eaten by the Indians and by the early European explorers. It was variously known as wild potato, Indian bread-root, and pomme-blanche. Attempts were even made to cultivate it – which it would seem desirable to continue. The stem grows from 4 to 16 inches tall. The plant is recognizable by its hairiness, by the five more or less lanceolate leaf-segments, and by the dense raceme of rather large blue flowers (up to ⅘ inch long).

May to July: in prairies and dry woods from Wisconsin to Alberta and southward to Missouri, Texas, and New Mexico. *Plate 177*.

P. cuspidata is very similar and was probably included among the "pommes blanches" by the early voyageurs. It is more western than *P. esculenta*, entering our range only in Minnesota.

SILVERLEAF SCURF-PEA, P. ARGOPHYLLA, is covered with silky white hairs. It grows up to 2 feet tall and has three or five narrow leaf-segments. The flowers are dark blue.

June to August: in prairies from Wisconsin to Saskatchewan and southward to Missouri and New Mexico. *Plate 177*. This species has caused poisoning of cattle.

PLATE 78

Medicago lupulina *Rickett*

Trifolium aureum *Johnson*

Medicago sativa *Gottscho*

Medicago lupulina *Rickett*

Trifolium campestre *Horne*

Melilotus officinalis *Rickett*

Melilotus albus *Scribner*

Melilotus officinalis *Johnson*

WILD-ALFALFA, P. TENUIFLORA, is badly named, being no relative of true alfalfa (which does grow wild). The botanical name refers to the slender (*tenui-*) flower-spikes, in which the small bluish flowers are scattered. The plants are much branched and up to 3 feet tall, somewhat grayish with close hairs. The leaf-segments number three or five; they are narrow ($\frac{1}{3}$ inch wide or less).

May to September: in prairies and open woodland from Indiana to Montana and southward to Missouri, Texas, and Arizona.

Several other species touch our boundaries: *P. lanceolata*, with white or violet-tinged flowers in short dense racemes and mostly narrow leaf-segments, has been reported from Iowa. *P. canescens*, hoary with very short hairs, very short main leaf-stalks, and some leaves with only one segment, and bluish flowers which turn green as they age, reaches Virginia.

THE SWEET-CLOVERS (MELILOTUS)

The sweet-clovers can be identified by their fragrance alone. Even the uninjured masses of foliage along roadsides give off, in hot sunlight, the odor of very sweet new-mown hay; this is still more intense when the plants are crushed or dried. The plants have scant resemblance to the clovers (*Trifolium*). They are tall and very bushy. The leaves are pinnately divided into three. The small flowers are in numerous spikes which grow from the axils. These plants are naturalized from the Old World.

WHITE SWEET-CLOVER, M. ALBUS, may be 10 feet tall, but the leaf-segments are only an inch long, and the white flowers not more than $\frac{1}{4}$ inch. The leaf-segments are toothed in the outer half.

May to October: on roadsides and in waste lands throughout the United States. *Plate 78.*

YELLOW SWEET-CLOVER, M. OFFICINALIS, is similar to the white, but not so large – 5 feet tall at most. The dimensions of leaf-segments and flowers are about the same as in *M. albus*. The flowers are bright yellow.

May to October: in waste ground and fields throughout the northern states, less common southward. *Plate 78. Officinalis*, a term found in many botanical names, means "of the shops," and refers to some economic use of the plants. Yellow sweet-clover was formerly used in medicine and for flavoring. Another yellow-flowered species, *M. indicus*, with smaller flowers, is established in the South and West and is occasionally found in our range. Still another species with yellow flowers, *M. altissimus*, occurs in our range; it differs from *M. officinalis* chiefly in having downy pods.

ALFALFA AND ITS RELATIVES (MEDICAGO)

The plants of *Medicago* are quite miscellaneous, embracing the important forage plant alfalfa and dooryard weeds. The leaves are pinnately divided into three segments – the end segment being stalked. The flowers are blue or yellow, in racemes. The chief distinguishing characteristic is in the pod, which is spirally coiled.

ALFALFA or LUCERNE, M. SATIVA, has stems which branch and form a bush up to 3 feet tall. The leaf-segments are narrow, toothed mostly at the end which is blunt. The flowers are blue-purple, in short racemes or heads, less than $\frac{1}{2}$ inch long.

May to October: naturalized along roadsides and in waste ground throughout the United States. *Plate 78.* This is a valuable crop-plant.

BLACK MEDICK or NONESUCH, M. LUPULINA, is a common weed, with stems that lie flat on the ground. The leaf-segments are widest towards their tips, which are blunt. The yellow flowers are in almost round heads.

March to December: in waste grounds and lawns and on roadsides throughout North America. *Plate 78.* Easily confused with the yellow-flowered clovers or hop trefoils. For the distinction see under *Trifolium*. Several other species of *Medicago* from the Old World are occasionally found in our range.

THE TICK-TREFOILS (DESMODIUM)

Everyone that has walked in fields and woods in late summer and autumn knows the tick-trefoils, even if he does not know the name. They are those plants that cover one's clothing with small oval or triangular "seeds" that cling despite brushing. They are not seeds but the sections of the pods (legumes) of *Desmodium;* each pod is formed of a number of joints which separate at a touch; they are covered with

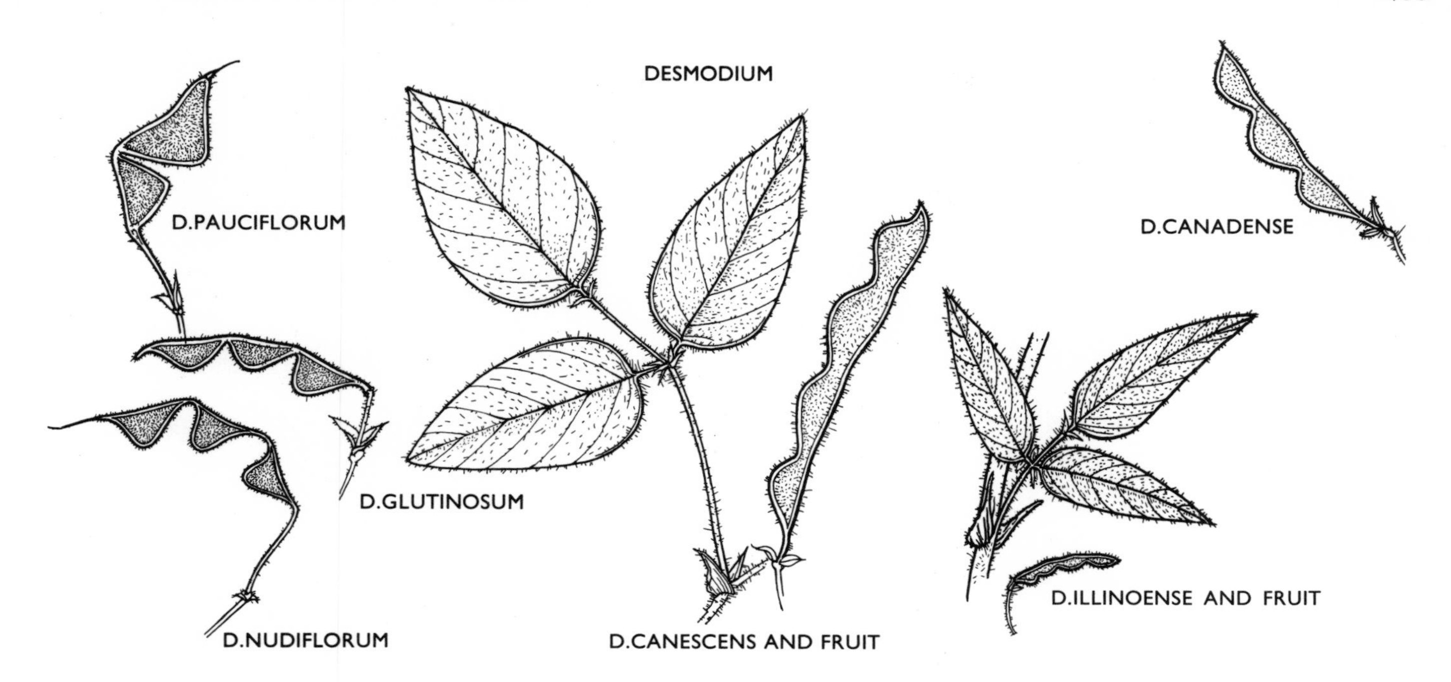

minute hooked hairs which account for their sticking to clothing.

The plants of the numerous species are rather weedy and straggling, lying on the ground or standing from 6 inches to 3 feet tall, with small purple or violet (or sometimes pale yellow or white or even green) flowers in long, often loose and branched racemes. The leaves are pinnately divided into three (the end segment at some distance along the midrib from the two at the side). To many persons the sticktights are weeds unworthy of inclusion among wild flowers. Some species, however, are conspicuous and decorative in flower.

Since the flowers are all much alike and quite small, identification of the species depends largely on the pods (see the drawings); as soon as the flowering season is well under way, both flowers and fruit are present together. The shape and size of the leaf-segments are also helpful, with the length of the leaf-stalk. Even so, many of the species are difficult to distinguish by easily seen characteristics, and for certain identification the technical books must be consulted. All flower in July and August except those otherwise noticed.

I. *Species whose pods have triangular joints connected only by their top corners; the whole pod standing on a stalk extending* above the calyx *at least three times as long as the calyx.*

D. NUDIFLORUM is easily distinguished from all other species: the flowers are on a leafless (*nudi-*) stem which grows directly from the ground, and taper to narrow points.

In woods from Maine to Minnesota and southward to Florida and Texas. *Plate 79.*

D. GLUTINOSUM has its leaves on the same stem as the flowers, all clustered just beneath the flowering part. The leaf-segments are like those of *D. nudiflorum.* The name refers to the flowering branches which bear numerous minute hooked (and therefore adhesive) hairs.

In woods from Maine to Minnesota and Nebraska and southward to Florida, Texas, and Mexico. *Plate 79.*

D. PAUCIFLORUM is named for the comparatively small number of flowers in its inflorescence. The stems may tend to lie on the ground with the tips turning up, or they may be wholly erect. The leaves are scattered along the stem. The leaf-segments are blunter than in the two preceding species.

June to August: in woods from western New York to Iowa (and perhaps farther north) and southward to Florida and Texas.

II. *Species the joints of whose pods are oval, or if triangular or flat on top are connected by a relatively broad joint; the stalk of the pod,* above the calyx, *rarely more than twice the length of the calyx.*

A. Among these we may first distinguish four species with stems that regularly lie on the ground, with or without tips that turn upward.

D. ROTUNDIFOLIUM is named for its round leaf-segments. The end segment is from 1 to 3 inches across.

July to September: in dry woods from Massachusetts to southern Ontario and Michigan and southward to Florida and Texas. *Plate 79.*

D. LINEATUM also has round leaf-segments, not much more than an inch across.

In sandy woods on the coastal plain from Maryland to Florida and Texas.

D. HUMIFUSUM has ovate leaf-segments. The pod joints number only three or four.

In dry, sandy woods from southern New England to Pennsylvania and southward to Maryland; also in Missouri.

D. OCHROLEUCUM is named for its pale yellow or white flowers. The leaf-segments are ovate. There are from two to four joints to a pod.

In sandy or loamy woods from Delaware to Georgia and Tennessee.

B. The remaining species of group II stand erect.

1. A number of these have pods typically with more than four joints.

a. The first two of these species have conspicuous stipules – ovate, indented at the base, ½ inch long.

D. CANESCENS has mostly hairy stems. The joints of the pod are up to ½ inch long, more deeply and sharply curved below than above.

In dry woods and fields from western Massachusetts to southern Ontario, Minnesota, and Nebraska and southward to Florida and Texas.

D. ILLINOENSE may have a downy but scarcely a hairy stem. The joints of the pod are about ¼ inch long, equally curved above and below.

In dry soil from southern Ontario to Wisconsin and Nebraska and southward to Ohio, Missouri, and Texas.

b. The remaining species with more than four joints to the pod have narrow, sharp-pointed stipules mostly less than ½ inch long.

The first two have rather conspicuous bracts adjacent to the flowers, up to ½ inch long. In the others the bracts are less conspicuous.

D. CANADENSE has a thickly flowered inflorescence which makes a good show. The leaf-segments are mostly lanceolate or narrowly ovate. The flowers are rose-purple at first, turning blue as they age. The joints of the pod are curved above and below (less sharply above), and are joined by a relatively wide connection.

In open woods and roadside thickets from Nova Scotia to Alberta and southward to Virginia, Illinois and Oklahoma; abundant in many places. *Plate 79.*

D. CUSPIDATUM has a less ample inflorescence than *D. canadense*. The leaf-segments are generally ovate and taper to a sharp point. The joints of the pod tend to be angled below and curved above.

In woods and thickets from New Hampshire to Minnesota and southward to Florida, Arkansas, and Texas.

D. PANICULATUM is distinguished in this group by its narrow leaf-segments, on a stalk an inch or more long. The joints of the pod are somewhat angular below, curved above.

In dry woods from Maine to Michigan and Nebraska and southward to Florida and Texas.

D. LAEVIGATUM has ovate leaf-segments. The joints of the pod are angular below and above (more obtusely above).

In dry woods from New York to Missouri and southward to Florida and Texas.

D. VIRIDIFLORUM is named for the color of its flowers; they are pink at first but turn green (*viridi-*). The leaf-segments are broad, ovate, and blunt. The whole plant is hairy or woolly. The joints of the pod are bluntly angular or round above and below.

In dry woods from New York to Missouri and southward to Florida and Texas. The northern plants, with round joints, have been separated as *D. nuttallii;* but they are hard to distinguish.

A group of similar plants, variously named *D. dillenii* (*Plate 79*), *D. glabellum*, and – appropriately – *D. perplexum*, offer no easy means of distinction; nor does *D. fernaldii*. All have more or less ovate or elliptic leaf-segments and somewhat triangular joints. The last-named is southern; the others widespread in our range.

2. The last of these erect species in group II have not more than four joints to a pod.

a. Among these, the first two can be immediately distinguished by their small leaf-segments – less than 2 inches long. The pods of both are somewhat curved, indented deeply below but scarcely above.

D. MARILANDICUM is a rather smooth plant with small blunt, broadly ovate or almost round leaf-segments, on a stalk from ⅜ to 1 inch long.

In dry woods from Massachusetts to southern Ontario and Michigan and southward to Florida and Texas.

D. CILIARE has small, blunt, ovate leaf-segments, on a stalk less than ½ inch long.

In dry, sandy soil from Massachusetts to Michigan and Missouri and southward to Florida and Texas.

b. The remaining species have leaf-segments more than 2 inches long.

Desmodium nudiflorum *Johnson*

Desmodium canadense *Gottscho*

Desmodium rotundifolium *McDowell*

Desmodium nudiflorum *Rickett*

Desmodium dillenii *Murray*

Desmodium glutinosum *Rickett*

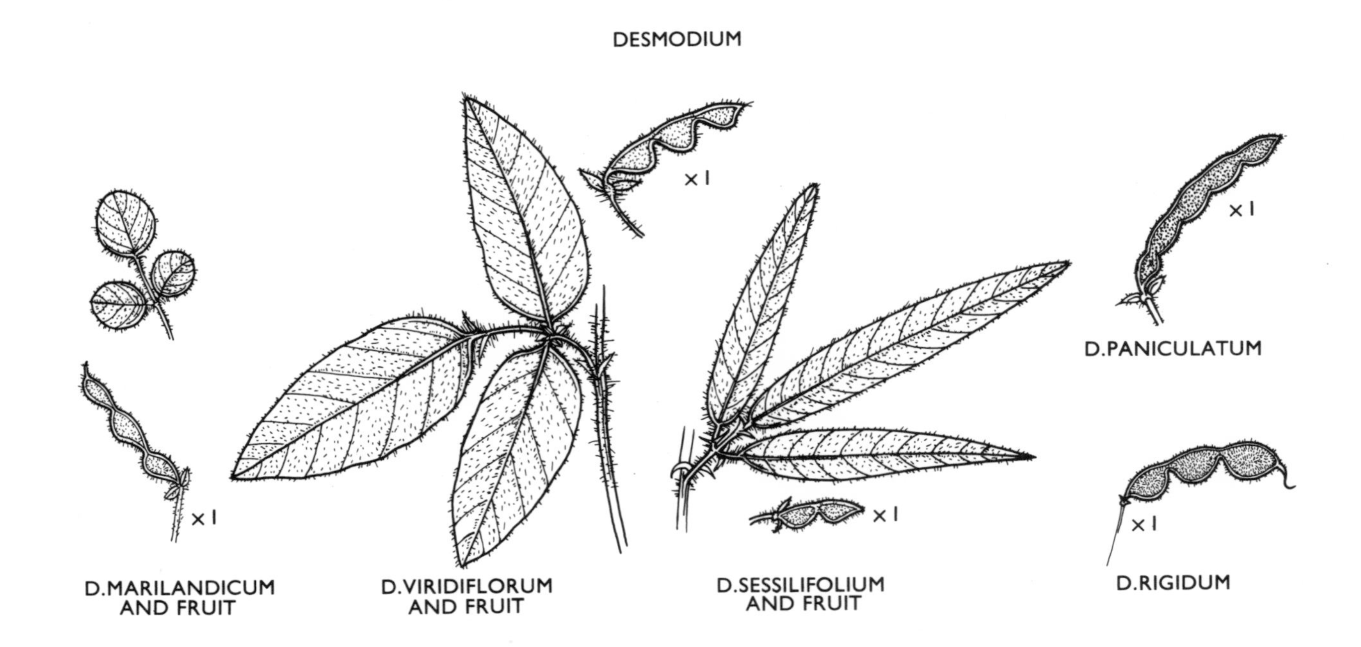

D. SESSILIFOLIUM has very narrow leaf-segments, on a leaf-stalk only $\frac{1}{5}$ inch long or even less (they are practically "sessile").

In dry soil from Massachusetts to southern Ontario, Michigan, and Kansas and southward to Pennsylvania, South Carolina, Louisiana, and Texas.

The southern *D. tenuifolium* also have very narrow leaf-segments, the leaf-stalk being about $\frac{1}{2}$ inch long. It grows in bogs and barrens from Virginia to Florida and Alabama. Another species of the southern coastal plain, *D. strictum* reaches up to the pine barrens of New Jersey. It has narrow leaf-segments on a stalk about $\frac{1}{2}$ inch long, and pods nearly straight along the top and indented below.

D. RIGIDUM has a finely downy stem and ovate leaf-segments on a short leaf-stalk (less than an inch long). The flowers are pale pink or almost white. The joints of the pods are curved above and below, more sharply below.

In dry woods from Massachusetts to Michigan and Kansas and southward to Florida and Texas.

THE BUSH-CLOVERS (LESPEDEZA)

The bush-clovers do not greatly resemble the true clovers, having small purplish or yellowish flowers in heads, spikes, or racemes, or singly in the axils. The leaves are pinnately divided into three segments which are generally blunt. Some species form pods in flowers that do not open. The pods of all species are short, roundish, with but a single seed. There is considerable hybridizing among some species, adding to their variability. Some are valuable crops.

We may separate the rather numerous species into two groups by the color of their flowers.

I. *Species with pale yellow or cream-colored flowers, with or without a purple spot on the standard; the stems are erect.*

L. CAPITATA is a common and variable species, recognizable by the dense heads of flowers in the axils of the upper leaves. The standard has a purple spot at the base. After the petals fall, the pods are concealed within the hairy calyx. The plants stand from 2 to 4 feet tall. The leaf-stalks are short and the leaf-segments generally rather narrow. In some forms the blades are silvery-silky underneath.

July to September: in dry open places from Quebec to Minnesota and Nebraska and southward to Florida and Texas. *Plate 80.*

L. HIRTA stands up to 5 feet tall. The leaves are short-stalked with leaf-segments typically broad and blunt; but there are forms with narrow segments which can scarcely be distinguished from *L. capitata.* The plants vary greatly in amount and kind of hairs. The flowers are in dense heads which are carried up from the leaf-axils on long stalks. The corolla is yellowish-white, the standard purple at the base. The pod about equals the calyx around it.

PLATE 80

Lespedeza hirta *Uttal*

Lespedeza capitata *Gottscho*

Lespedeza virginica *Ryker*

Lespedeza intermedia *Scribner*

Lespedeza procumbens *Clewell*

Lespedeza hirta *Rhein*

Stylosanthes biflora *Johnson*

July to October: in dry places from Maine to southern Ontario and Wisconsin and southward to Florida and Texas. *Plate 80.*

L. ANGUSTIFOLIA is quite similar to *L. hirta* but the leaf-segments are only about $\frac{1}{4}$ inch wide.

August and September: in sandy soils of the coastal plain from Massachusetts to Florida and Louisiana; also in Tennessee.

L. LEPTOSTACHYA is another species with narrow leaf-segments. The flowers are loosely arranged in slim spikes that grow from the upper axils.

July and August: in prairies from Illinois to Minnesota and Iowa.

L. CUNEATA is a slender plant up to 5 feet tall, with short-stalked leaves divided into narrow segments only an inch or less long. The flowers are borne singly or in small clusters in the axils of the leaves. They are white with purple veins. The pod is about as long as the calyx.

September and October: cultivated in the South and escaped from Pennsylvania to Michigan and Missouri and southward.

II. *Species with pinkish or purplish flowers, their stems reclining on the ground. The stipules are broad and rather papery.*

JAPANESE-CLOVER, L. STRIATA, has half-inch leaf-segments on a very short leaf-stalk. The stipules are relatively conspicuous, longer than the leaf-stalk on which they grow, becoming dry and brownish and marked lengthwise by fine ridges. The pinkish flowers are in small clusters on short stalks from the axils.

July to October: on roadsides and in dry soil from New Jersey to Kansas and southward to the Gulf; a native of Asia, cultivated as forage in the South.

KOREAN-CLOVER, L. STIPULACEA, is similar to *L. striata* in general aspect, but the flowers are more numerous in each cluster, and the clusters project from the axils. The leaf-segments are broader.

July to October: in dry open places from Pennsylvania to Iowa and Kansas and southward.

III. *Species with violet or purple flowers.*

A. The first five of these stand erect.

L. VIRGINICA has a generally unbranched stem from 1 to 4 feet tall. The leaves are crowded, with rather long stalks bearing the three narrow segments. The flowers also are crowded on very short branches in the upper axils. The petals are purple.

July to September: in dry open woods and thickets from New Hampshire to Wisconsin and Kansas and southward to Florida and Texas. *Plate 80.* The various plants called *L. simulata* are probably hybrids between this and other species.

L. INTERMEDIA is very like *L. virginica*, but has longer and broader leaf-segments (up to $\frac{2}{5}$ as against $\frac{1}{4}$ inch wide). The flowering branches are often somewhat longer. It has about the same range as *L. virginica. Plate 80.*

L. NUTTALLII grows from 2 to 4 feet tall, with a hairy stem. The leaf-segments are elliptic, on a fairly long leaf-stalk. The flowers (pink or purple) are crowded in a cluster on a stalk longer than the adjacent leaf.

August to October: in dry woods from New Hampshire to Michigan and southward to South Carolina, Alabama, Missouri, and Kansas. The plants known as *L. brittonii*, which are covered with a grayish velvety wool, are probably hybrids between *L. nuttallii* and other species.

L. VIOLACEA has a stem from 8 inches to nearly 3 feet tall, with spreading branches. The leaves are distinctive: the three leaf-segments are elliptic; scarcely 2 inches long and half as broad; on the branches they are often much smaller than on the main stem. The flowers are purple, in small, rather loose clusters on long stalks.

July to September: in dry woods and thickets from New Hampshire to Wisconsin and Kansas and southward to Florida and Texas.

L. STUEVEI is from 1 to 4 feet tall, downy all over. The leaves are crowded, with three elliptic segments on a short leaf-stalk. The flowers are in dense heads which are seated in the axils on short stalks. The petals are purple.

August and September: in dry open woods from Massachusetts to southern Vermont and Kansas and southward to Florida and Texas.

B. Two purple-flowered species have creeping stems with upright flowering branches.

L. PROCUMBENS is softly hairy. The leaf-segments are only an inch long or less, and about half as wide. The flower-clusters are on very long stalks.

August to October: in sandy or rocky soil from New Hampshire to Wisconsin and Kansas and southward to Florida and Texas. *Plate 80.* Some botanists would assign this to the following species, as a variety; the two are certainly much alike. There is also a narrow-leafed form of *L. procumbens.*

L. REPENS is like *L. procumbens* but smooth or almost so. The leaf-segments are more round, often finely silky underneath.

May to September: in sandy or rocky woodland from Connecticut to southern Wisconsin and Kansas and southward to Florida and Texas.

STYLOSANTHES

Only two species of *Stylosanthes* occurs in the northeastern United States. Many others are tropical. The names, both English and botanical, refer to the stalk-like part of the flower at the summit of which the small yellow petals are borne. This "pencil" or "stylus" is a hollow tube or receptacle, the pistil being seated within it (the ovary is not inferior since its *sides are not joined* to this receptacle). The leaves are pinnately divided into three segments (the end segment having a longer stalk and the arrangement therefore not palmate). The flowers are in small clusters at the tips of the stems. The fruit usually has two joints, of which one commonly contains a seed.

PENCIL-FLOWER, S. BIFLORA, stands more or less erect to a height of from 6 inches to nearly 2 feet. The leaf-segments are mostly elliptic or lanceolate, with a minute spine at the tip. The uppermost leaf-stalks are surrounded by small bristly tubes (made of the stipules).

June to September: in dry woods and thickets from southern New York to Kansas and southward to Florida and Texas. In one variety the stem is bristly all over. *Plate 80*.

S. RIPARIA tends to spread on the surface. The leaf-segments are commonly shorter and broader than those of *S. biflora*, not more than an inch long. The small tubes around the leaf-stalks are smooth.

June to September: in dry woods from New Jersey to Missouri and southward to Florida and Texas.

RHYNCHOSIA

Three species of this genus are found barely within the southern boundaries of our range. The genus is characterized by leaves pinnately divided into three segments (i.e. the end segment with longer stalks), yellow flowers in short clusters on stalks from the axils, and flat pods with one or two seeds. They grow in dry sandy places, some species flowering from June to August, others until September.

R. TOMENTOSA is downy with hairs that point upwards. The stem is erect, from 6 inches to 3 feet tall. The flower-clusters are shorter than the adjacent leaf-stalks. It grows from Delaware to Tennessee and southward to Florida and Texas.

R. DIFFORMIS is a trailer or twiner. The stems bear hairs that point downward. The first leaves are undivided. The flower-clusters do not project beyond the adjacent leaves. This occurs in Virginia and Missouri and southward to Florida and Texas.

R. LATIFOLIA also has a trailing or twining stem, covered with short hairs. The flower-clusters are on stalks longer than the adjacent leaves. It is found from Missouri and Oklahoma to Louisiana and Texas.

THE PRAIRIE-CLOVERS (PETALOSTEMON)

The prairie-clovers do not conform to the papilionaceous flower-pattern of the family. They have a standard but no wings or keel. There are only five stamens with pollen. Alternating with these are four narrow petal-like bodies which may be regarded as transformed stamens; they bear no pollen. The leaves are divided pinnately into from three to many narrow segments. The pod, which does not open, contains only one or two seeds, and is surrounded by the calyx.

P. PURPUREUS has a stem from 1 to 3 feet tall, bearing leaves divided into three or five very narrow, almost hairlike segments less than an inch long. The flowers are pink or crimson (not purple, in spite of the name).

June to September: in prairies from Indiana to Alberta and southward to Alabama, Arkansas, Texas, and New Mexico; occasionally found farther east. *Plate 81*.

P. CANDIDUS is named for its white flowers. The stems are from 1 to 2 feet tall, or taller. There are five or seven narrow leaf-segments ranging from 2/5 inch to over 1 inch in length.

June and July: in prairies from Indiana to Minnesota and Saskatchewan and southward to Alabama and Texas. *Plate 81*.

P. FOLIOSUS is distinguished by its numerous leaf-segments, from eleven to thirty-one; they are not more than 1/2 inch long. The flowers are rose.

July to September: on river-banks and rocky hills and in glades from Ohio to Illinois and southward to Tennessee.

Besides these, several other species just enter our borders. The southwestern *P. multiflorus*, with from three to nine narrow leaf-segments, very short flower-clusters, and white flowers, is found in western Iowa. The western *P. occidentalis*, also with white flowers, like *P. candidus*, but with narrow leaf-segments not more than ½ inch long, occurs in Minnesota. Another westerner, *P. villosus*, is named for its soft or silky hairs; it has from thirteen to nineteen leaf-segments; the flowers are rose; it is found from Michigan westward and southwestward. *Plate 81*. *P. pulcherrimus*, with from three to seven leaf-segments about an inch long or longer, and rose flowers, grows in Missouri.

FALSE INDIGOES AND LEADPLANTS (AMORPHA)

The botanical name signifies "formless," and refers to the flower which has only one petal (the standard). These are somewhat shrubby plants with leaves divided pinnately into numerous segments. The purple flowers are in long, dense, tapering spikes at and near the top of the plant, their yellow stamens projecting beyond the perianth. The pod has only one or two seeds; it projects from the calyx when ripe. For other false indigoes see *Baptisia*.

LEADPLANT, A. CANESCENS, may be recognized by the usually dense covering of white hairs on its leaves. It grows from 20 to 40 inches tall. There are from fifteen to fifty-one segments in a leaf, each from ⅓ to ½ inch long.

May to August: in prairies and dry woodland from Michigan to Saskatchewan and southward to Indiana, Arkansas, Texas, and New Mexico. *Plate 81*.

FALSE INDIGO, A. FRUTICOSA, is almost smooth. This is generally a shrub and may reach a height of 15 feet. The leaves have from eleven to thirty-five or more segments, from less than an inch to more than 2 inches long.

May and June: in woods and thickets and on stream-banks from Pennsylvania to Minnesota and Saskatchewan and southward to Florida, Texas, and Arizona. *Plate 81*.

FRAGRANT FALSE INDIGO, A. NANA, is somewhat shrubby, from 1 to 3 feet tall, smooth, with from twenty-one to forty-one leaf segments per leaf, each less than ½ inch long. The spikes of flowers are only about 3 inches long.

June and July: a western species, growing in prairies from Minnesota to Saskatchewan and southward to Iowa and Colorado.

Other species may be found within the borders of our range. *A. brachycarpa* is a slender, smooth shrub of southwestern Missouri. *A. nitens* usually has only one inflorescence, 10 inches long, and relatively few and large leaf-segments; it is southern, extending to southern Illinois.

THE MILK-VETCHES (ASTRAGALUS)

The milk-vetches are herbs which grow from a perennial root or underground stem. They have leaves pinnately divided into many segments. The flowers are in short or long clusters (heads and racemes) on stalks which arise in the axils of leaves. The petals are white, yellowish, or purple. The pods vary greatly and are the best guides for identification: they may be flat or plump, with thin or thick, woody or almost succulent walls. This is an enormous genus in the West, where some species are called loco-weeds because, when eaten by animals, they cause them to act as if mad (*loco*) and often lead to death. Other species accumulate the deadly poisonous element selenium from the soil, without harm to themselves but with fatal results to any animal grazing them.

We have five species that seem at home in our range, though a number of others just cross our western, northern, or southwestern boundaries. To distinguish even our few species it is generally necessary to count leaf-segments and observe the fruits.

GROUND-PLUM, A. CRASSICARPUS, is named for the plump pod, nearly an inch through; it is thick-walled and does not split open. The plant usually has several stems up to 20 inches tall. The leaf-blades have from fifteen to twenty-three segments. The flowers are either purple or yellowish, less than an inch long, in a loose head.

April and May: in prairies and plains from Minnesota to Alberta and southward to Tennessee, Oklahoma, and Texas.

A. CANADENSIS has a pod from ⅖ to ⅘ inch long and about ⅕ inch thick. The stem grows up to 5 feet tall. The leaf-blades are divided into from thirteen to twenty-nine segments. The flowers are white or yellowish, about ½ inch long, in a cluster (raceme) up to 6 inches long.

July and August: in open woodland and on riverbanks from Quebec to British Columbia and southward to Georgia, Texas, and Utah. *Plate 177*.

PLATE 81

Petalostemon villosus *Johnson*

Petalostemon purpureus *Johnson*

Amorpha fruticosa *Phelps*

Astragalus distortus *Leeson*

Amorpha canescens *Johnson*

Oxytropis lambertii *Rickett*

Petalostemon candidus *Johnson*

A. COOPERI has a pod from $\frac{2}{5}$ to $\frac{4}{5}$ inch long, up to $\frac{2}{5}$ inch thick. The stem is from 16 to 28 inches tall. The leaves have from eleven to seventeen segments. The flowers are white; they are in short clusters (racemes) which their stalks scarcely carry above the leaves.

June: on river-banks and lake-shores from New York to South Dakota.

A. DISTORTUS has a curved pod – half-moon-shaped – about an inch long and $\frac{1}{6}$ inch thick. The stems spread in all directions to a length of from 4 to 12 inches. The flowers are purple, about $\frac{1}{2}$ inch long; they are in short but loose clusters (racemes) about an inch long.

May and June: in rocky barrens, dry prairies, and open woodland from Virginia and Maryland to West Virginia and from Illinois to Iowa and Kansas and southward to Mississippi and Texas. *Plate 81.*

A. ALPINUS has a pod directed down along the stem; it is inclined to be sickle-shaped, flattish, grooved on the lower side, about $\frac{1}{2}$ inch long. The flowers are purplish, about $\frac{2}{5}$ inch long; they are in loose, long-stalked clusters.

June to August: on shores and gravel in the far north of America, Europe, and Asia, extending southward in North America to Maine, Vermont, and Colorado.

OXYTROPIS

The species of *Oxytropis* closely resemble *Astragalus*, the difference being in their keel, which ends in a sort of beak. The pods are cylindric, tipped with a narrow beak. These plants barely enter our territory. Loco-weed, *O. lambertii*, with stem and leaves whitened with close hairs, is found in Minnesota and Missouri. The flowers are purple. *Plate 81. O. johannensis*, with silky hairs and purple flowers, enters Maine and Wisconsin. *O. viscida* is found in Minnesota; it is sticky (viscid).

DALEA

This is a genus mostly of the southwest, of Mexico, and of South America. It is characterized by petals that taper to a stalk-like base, and by the union of wings and keel with the stamens. The pod is short and contains only one seed; it does not split open, and is enclosed in the calyx. The leaves are pinnately divided into numerous segments.

We have two species in the western part of our range. *D. alopecuroides* is from 1 to 2 feet tall, or taller. The flowers are in a dense cylindric inflorescence which terminates the stem (this somewhat resembles the fox-tail grass, *Alopecurus;* whence the second half of the name). It grows from Indiana to Minnesota and Colorado and southward to Alabama and New Mexico, and is sometimes found farther east. It flowers from July to September. *D. enneandra* may reach 4 feet in height, with fewer segments (from five to eleven) to a leaf. The flowers are scattered in a loose spike. The calyx is remarkable, the teeth prolonged in long feathery parts. This is found from Iowa and Minnesota to Saskatchewan and southward to Mississippi and Texas, flowering from May to August.

HOARY-PEAS (TEPHROSIA)

Our two species of this genus (there are many in tropical Africa and America) earn their English name by being hairy – the first with white hairs, the second with a somewhat rusty color. They have leaves pinnately divided into many segments, and attractive flowers in racemes at the tip of the stem. The pods are narrow, round in cross section, with several seeds.

GOAT'S-RUE or RABBIT'S-PEA, T. VIRGINIANA, has a covering of white hairs, sometimes silvery-silky (but there is a smooth form). There may be up to twenty-nine segments to a leaf. The flowers are in a dense cluster. The standard is yellow, the other petals pink or lavender. The plant is from 1 to 2 feet tall.

May to August: in sandy soil, in woods and other dry places from New Hampshire to southern Ontario and Minnesota and southward to Florida and Texas. *Plate 82.*

T. SPICATA has somewhat zigzag stems covered with rusty hairs, from 1 to 2 feet tall but often not standing erect. There are up to fifteen leaf-segments. The flowers are $\frac{1}{2}$ inch long, purple (sometimes white).

May to September: in sandy fields and woods from Virginia to Kentucky and southward to Florida and Louisiana.

LICORICE (GLYCYRRHIZA)

We have only one species in this genus.

WILD LICORICE, G. LEPIDOTA, grows up to 3 feet tall. There are numerous leaf-segments sprinkled when young with small scales (*lepidota*) which become minute dots as the plant ages. The calyx has four teeth. The petals are nearly white. The pod is only half an inch long; it is beset with hooked prickles.

May to August: in prairies and waste places from Ontario to Washington and southward to Arkansas, Texas, California, and Mexico; also in Virginia. *Plate 177*. Commercial licorice is obtained from the root of another species of this genus (the botanical name means "sweet root").

BIRD'S-FOOT TREFOILS (LOTUS)

The Greek name *Lotos* was applied to a number of different plants, including the water-lily of Egypt, the plant used by the legendary lotus-eaters to induce dreams and oblivion, and certain trees. None of these is the botanical *Lotus*, which is a genus of small plants with papilionaceous flowers in long-stalked umbels. Only one species is at all widespread in our range, though a second, western species, just touches our western borders.

L. CORNICULATUS has peculiar leaves with five segments, two at the base like large stipules, the other three close together at the end of the midrib. The flowers are yellow or reddish.

June to September: an European plant naturalized in America on roadsides and in fields and waste land from Newfoundland to Minnesota and southward to Virginia and Ohio; also on the Pacific Coast. *Plate 82*. The English name means "three leaves" and refers doubtless to the three end segments of each leaf. Geoffrey Grigson records more than seventy other English names for this diminutive plant! Some of these – Hop-o'my-thumb, Tom-thumb's-fingers-and-thumbs – are associated with a well-known goblin, and the group of pods suggested fingers or claws, whence cat's-claws, devil's-claws, five-fingers, etc. The pods also provide the "bird's foot." It was evidently a fairy plant, to be treated with respect.

The western species, *L. purshianus*, is found from western Minnesota to western Missouri. It has three leaf-segments on a very short stalk, and short-stalked clusters of pinkish flowers.

CORONILLA

One species of this Old-World genus has become established in North America.

CROWN-VETCH, C. VARIA, is a bushy plant sometimes attaining a height of several feet, with leaves divided pinnately into from eleven to twenty-five segments, and long-stalked clusters (umbels) of small pink flowers.

May to September: on roadsides and in waste land from New England to South Dakota and southward to Virginia, Kentucky, and Missouri. *Plate 82*.

ANTHYLLIS

One species of this Old-World genus has become established in a few places in the United States.

LADY'S-FINGERS or KIDNEY-VETCH, A. VULNERARIA, has several downy stems standing about a foot tall. The leaves have from five to thirteen segments (but some of the lower leaves may have only one!). The flowers are in a long-stalked head, surrounded by conspicuous bracts. The petals vary from red to yellow.

June to August: in fields and waste land scattered throughout our range (reported from Quebec and Ontario, Michigan, North Dakota, New England, Pennsylvania, Ohio, and Missouri). *Plate 177*.

THE TOUCH-ME-NOT FAMILY (BALSAMINACEAE)

The *Balsaminaceae* offer no problems of identification; their curious flower is like no other. There are (in our species) three sepals, one very large and saclike or funnel-shaped, and not green; the flower hangs from its stalk, and this large sepal is the lowest. From its mouth issue the three petals (the lower two are lobed; probably each is two petals joined). There are five stamens. The fruit is a capsule; it opens at a touch, the five pieces coiling up in opposite directions, almost explosively ejecting the seeds.

The juice of these plants is supposed to afford protection against poison-ivy, or even to remedy the effects. It has also been used against warts, for which reason the plants have been called celandine, with *Chelidonium majus*. They are annual plants coming up in great numbers in moist ravines, along streams, and in similar places. The paired leaves of the seedlings, pale green and round, are familiar in such places in spring.

There is only one genus native in North America. Another is found in Malaysia.

THE TOUCH-ME-NOTS, JEWELWEEDS, OR SNAPWEEDS (IMPATIENS)

These plants have the characteristics described above. The flowers of our wild species are yellow or orange, variously marked and mottled.

ORANGE TOUCH-ME-NOT or JEWELWEED, I. CAPENSIS, grows from 2 to 5 feet tall. The orange flowers are usually marked with reddish-brown.

June to September: in damp woods, bottomlands, ravines, etc. from Newfoundland to Saskatchewan and southward to South Carolina, Alabama, and Oklahoma. *Plate 82*. These plants have become naturalized in England, where they are known as American jewelweed. There are various color-forms. The botanical name is due to an error, its author believing that it came from the Cape of Good Hope. It seems absurd to retain the error, but international rules forbid us to change the original name.

YELLOW TOUCH-ME-NOT or JEWELWEED, I. PALLIDA, is somewhat taller than *I. capensis*. The flowers are lemon-yellow, often with red markings.

June to September: in damp woods and thickets from Quebec to Saskatchewan and southward to North Carolina, Tennessee, and Missouri. *Plate 82*. Like the preceding species, this one has several color-forms. Notice the spur on the lower sepal; in this species it points straight down.

I. balsamina is the familiar garden balsam, a native of Asia; it is occasionally found wild. Two other exotic species, *I. parviflora* and *I. glandulifera*, may be found.

THE VIOLET FAMILY (VIOLACEAE)

Essentially, this family in North America is the genus *Viola*, the violets. There is, however, one species, with small green flowers, in another genus named *Hybanthus*.

VIOLETS (VIOLA)

The flowers of the violets have a flat lower petal which provides a landing-place for insects, two side petals or "wings," and two upper petals. The lower petal is generally prolonged backward into a hollow sac or tube, a "spur." The five stamens closely surround the pistil, only the short style and stigma generally being visible. The two lower stamens bear nectaries, bodies that exude nectar, which extend into the spur. The whole arrangement makes self-fertilization almost impossible but favors cross-fertilization. Many violets also form flowers that fertilize themselves without opening, and these may be responsible for most of the seeds. The fruit is a small capsule which splits into three parts, each bearing a row of seeds.

The species are numerous and some of them, which are distinguished by only minor details, hybridize freely in nature, making identification almost impossible; this is particularly true of the "stemless blue

PLATE 82

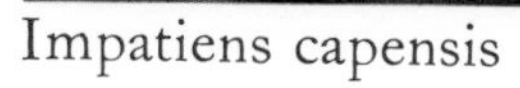

Impatiens capensis *Rickett*

Tephrosia virginiana *Ryker*

Lotus corniculatus *Ryker*

Coronilla varia *Rickett*

Viola cucullata *Rickett*

Viola papilionacea *Rickett*

Impatiens pallida *Johnson*

violets" (see below). In fact it has been said that there are no true species in this group, or that they all form one vast and heterogeneous species. However, the descriptions below will enable the reader to name at least some of the plants he finds; perhaps most of them.

All the violets are spring-flowering. If, however, frosts are long delayed in the autumn until the days are again as short as in early spring, flowers may be seen in this season also.

It is easy to classify all the violets in two groups according to the type of stem from which leaves and flowers grow.

I. *The so-called stemless violets have a short, horizontal, underground stem (rhizome), from the tip of which leaves and flowers grow in the spring. The only stem visible above ground is that of the flower. (Compare II; some species of II may seem to belong here when young.) Some of these species may be separated by the color of the flowers, but to distinguish some of them it is necessary to look at the hairs on the leaves and on the petals, and other minute details.*

A. Species with blue, violet, or white flowers. These may be further divided into those with no lobed or cleft leaves and those with lobed or cleft leaves. (This refers to the leaves present at flowering time; see the note under *V. triloba.*)

1. Blue, violet, or white flowers with no lobed or cleft leaves.

COMMON BLUE VIOLET, V. PAPILIONACEA, is perhaps the commonest violet in the Northeast. It has typically heart-shaped leaves with bluntly toothed margins and smooth stalks, but many plants will be found intermediate in these respects between this and other species. The side petals wear "beards," tufts of white hairs. The flowers vary from light to deep violet or white, and a well-known form, the confederate violet, has a gray-white center.

In all sorts of situations, mostly moist, from Massachusetts to Minnesota and southward to Georgia and Oklahoma. *Plates 82, 83.*

WOOLLY BLUE VIOLET, V. SORORIA, is much like *V. papilionacea* except for its hairy leaf-stalks, leaf-blades, and flower-stalks. The petals are apt to be white at the base.

In moist meadows and woods from Quebec to Minnesota and southward to North Carolina and Oklahoma. *Plate 83.*

MARSH BLUE VIOLET, V. CUCULLATA, is distinguished by the long flower-stalks which carry the flowers well above the leaves. The petals are typically rather pale or even white, with darker veins. A hand lens will reveal that the hairs or "beard" on the side petals end in round knobs.

In wet places, often in swamps, from Quebec to Ontario and Minnesota and southward to Virginia, the mountains of Georgia, Arkansas, and Nebraska. *Plate 82.*

V. AFFINIS resembles *V. papilionacea* but the lower petal bears a tuft of hairs, or "beard," like that of the side petals. The leaf-blades are somewhat narrower and more tapering.

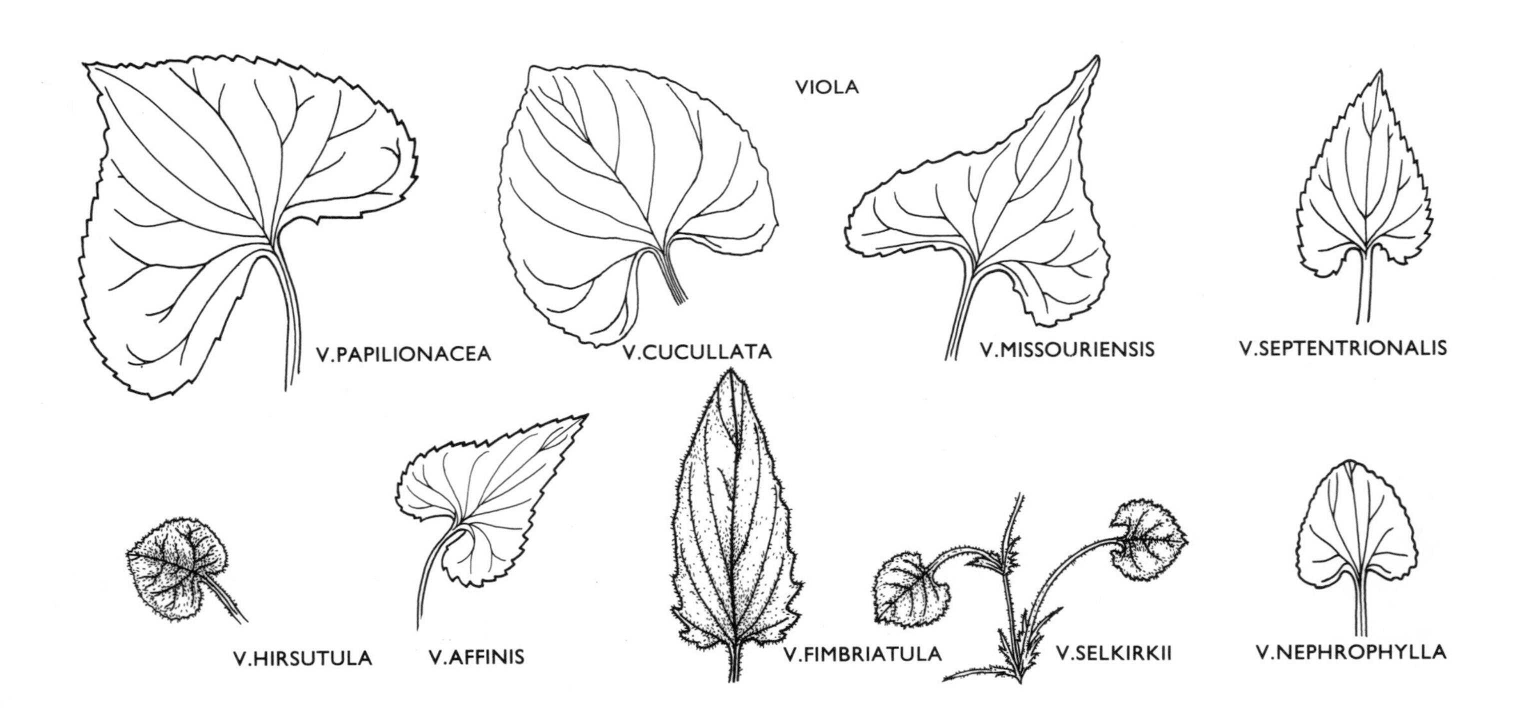

PLATE 83

Viola nephrophylla *Rhein*

Viola pedata *Johnson*

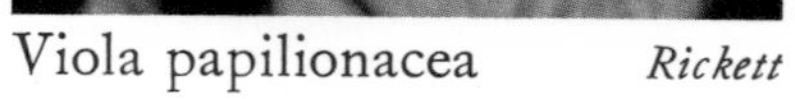
Viola papilionacea *Rickett*

Viola selkirkii *Elbert*

Viola sororia *Donahue*

Viola pedata *Rickett*

Viola hirsutula *Elbert*

Viola fimbriatula *Johnson*

In moist meadows and woodland from Quebec and Vermont to Wisconsin and southward to Georgia, Alabama, and Arkansas.

V. HIRSUTULA is a small plant, with leaf-blades not more than 2 inches wide, the under surface purplish and smooth, the upper silvery with a fine down, and with usually purple veins.

In dry woods from Connecticut to Indiana and southward to Alabama and Georgia. *Plate 83.*

MISSOURI VIOLET, V. MISSOURIENSIS, differs from all the foregoing species in its triangular leaves and pale violet flowers with a white center.

In bottomlands and low woods from Indiana to South Dakota and southward to Kentucky, Louisiana, and Texas. It should be borne in mind that all the region west of the Mississippi was once known as "Missouri territory."

V. FIMBRIATULA is a small plant with lanceolate leaf-blades indented at the base (i.e. somewhat heart-shaped), often with long teeth or even lobes near the base.

In dry woods and on open slopes from Nova Scotia to Minnesota and southward to Florida, Louisiana, and Oklahoma. *Plate 83.*

V. NEPHROPHYLLA is a smooth plant with the lower petal hairy, growing in cold bogs and along streams from Newfoundland to British Columbia and southward to Connecticut, Wisconsin, New Mexico, and California. *Plate 83. V. novae-angliae*, hairy, with narrow, triangular leaf-blades, and lower petal hairy at the base, is found along shores from New Brunswick to Minnesota.

V. SEPTENTRIONALIS is downy, much like *V. papilionacea* but with the lower petal hairy at the base; it grows in open woodland from Newfoundland to British Columbia and southward to Connecticut, the mountains of Virginia, Tennessee, Wisconsin, Nebraska, and Washington.

GREAT-SPURRED VIOLET, V. SELKIRKII, has pale violet, smooth petals, the spur on the lower one comparatively large and blunt; it grows from New Brunswick to Alberta and extends southward to Pennsylvania. *Plate 83. V. latiuscula* is smooth, with leaves tinged with purple on the under surface, and the lowest petal slightly hairy; it is found in dry woodland from Vermont and New York to Virginia.

NORTHERN MARSH VIOLET, V. PALUSTRIS, is a small plant with pale lilac, smooth petals, with darker veins, and short-spurred; it is found in wet places across Canada and extends along mountain ranges into the United States. *Plate 84.*

2. Stemless blue violets with leaf-blades (at flowering time) cleft or lobed or deeply toothed or even palmately divided (compare 3).

BIRD'S-FOOT VIOLET, V. PEDATA, is perhaps the queen of all violets and indeed one of the loveliest of wild flowers. It exists in two forms. In one, locally called velvets, the two upper petals are a deep

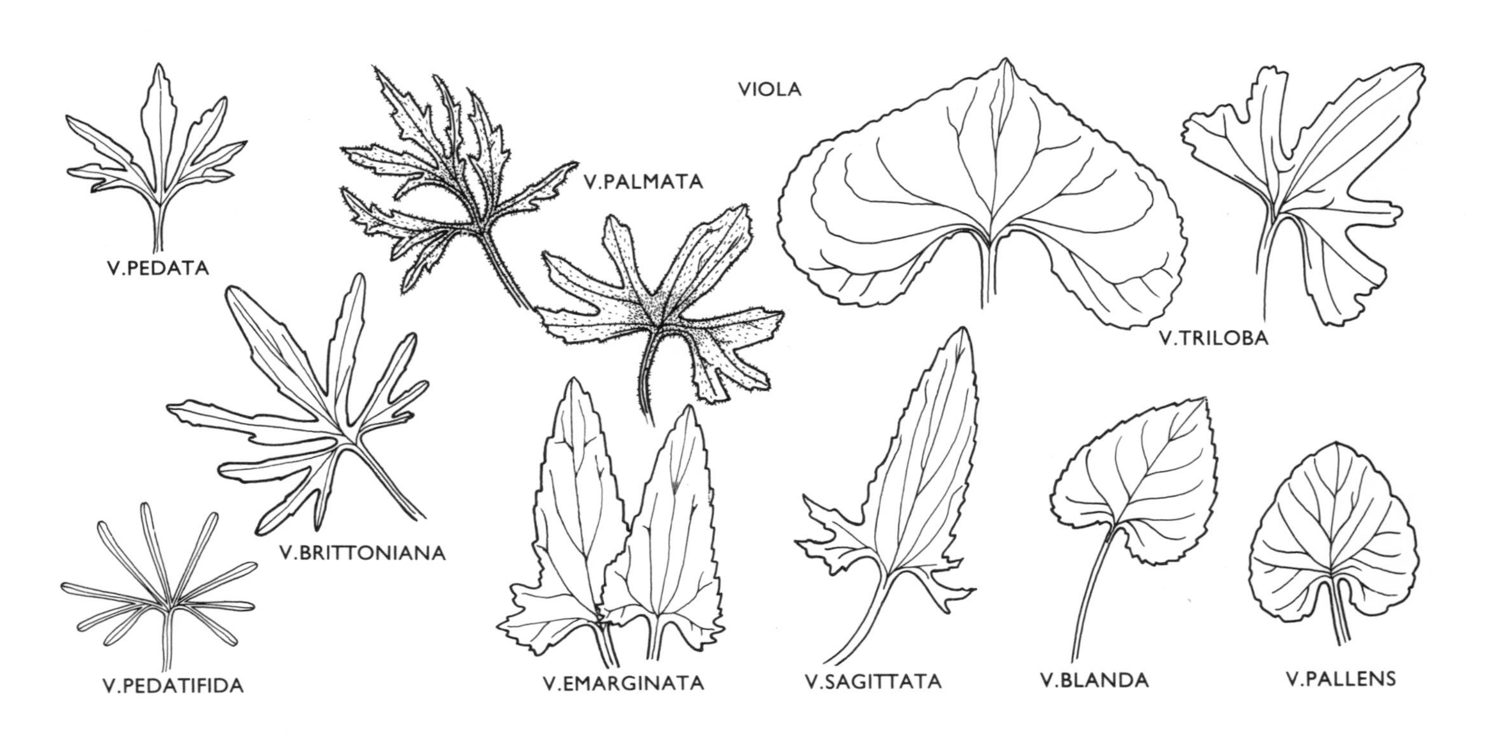

PLATE 84

Viola palustris *Gottscho*

Viola palmata *Johnson*

Viola emarginata *Johnson*

Viola brittoniana *Elbert*

Viola triloba *Elbert*

Viola pedatifida *Johnson*

Viola sagittata *Johnson*

purple, the three lower lilac. In the other all petals are lilac. There are intermediate forms, some with the lower petals blotched with purple. All the upper petals flare backwards, and the orange tips of the stamens are conspicuous in the middle. The leaf-blades are deeply cleft or divided into narrow parts; these are the "bird's feet."

In dry soil, often rocky, in open woodlands or full sun, from Maine to Minnesota and Kansas and southward to Florida and Texas. *Plate 83*. The form with all petals lilac is commoner eastward and northward; in the south central states both forms are equally abundant. Both are variously called wild pansies and (with other species) Johnny-jump-up.

WOOD VIOLET, V. PALMATA, is a hairy plant with leaf-blades cleft or lobed, the lobes numbering from five to eleven, the middle one rather broader than the others. It is almost impossible to distinguish plants of this species with five or seven lobes to a leaf from some forms of *V. triloba;* it is necessary to see the plants before and after flowering. The lowest petal of both species has a tuft of hairs at the base.

In woods from New Hampshire to Minnesota and southward to Florida and Mississippi. *Plate 84*.

V. TRILOBA is also hairy, with leaf-blades at flowering time cleft or lobed into from three to seven parts, the middle one wider than the others, and the basal ones usually somewhat wider than those just above. The very earliest leaves, however, are unlobed, as are those formed after flowering and lasting into summer. In summer, indeed, the plants have been mistaken even by expert botanists for *V. sororia*. The suggestion has been made that these two species, and *V. palmata*, are really all one species.

In woods from Massachusetts and Vermont to Missouri and Oklahoma and southward to Florida and Texas. *Plate 84*.

V. STONEANA, with a more restricted range, differs from the preceding species chiefly by its smoothness and by the insignificant characteristic of having no "beard" on its lowest petal!

In moist woodlands from New Jersey, Pennsylvania, and Kentucky to Maryland and Virginia.

PRAIRIE VIOLET, V. PEDATIFIDA, has leaves deeply cleft into three parts and these again cleft nearly as deeply into narrow lobes which are often lobed at their tips. The flowers are generally pale violet, with hairs ("beards") on all three lower petals.

In prairies and other dry open places from Ohio to Alberta and southward to Missouri, Texas, and Arizona. *Plate 84*.

V. BRITTONIANA is rather similar to *V. pedatifida* but has an entirely different range. Its smoothness distinguishes it from *V. palmata* and *V. triloba*. The flower has a conspicuous white center.

In moist sandy or peaty soil from Maine to North Carolina, mostly near the coast but to the mountains at the south. *Plate 84*.

Several other species with somewhat similar leaves, which barely enter our range, may be mentioned here but not described: *V. viarum* and *V. lovelliana* in Missouri, *V. egglestonii* in Kentucky, *V. septemloba* and *V. villosa* in southeastern Virginia.

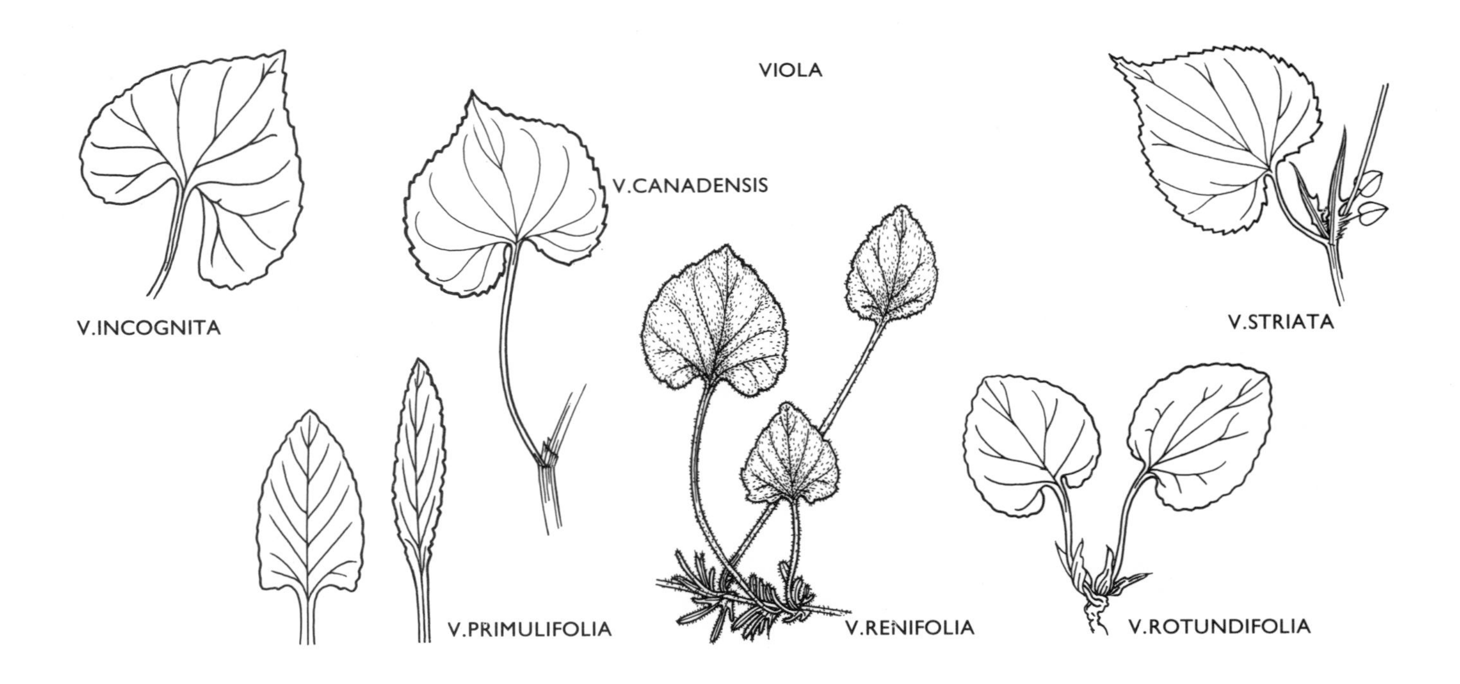

3. Stemless blue violets with leaf-blades more or less lanceolate in shape, often with prominent teeth or lobes near the base. (Some leaves of some plants may lack these projections. *V. fimbriatula* might qualify for a place here; see under 1.)

ARROW-LEAVED VIOLET, V. SAGITTATA, has relatively long leaf-stalks (longer than the blades), and flower-stalks that carry the flowers up as high as or higher than the leaves. The flowers are purple with a white center, and darker veins on the lower petals. Compare the following species.

In meadows and open woodland from Massachusetts to Minnesota and southward to Georgia and Texas. *Plate 84*.

V. EMARGINATA derives its name from the petals, which may be notched ("emarginate"). It is very similar to *V. sagittata*. The leaves are very variable, sometimes deeply lobed at the base, sometimes not lobed or even prominently toothed (in which case it may be mistaken for *V. fimbriatula*).

In woods and meadows from Massachusetts to Kansas and southward to Georgia and Texas. *Plate 84*.

4. Stemless violets always with white flowers. (These may be confused with the numerous white-flowered forms of blue violets. The chief distinguishing feature of this group – except *V. renifolia* – is the presence of runners – long stems creeping over the surface of the soil.) The leaves are various in shape.

SWEET WHITE VIOLET, V. BLANDA, is well named for its fragrant flowers. The plants are often tiny, the flowers not more than $\frac{1}{2}$ inch across, the heart-shaped leaf-blades only about an inch wide at flowering time, on stalks not much longer. The upper petals are narrow and apt to be bent back and twisted. The lower petals have brownish veins. All the petals lack hairs.

In cool, moist woods and on wet slopes from Quebec to Minnesota and southward to Maryland, the mountains of Georgia, Tennessee, and Wisconsin. *Plate 85*.

V. PALLENS is very similar to *V. blanda* but scarcely fragrant, and the upper petals are not twisted. The plant and the flowers are even smaller.

In wet places (sometimes in shallow water) in woods and in the open from Labrador to Alaska and southward to Delaware, North Carolina, Alabama, Iowa, and Montana, chiefly in mountains. *Plate 85*. Authorities differ greatly on the range, doubtless because of the difficulty of separating these species (especially dried specimens!).

V. INCOGNITA, the "unknown" violet, differs from the foregoing two species in being downy; also it is rather larger. The side petals have tufts of hair ("beards").

In woods and on mountain slopes from Labrador to North Dakota and southward to Long Island, Delaware, the mountains of North Carolina, and Tennessee. *Plate 85*.

KIDNEY-LEAVED VIOLET, V. RENIFOLIA, is a northern species usually without runners. The leaves are round or even broader than long, distinctly scalloped,

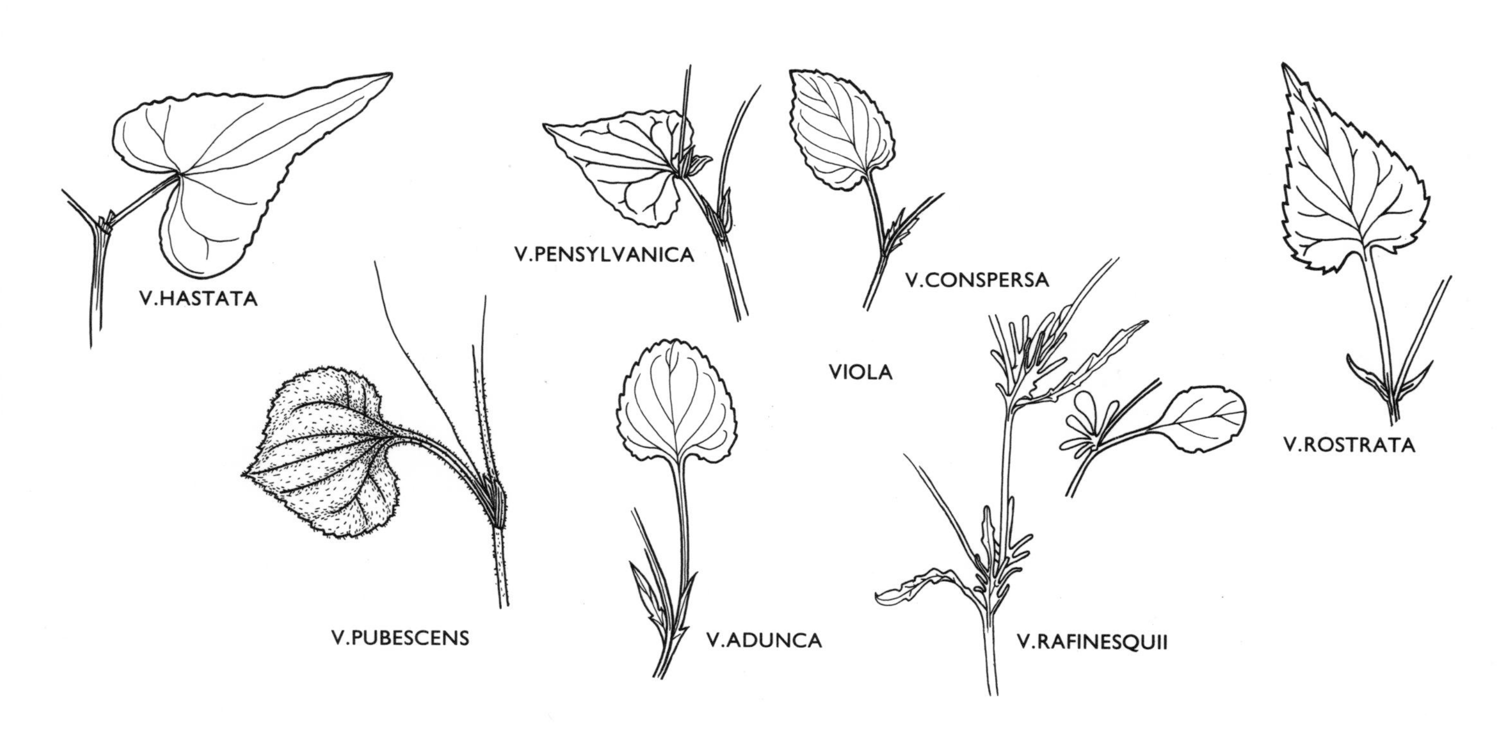

not pointed, often downy with soft white hairs. The petals lack hairs; the three lowest have brown-purple veins.

In wooded swamps and cool woods from Labrador to Alaska and southward to Connecticut, Michigan, South Dakota, Colorado, and British Columbia. *Plate 85*.

PRIMROSE-LEAVED VIOLET V. PRIMULIFOLIA, has oblong or ovate leaf-blades which taper at the base to the broad, thin stalk. The three lower petals are marked with brown-purple veins; those at the side may have a few hairs, or none.

In meadows and moist open woods from New Brunswick to Minnesota and southward to Florida and Texas. *Plate 85*. There is a hairy form.

LANCE-LEAVED VIOLET, V. LANCEOLATA, is known by its characteristic narrow leaves, which may stand 2 inches or more tall. The flower-stalks are usually even longer. The petals all lack "beards"; the three lowest are marked with brown-purple veins. A southern variety (sometimes considered a distinct species, *V. vittata*) has narrower leaves and is more likely to have the upper petals twisted.

In wet meadows, bogs, and shores from Nova Scotia to Minnesota and southward to Florida and Texas. *Plate 85*.

B. Stemless violet with yellow flowers.

ROUND-LEAVED or EARLY YELLOW VIOLET, V. ROTUNDIFOLIA, cannot be mistaken for any other violet. The leaves are not quite "round" (*rotundi-*), but rather ovate and heart-shaped, often bearing minute white hairs. The flowers are bright yellow, the three lowest petals with brown veins, the two at the sides with "beards."

In woods from Quebec to southern Ontario and southward to Delaware, the mountains of Georgia, and Tennessee. *Plate 86*.

II. *Violets with a stem above ground which bears both leaves and flowers, the flowers borne singly in the axils. (Caution: in some of these species young plants may have scarcely any erect stem.)*

A. Some of these species have blue, violet, or white flowers. The first three below have white, or nearly white, flowers.

CANADA VIOLET, V. CANADENSIS, may grow more than a foot tall, with leaves from 2 to 4 inches long. The flowers are on rather short stalks. The side petals have tufts of hairs. The lowest petal is yellow at the base and marked with brown-purple lines.

In woods from Newfoundland to Saskatchewan and southward to Maryland, the mountains of South Carolina, Alabama, Iowa, New Mexico, and Arizona. *Plate 86*.

V. RUGULOSA differs from *V. canadensis* only in certain subterranean details: chiefly the formation of many branching underground stems. The leaves are more wrinkled ("rugulose") and hairy.

In woods from Minnesota to British Columbia and southward to Wisconsin, Iowa, New Mexico, and Arizona; also reported from Virginia, Tennessee, and North Carolina.

PALE or CREAM VIOLET, V. STRIATA, may be recognized by its stipules, which are toothed. The stem is up to a foot tall at flowering time, with leaves from 1 to 3 inches long. The flowers are cream-colored with brown-purple lines; the side petals have tufts of hairs.

In woods from New York and southern Ontario to Minnesota and southward to Georgia and Arkansas. *Plate 86*.

AMERICAN DOG VIOLET, V. CONSPERSA, is a small violet, often with a stem only an inch tall at flowering time, becoming 6 or 8 inches tall later. The stipules are deeply toothed. The small flowers are pale violet; the side petals bear tufts of hairs.

In meadows, bottomlands, borders of woods, etc. from Quebec to Minnesota and southward to Georgia, Alabama, and Missouri. *Plate 86*. A form is known with white petals.

V. ADUNCA resembles *V. conspersa*, but is smaller and tends to spread over the surface. It grows in rocky open places and dry woodlands from Quebec to Alaska and southward to New York, Michigan, South Dakota, Colorado, and California. *Plate 86*. The southern *V. walteri* also resembles *V. conspersa*, but forms runners which develop tufts of leaves at their tips. The leaves are almost round. This extends from Florida to Texas and northward to West Virginia and Ohio.

LONG-SPURRED VIOLET, V. ROSTRATA, resembles *V. conspersa* in general aspect but has a spur – hollow, narrow extension – extending back about ½ inch from the lowest petal and curving upward. The side petals are without hairs.

In woods from Quebec to Wisconsin and southward to New Jersey, the mountains of Georgia, and Alabama. *Plate 86*.

JOHNNY-JUMP-UP, V. RAFINESQUII, is from 2 inches to a foot tall. The leaves are distinctive, having blades that taper gradually into the stalk, and comparatively large stipules cleft into narrow lobes like a cock's-comb. The flowers are small, varying greatly in color from cream-colored to blue-violet and often marked with yellow.

In fields and open woodland from New York to Michigan and Colorado and southward to Georgia and Texas, often very abundant. *Plate 87*. This is often called wild pansy in parts of the country where *V. pedata* is not so named.

PLATE 85

Viola primulifolia *Gottscho*

Viola blanda *Johnson*

Viola pallens *Johnson*

Viola incognita *Elbert*

Viola renifolia *Johnson*

Viola lanceolata *Rickett*

V. TRICOLOR and V. ARVENSIS are two European species somewhat resembling *V. rafinesquii* but with scalloped leaf-blades. They interbreed in their native lands, yielding an almost infinite number of combinations. In this country they appear in cultivated ground. In general *V. arvensis* has yellow petals more or less marked with lavender, while *V. tricolor* often has the two upper petals purple and the others variously colored; it is one of the ancestors of the pansies. *Plate 87.*

B. Species with leafy erect stems that bear yellow flowers. (See also *V. tricolor* and *V. arvensis* above.)

SMOOTH YELLOW VIOLET, V. PENSYLVANICA, has from one to three long-stalked leaves at the base with broad, heart-shaped blades. The petals are bright yellow with purplish veins. The side veins bear tufts of hairs. One variety has pods covered with white wool.

In moist woods and bottomlands from Quebec to Manitoba and southward to Maryland, the mountains of Georgia, Tennessee, and Texas. *Plate 87.*

DOWNY YELLOW VIOLET, V. PUBESCENS, differs from the preceding species in having softly hairy stems and not more than one long-stalked leaf at the base – usually none.

In woods from Maine to North Dakota and southward to Georgia, Mississippi, Oklahoma, and Nebraska. *Plate 87.*

HALBERD-LEAVED YELLOW VIOLET, V. HASTATA, has a smooth stem up to a foot tall. There are from two to four leaves at the summit of the stem, their blades ovate or triangular with projecting lobes. They are apt to be marked with silvery blotches. The stipules may be toothed.

In woods from Pennsylvania and Ohio southward to Florida and Alabama. *Plate 87. V. tripartita* is similar in aspect to *V. hastata* but has some leaves deeply cleft into three narrow lobes. It grows from Florida and Alabama northward to Ohio. The prairie yellow violet, *V. nuttallii*, with narrow leaf-blades which taper to long stalks, is found from Minnesota to Missouri and westward.

HYBANTHUS

There is one species of *Hybanthus* in our range.

GREEN-VIOLET, H. CONCOLOR, is a plant from 1 to 3 feet tall, with ovate or elliptic leaves. The small green flowers hang on drooping stalks singly or in small clusters from the axils. The petals and sepals are alike in color and about equal in length, but the lowest petal is sac-like, like that of a violet. The pistil is much like that of a violet, and the stamens surround the ovary in much the same way.

April to June: in woods and bottomlands from New York and southern Ontario to Michigan and Kansas and southward to Georgia, Mississippi, and Arkansas. *Plate 87.*

PLATE 86

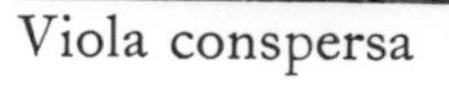
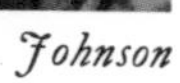
Viola conspersa *Johnson*

Viola rotundifolia *Johnson*

Viola adunca *Johnson*

Viola rostrata *Elbert*

Viola conspersa *Rickett*

Viola canadensis *Johnson*

Viola striata *Johnson*

GROUP X

SEPALS four or five, petals four or five; petals joined, radially symmetric. Stamens as many as the petals or twice as many. Leaves all crowded, small and narrow, or all at the base of the stem, or in pairs or circles, generally unlobed and undivided.

Exceptions: A species of *Sabatia* has more than five petals. In *Limonium* the petals may be only slightly joined or even not at all. The leaves are borne singly in two genera of the primrose family, in one genus of the dogbane family, in some species of milkweeds, in three small genera of the gentian family, and in two genera of the phlox family; they are divided in one genus of the gentian family and in one of the phlox family.

Triosteum in the honeysuckle family (XII) may seem to belong here because of its five stamens.

I. *Flowers with twice as many stamens as petals:* heath family.

II. *Plants with matted stems covered with crowded small leaves:* pyxie family.

III. *Plants with all leaves at the base of an erect flowering stem:* leadwort family; *one genus in the* pyxie family; *some genera of the* primrose family.

IV. *Plants with leaves in pairs or circles on a generally erect stem (creeping in some species). In this group the head of a stamen either (A) stands opposite the middle of a petal; or (B) coincides with the junction of two adjacent petals.**

A. Stamens opposite the middle of the petals: primrose family.

B. Stamens opposite the line of junction of adjacent petals.

1. Flowers on one side of the stem, bright red and yellow: logania family.
2. Plants with colorless watery juice; style tipped with a single stigma or cleft into two at the summit: gentian family.
3. Plants with milky juice; two ovaries with one style and one stigma: dogbane family.
4. Plants with milky juice (except one species): two ovaries with a style apiece but only one stigma which is joined to the tips of the stamens: milkweed family.
5. Plants with colorless juice: style with three branches at the tip: phlox family.

THE HEATH FAMILY (ERICACEAE)

The heaths are all trees and shrubs, and must therefore, in spite of the beauty and interest of many of them, be regretfully left for some other author to describe. However, a few are such low shrubs as to merit treatment here. These are all characterized by mostly five joined petals and five or ten stamens. The stamens discharge their pollen through tubes or pores as in the shinleaf family, not by splitting lengthwise. The leaves are mostly evergreen (lasting through winter).

*When the stamens are hidden inside the tube of the corolla, one must, as it were, prolong them to see whether they end opposite the middle of a corolla-lobe or in the gap between two adjacent lobes.

PLATE 87

Viola rafinesquii *D. Richards*

Viola pensylvanica *Johnson*

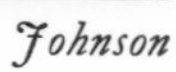

Viola arvensis *Elbert*

Viola tricolor *Rickett*

Viola hastata *Elbert*

Hybanthus concolor *Johnson*

Viola pubescens *Johnson*

EPIGAEA

There are only two species of *Epigaea*, and only one in the New World.

TRAILING ARBUTUS or MAYFLOWER, E. REPENS, is a creeping plant with rather leathery and veiny leaves (borne singly). In the axils of the leaves and at the tips of branches are clusters of small, waxy, pink or white flowers known for their exquisite fragrance.

March to May: on sandy or peaty woodland banks from Labrador to Saskatchewan and southward to Florida, Mississippi, and Iowa. *Plate 88*. These plants are so attractive that the species has been exterminated in many places; it is extremely hard to transplant and cultivate successfully. In many states it is protected by law, and should be respected by all lovers of nature. It is said that this mayflower was abundant around Plymouth, Massachusetts, and was the first flower to assure the pilgrim fathers, after their first terrible winter, that spring was really on the way. The stamens and pistil are often in separate flowers.

ARCTOSTAPHYLOS

This is a big genus in the West, with many species known as manzanita. In the northeastern United States there is only one species at all widespread, with an arctic species reaching the high mountains of New England.

BEARBERRY or KINNIKINICK, A. UVA-URSI, has woody stems that lie on the ground, forming mats. They bear small, thick leaves with round ends, attached singly. The petals form a hanging, narrow-mouthed bell, white, pink, or pink-tipped. The flowers are followed by red berries, tasteless but presumably relished by bears. (*Arcto-* is Greek for "bear," *ursus* is the Latin for the same animal. *Staphylos* is, in Greek, a "cluster" of berries; *uva* is Latin for "grape." So the botanical name means "bear-cluster, the grape of a bear"!)

May to July: in sandy and rocky woods and open places from Labrador to Alaska and southward to Virginia, northern Indiana and Illinois, Minnesota, New Mexico, and California. *Plate 88*.

Kinnikinick is an Indian word used for any substitute for tobacco. Bearberry leaves were so used.

ALPINE BEARBERRY, A. ALPINA, is similar, with narrower leaves and black berries. It occurs throughout Arctic Canada and extends south to the high mountains of Maine and New Hampshire.

GAULTHERIA

Of this large genus two species are known in the northeastern United States, one with five petals and stamens, the other with four petals and eight stamens. Both have creeping woody stems and small leathery leaves, with white, bell-shaped flowers in the axils. The first species, below, has more or less erect flowering stems.

WINTERGREEN or CHECKERBERRY, G. PROCUMBENS, has leaves and berries with the familiar odor and taste; this is the true wintergreen (though many other species are green in winter). It has a corolla of five petals, joined to make the hanging bell, and five stamens within the bell. The fruit is red. It is not a true berry, but is formed from the calyx, the ovary becoming a sort of pod within. It matures during the winter, being more succulent in spring.

July and August: in woods from Newfoundland to Manitoba and southward to Georgia, Alabama, and Minnesota. *Plate 88*. The leaves have often been used as a substitute for tea.

CREEPING SNOWBERRY, G. HISPIDULA, forms mats of delicate stems with leaves less than ½ inch long. The corolla is formed of four petals and encloses eight stamens. The berries are white, formed as in *G. procumbens*, and having something of the same taste. They are said to be delicious with sugar and cream; and an excellent preserve is made from them.

April and May (to August in the mountains): in mossy, mostly evergreen woods from Labrador to British Columbia and southward to New Jersey, the mountains of North Carolina, Michigan, Wisconsin, and Minnesota.

Gaultheria procumbens *D. Richards*

Epigaea repens *V. Richard*

Kalmia polifolia *Scribner*

Gaultheria procumbens *Rhein*

Calluna vulgaris *Gottscho*

Arctostaphylos uva-ursi *V. Richard*

CALLUNA

The single species of *Calluna* is native in Europe and is found established in a few places in North America.

HEATHER, C. VULGARIS, is a small shrub usually a foot or two tall, the branching stems covered with many small leaves (scarcely more than $\frac{1}{8}$ inch long) in pairs. There are four petals which form a tube with four teeth.

July to September: in sand near the coast from Newfoundland to New Jersey; inland to Michigan and West Virginia. *Plate 88*. The name *Calluna* is from the Greek for "broom," the wiry, much-branched stems having been used in Europe to make brooms.

KALMIA

The genus *Kalmia* includes the mountain-laurel (state flower of Connecticut), sheep-laurel, and swamp-laurel. Only the last is here considered, but the flowers of all three are similar. These are not the classical laurels with which Greek heroes were crowned; those were *Laurus* in the laurel family. Some shrubs in the rose family are also known in England as laurels.

SWAMP-LAUREL, K. POLIFOLIA, is a little shrub not exceeding 3 feet in height, with narrow, mostly paired leaves. The flowers are in a cluster at the tip of the stem. Each has a pink or crimson, saucer-shaped corolla of five joined petals with ten small pouches in which the heads of the ten stamens are caught. At a touch these are released and fly up, scattering their pollen over the intruding object.

May to July: in peat and bogs from Labrador to Alaska and southward to New Jersey, Pennsylvania, Michigan, Minnesota, Idaho, and Oregon. *Plate 88*. The leaves of *Kalmia* are poisonous – another species being known as lambkill. *K. polifolia*, however, seldom grows where sheep graze, and the wild animals seem to know about it.

THE DIAPENSIA FAMILY (DIAPENSIACEAE)

This is a very small but distinct family of northern latitudes. The plants are low, with crowded, evergreen, undivided and unlobed leaves. The parts of the flower are in fives, but the pistil bears a style with a three-lobed stigma.

This is a small genus of a few arctic and alpine species; one of these extends southward into the United States.

DIAPENSIA

D. LAPPONICA forms low mats of woody stems (it is, therefore, technically a shrub). The leaves are flat, blunt, a little wider at the tip than at the base, only $\frac{1}{2}$ inch long or less. The pretty white flowers grow singly on leafless stalks that emerge from the tips of the branches. The corolla is an erect bell with a margin of five round lobes.

June and July: on rocks in arctic regions around the world, and southward in our range to the high mountains of New England and New York. *Plate 89*.

PYXIDANTHERA

We have one species of this genus.

PYXIE or FLOWERING-MOSS, P. BARBULATA, has branched, creeping, woody stems covered with sharp-pointed leaves not more than $\frac{1}{3}$ inch long. In season these mats are covered with white or pink, stalkless flowers each about $\frac{1}{4}$ inch across. The corolla is an upright bell with five spreading lobes (the lobes nearly triangular with the points inward).

March to May: in sandy pine-barrens from New Jersey to South Carolina. *Plate 89*. The name does not refer to the fairies (pixies) but to the stamens.

PLATE 89

Allen

Limonium carolinianum

Limonium carolinianum *Uttal*

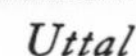

Dodecatheon amethystinum *Rhein*

Pyxidanthera barbulata *Elbert*

Dodecatheon meadia *Johnson*

Galax rotundifolia *Scribner*

Diapensia lapponica *Johnson*

GALAX

There is only one species, mainly southern.

G. ROTUNDIFOLIA sends up, from a mat of underground stems, long-stalked leaves in a cluster, each with a heart-shaped, evergreen blade from 2 to 6 inches across. From the midst of the cluster of leaves rises a leafless stem up to 2 feet tall or more bearing numerous small white flowers along its sides. The corolla – less than $\frac{1}{4}$ inch long – is divided nearly to the base into five lobes.

May to July: in open woods, chiefly in the mountains, from Virginia to Kentucky and southward to Georgia and Alabama. *Plate 89*. The plant has long been known as *G. aphylla*, which means "leafless" – referring to the leafless flowering stem. It has recently been shown that Linnaeus was confusing two species, the *other* one being "aphylla."

THE LEADWORT FAMILY (PLUMBAGINACEAE)

The leadworts and their numerous relatives are plants chiefly of alkaline deserts or salt marshes. *Plumbago* itself (*plumbus* means "lead") is a genus much cultivated in the warmer parts of the world. The family is represented in our range by only one small genus.

SEA-LAVENDERS (LIMONIUM)

The sea-lavenders are plants of salt marshes along the coast. They have a cluster of thick leaves at the base of the stem. The numerous small lavender flowers are borne along the numerous branches of the stem. The five petals are nearly (or quite) separate. There are five styles on the single ovary. The small clusters of flowers are enveloped by bracts. We have two species, differing only by the hairiness of the calyx, as far as easily visible characteristics are concerned. Both flower from July to October. They are not related to lavender (*Lavandula*) in the mint family.

L. NASHII grows from Quebec to Florida and northeastern Mexico. This has hairs on the calyx, especially on the ribs.

L. CAROLINIANUM does not extend northward of New York. The calyx is quite without hairs. *Plate 89*.

THE PRIMROSE FAMILY (PRIMULACEAE)

There are few true primroses (genus *Primula*) in the United States, but many species in the family. They hang together by several technical characteristics, of which one is fairly easily seen with a hand magnifier. In most families whose petals and stamens are of the same number the stamens stand opposite the gaps between adjacent petals; or, if the petals are joined to make a cup or tube with teeth or lobes at the margin, a line prolonging the stamen will run to the notch between adjacent teeth or lobes. In the primrose family and a few others, such a line will run along the middle of a tooth or lobe to its apex; each stamen is opposite the center of a petal.

Almost all our species have unlobed and undivided leaves, all at the base of the stem or in pairs or circles on the stem. There are generally five joined petals (in some genera *nearly* separate) and five stamens. The pistil has a single style. A further characteristic, difficult to see, is that the ovules (rudimentary seeds) are borne on a central column that rises from the base of the ovary and is not attached at the tip. Either the stamens or the style (not both) projects from the corolla.

Guide to the Genera of Primulaceae

I. *In several genera the leaves are all at the base of the stem:* Dodecatheon (*petals joined only at the base, bent back sharply*); Primula (*petals forming a distinct tube with flaring lobes*); Androsace (*petals shorter than sepals*); Hottonia (*aquatic; leaves pinnately divided, stem very thick*).

II. *In other genera the leaves are borne in pairs or circles on the stem (perhaps with some at the base also):* Trientalis (*leaves in one circle*); Lysimachia *and* Steironema (*leaves in several pairs or circles; petals yellow, joined only at the base*); Naumburgia (*flowers yellow, in dense clusters on stalks from the axils*); Anagallis (*leaves in pairs; petals not yellow*); Glaux (*succulent; flowers almost without stalks in the axils; petals lacking*).

III. *In a few genera the leaves are borne singly, at least in part:* Centunculus (*small flowers in axils almost without stalks; their parts in fours*); Samolus (*small flowers in racemes*).

SHOOTING-STARS (DODECATHEON)

For the characteristics of this genus see under the first species below.

SHOOTING-STAR, D. MEADIA, has a tuft of smooth, narrow leaves, commonly reddish at the base, from among which rises a leafless flowering stem from 6 to 20 inches tall. At the summit of this stem is a cluster (an umbel) of flowers; the buds stand erect but as the flowers open they hang down on curved stalks. The five lilac or crimson petals bend sharply back, exposing a sort of beak – the "star" – made of the tips of the stamens.

April to June: in open woods, on cliff edges, and in meadows from Pennsylvania to Wisconsin, and southward to Georgia and Texas. This is sometimes called American cowslip. The English cowslip is in this family, being a species of *Primula*, but bears no resemblance to this genus. *Plates 89, 90.*

JEWEL SHOOTING-STAR, D. AMETHYSTINUM, is a smaller plant, not more than a foot tall, with smaller, deep crimson flowers.

April to June: on damp, shaded cliffs and riverbanks from Pennsylvania to Kentucky, Wisconsin, Minnesota, and northern Missouri; a closely related species occurs in the Rocky Mountains. *Plate 89.*

PRIMROSES (PRIMULA)

The true primroses grow mostly in Asia, but some are found around the world in the higher latitudes. Many are known in cultivation. We have only two species in the Northeast, and those mostly along our northern borders. The flowers cannot be mistaken for those of any other genus. They grow in a tight umbel at the tip; the petals form a tube which projects far beyond the calyx, flaring into five notched lobes. The stamens are attached to the inside of the corolla-tube.

BIRD'S-EYE PRIMROSE, P. MISTASSINICA, grows at most 8 or 10 inches tall. The leaves are from 1 to 3 inches long, widest near the tip, rather blunt, with or without minute teeth on the edges. The flowers vary greatly in color, from white to pink, lilac, or purplish, with a yellow "eye" in the center which gives the plant its English name. The botanical name is from Lake Mistassini in central Quebec.

May to August: on rock ledges and in meadows from Labrador westward, extending southward to Maine, New York, Michigan, northern Illinois, and Iowa. *Plate 90.*

Another northern primrose, *P. laurentiana*, may be found in Maine. It is a stouter plant, up to 18 inches tall, but with the corolla up to ½ inch across (as against nearly an inch in *P. mistassinica*), and with a mealy under surface to the leaves. It grows in eastern Canada. Plants with a mealy under surface to the leaves, but with the dimensions of *P. mistassinica*, are found in northern Michigan and Minnesota; they have been called *P. intercedens.*

ANDROSACE

We have one species of this genus. Many others occur in the West and in Asia.

A. OCCIDENTALIS is a little plant, scarcely 3 inches tall at most. The basal leaves form a rosette less than 2 inches across. The tiny flowers are in an umbel, the white petals shorter than the sepals.

March to May: in sandy soil and on dry rock ledges from Ontario to British Columbia and southward to Indiana, Arkansas, Texas, and Arizona.

HOTTONIA

One species of *Hottonia* occurs in America, another in the Old World.

FEATHERFOIL or WATER-VIOLET, H. INFLATA, is an extraordinary aquatic plant with thick, hollow ("inflated") stems and leaves all divided pinnately into numerous very narrow segments. The flowers grow at the joints of the stem, in circles. The white corolla is shorter than the calyx, which is less than ½ inch long.

April to August: in shallow pools and ditches from Maine to Florida and Texas and northward to Ohio, Indiana, Illinois, and Missouri. *Plate 90*.

TRIENTALIS

We have one species of this genus.

STAR-FLOWER, T. BOREALIS, has a stem from 2 to 10 inches tall, bearing a circle of lanceolate leaves at its summit, and, above these, one or several white flowers on delicate stalks. The petals number from five to nine but are most often seven; they are white, narrow, and sharp-pointed.

May to August (at high altitudes): in woodlands and on mountain slopes from Labrador to Saskatchewan and southward to Virginia, West Virginia, Indiana, and Minnesota. The fanciful name chickweed-wintergreen is also given – the plant being neither a chickweed nor a wintergreen, nor resembling either. *Plate 90*.

THE LOOSESTRIFES (LYSIMACHIA AND STEIRONEMA)

These two genera are treated together because they are hard to tell apart. Indeed some authors unite them into one, under the first name. There *is* a difference, however, which may be seen with a hand magnifier. In *Lysimachia* the stalks of the stamens are joined at the base to make a sleeve around the ovary. In *Steironema* the stamens are quite separate, and in the bud each lobe of the corolla is wrapped around a stamen. In both genera the petals are yellow and joined only at the base. The leaves are in pairs or circles.

The name loosestrife is also, unfortunately, used for species of *Lythrum*. For the origin of the name see under that genus.

I. *Species with flowers in a cluster longer than wide (a raceme) at the summit of the stem.*

GARDEN LOOSESTRIFE, L. VULGARIS, has flowers in a branched raceme. The petals spread to a width of an inch. The stem, which is slightly sticky-hairy, grows up to 4 feet tall. The leaves are in pairs or circles.

June to September: an Old-World plant, escaped from cultivation on roadsides and in wet places from Quebec to Ontario and southward to Maryland, Ohio, and Illinois. *Plate 90*.

SWAMP-CANDLES, L. TERRESTRIS, has a tall, narrow inflorescence of small flowers about ½ inch across. The stem may reach a height of 3 feet, bearing paired leaves. The petals may have dark dots or streaks.

June to August: in wet meadows and swamps from Newfoundland to Manitoba and southward to Georgia, Kentucky, and Iowa. *Plate 91*.

II. *Species with flowers singly or in small clusters in the axils of leaves or at the tips of axillary branches – not in racemes.*

A. Of these, several species have flowers on leafless stalks arising directly from the axils of leaves (compare B).

MONEYWORT or CREEPING-JENNY, L. NUMMULARIA, is a creeping plant with almost round leaves (the "money") in pairs. The yellow petals are marked with dark dots.

PLATE 90

Primula mistassinica *Johnson*

Dodecatheon meadia *Johnson*

Lysimachia vulgaris *Mayer*

Trientalis borealis *Rickett*

Lysimachia nummularia *D. Richards*

Hottonia inflata *Phelps*

June to August: an European plant now naturalized in damp places and in gardens from Newfoundland to Ontario and beyond and southward to Georgia and Kansas; also on the Pacific Coast. *Plate 90.*

WHORLED LOOSESTRIFE, L. QUADRIFOLIA, has leaves in circles of from three to six (or sometimes of other numbers) on a stem from 1 to 3 feet tall. The flowers are on long and delicate stems growing from the axils, usually as many as there are leaves at any one level. The petals are marked with dark dots or streaks.

May to August: in woods and open land from Maine to Wisconsin and southward to Georgia, Alabama, and Tennessee. *Plate 91.* This species crosses with *L. terrestris.*

GARDEN LOOSESTRIFE, L. PUNCTATA, has leaves chiefly in threes and fours, and flowers more numerous than the leaves at any level. The upper part of the stem, which may be 3 feet tall, may approach a raceme. The corolla is ½ inch or more across.

June to September: an European plant now established in waste ground from Newfoundland to Illinois and southward to New Jersey and Pennsylvania. *Plate 91.*

B. In other species with axillary flowers, the flowers are clustered at the ends of branches which usually bear leaves also. The first three of these species have ovate or lanceolate leaf-blades on distinct stalks.

FRINGED LOOSESTRIFE, S. CILIATUM, is usually much branched and from 1 to 4 feet tall. The leaf-stalks are fairly long and fringed with large hairs. The flowers are from ½ inch to over an inch across, with broad petals that are toothed and pointed on their outer margins. The flowers characteristically face outwards and even downwards on the end of their long stalks.

June to August: in wet meadows and thickets from Quebec to British Columbia and southward to Florida, Texas, and Arizona. *Plate 91.*

S. TONSUM is a southern species which closely resembles *S. ciliatum* but lacks the fringed leaf-stalks. In our range it is found in dry woods in Virginia and Kentucky.

S. RADICANS gets its name from its habit of lying down and forming roots (*radicans* means "rooting") at the points on the stem where the leaves are attached. The stem is long (up to 3 feet) and slender. The flowers are about ½ inch across. The petals are toothed and pointed on their outer edges.

June to August: in wet woods and swamps in Virginia and Missouri, and in the South from Tennessee and Mississippi to Texas.

S. QUADRIFLORUM (not to be confused with *L. quadrifolia;* neither name can be literally applied) has a stem from 1 to 3 feet tall bearing very narrow leaves without stalks. Bunches of smaller leaves may be seen in the lower axils, the flowering branches springing mostly from the upper axils. The flowers are nearly an inch across, with pointed petals.

July and August: in bogs and wet open places from western New York and southern Ontario to Manitoba and southward to Virginia, Kentucky, and Missouri. *Plate 91.*

S. LANCEOLATUM reaches a height of 2 feet or more; the stem forms long runners at the base, by which the species spreads. The leaves are lanceolate or narrow with parallel sides, and have no stalks. The flowering branches, in the upper axils, are short. The flowers are about ¾ inch across; the petals are toothed and pointed at their outer edges.

June to August: in open woodland from Pennsylvania to Wisconsin and southward to Florida and Louisiana. *Plate 91.*

S. HYBRIDUM resembles *S. lanceolatum* but the leaf-blades taper downward into short stalks.

July and August: in swamps, wet prairies, and on shores from Quebec to Ontario and North Dakota and southward to Florida, Mississippi, and Texas; and perhaps farther west. *Plate 91.*

NAUMBURGIA

There is only one species of this genus – by some botanists placed in *Lysimachia.*

TUFTED-LOOSESTRIFE, N. THYRSIFLORA, is from 8 to 30 inches tall, with leaves in pairs. The leaves are narrowly lanceolate or elliptic, without stalks. The flowers are in dense tufts in axils in the middle part of the stem. The yellow petals are very narrow and separate nearly to their base; they vary in number but are commonly six or seven. The stamens are of the same number.

May to July: in swamps and bogs from Quebec to Alaska and southward to New Jersey, Indiana, Missouri, Colorado, and California. *Plate 92.*

PLATE 91

Lysimachia quadrifolia *Gottscho*

Lysimachia terrestris *Johnson*

Steironema hybridum *V. Richard*

Steironema quadriflorum *D. Richards*

Lysimachia punctata *Scribner*

Steironema ciliatum *Rickett*

Steironema lanceolatum *Becker*

ANAGALLIS

We have only one species of this Old-World genus, a well-established immigrant.

PIMPERNEL, A. ARVENSIS is a small, branched plant, often spreading horizontally, with pairs of small leaves and single flowers in their axils. The flowers are commonly scarlet, and the species is often known as scarlet pimpernel; but races are quite common with blue or white petals. The name pimpernel is a corruption of a Latin word meaning "featherlike," or, botanically, pinnately divided; it has been applied to a number of species (e.g. *Hottonia inflata*, *Samolus parviflorus*, *Taenidia integerrima*, *Lindernia dubia*), with some prefixed word such as "yellow" or "false", and apparently without reference to divided leaves.

May to August: on roadsides and in gardens and waste places, practically throughout North America. *Plate 92*. In England this little plant is known as poor-man's-weather-glass, from its habit of closing its flowers at the approach of bad weather. As might be expected, it is a magical plant. In Ireland one only has to hold it to understand the language of birds. It had also various reputed medicinal properties, being a cure for melancholy among other afflictions. Actually the leaves are known to cause a rash on the skin of susceptible persons.

GLAUX

There is but one species of *Glaux*.

SEA-MILKWORT, G. MARITIMA, is a small succulent plant of salty places around the world. The small, narrow leaves are paired. The minute flowers are borne singly in the axils, almost without stalks. There are no petals, but the sepals are variously colored from white to crimson. The stamens, which in this family stand opposite the middle of the petals, in this species are opposite the gaps between adjacent sepals.

June and July: on seashores and on the borders of salt marshes, and in alkaline soil from Quebec to Virginia, and from Saskatchewan to British Columbia and southward to New Mexico and California; also in the Old World. The name milkwort is more properly applied to *Polygala*.

CENTUNCULUS

This genus has only one species.

CHAFFWEED, C. MINIMUS, is an insignificant plant rarely exceeding 4 inches in height. The small oblong leaves, which have no stalks, are mostly borne singly on the branching stems. The minute flowers are in their axils, practically without stalks. The petals are pink; they mostly number four.

April to September: in damp soil in scattered places in Nova Scotia and Delaware and from Ohio to Minnesota and British Columbia; throughout the South and in Mexico; and in the Old World.

SAMOLUS

One species of *Samolus* occurs in the northeastern United States.

WATER PIMPERNEL, S. PARVIFLORUS, grows up to 2 feet tall. The leaves, which are borne singly, have blades widest towards their tips, tapering downward into stalks. The small flowers are on leafless stems (racemes) at the summits of the branches; they are only about $\frac{1}{8}$ inch across. The petals are white.

May to September: in shallow water, mud, and wet sand practically throughout the United States. *Plate 92*.

THE LOGANIA FAMILY (LOGANIACEAE)

The *Loganiaceae* are a tropical family, with a few species in the temperate parts of North America. Their best-known product is strychnine, from a South American genus; other genera also contain poisons. The shrub *Buddleia* is widely cultivated in gardens.

PLATE 92

Gentiana amarella *Scribner*

Spigelia marilandica *Johnson*

Naumburgia thyrsiflora *Houseknecht*

Anagallis arvensis *D. Richards*

Samolus parviflorus *D. Richards*

Gentiana procera *Johnson*

Gentiana crinita *Gottscho*

SPIGELIA

One species of *Spigelia* extends from the South into the southern states of our range.

PINKROOT, WORM-GRASS, or INDIAN PINK, S. MARILANDICA, has striking trumpet-shaped flowers, scarlet outside and yellow inside, up to 2 inches long. The trumpet flares into five short teeth which display the yellow inside. The flowers are borne close together on one side of the stem at and near its tip, the upper flowers being the youngest.

May and June: in moist woods from Maryland to Indiana, Missouri, and Oklahoma and southward to Florida and Texas. *Plate 92*. This flower suggests the tropics by its very colors. Like others in the family, the plant contains a poisonous alkaloid, which has been used in medicine for destroying parasitic worms.

Other southern species that enter our range are mitrewort, *Cynoctonum mitreola*, with minute white flowers in a branched inflorescence (southeastern Virginia); yellow jessamine, *Gelsemium sempervirens*, a twining woody vine with fragrant yellow flowers (southeastern Virginia); and *Polypremum procumbens*, a foot-high plant with narrow leaves and many small white flowers in a branching inflorescence (Long Island, New Jersey, Pennsylvania, Kentucky, and southeastern Missouri).

THE GENTIAN FAMILY (GENTIANACEAE)

The gentian family is noted for handsome, brightly colored flowers (but includes some species with small, inconspicuous flowers). Our species mostly have a corolla of four or five joined petals. Most of them also have leaves in pairs or circles; the leaves lack stalks and except in one genus are undivided and unlobed.* The family is characterized by an ovary (and capsule) with one chamber containing numerous ovules (which become seeds).

I. *Genera with petals that form a distinct tube or funnel, in some species flaring at the end into four or five lobes, in others merely with four or five teeth, or even closed at the tip; leaves paired, unlobed and undivided:* Gentiana (*comparatively large flowers, mostly blue, with a tubular calyx*); Centaurium (*plants small with pink or white flowers less than ½ inch across the teeth of the corolla*); Halenia (*four purplish-green petals prolonged at the base into narrow hollow sacs – "spurs"*); Obolaria (*flowers small with four white petals and two sepals*).

II. *Genera with four, five, or more petals that at first glance may seem separate, being joined only at the base; leaves in pairs or circles except in one species and undivided and unlobed:* Sabatia (*handsome flowers with comparatively broad rose, lilac, or white petals, four or five in most species, up to twelve in two*); Swertia (*four greenish-yellow petals with a spot – gland – in the middle of each surrounded by a long erect fringe*); Lomatogonium (*five blue petals, the flowers an inch across or less*); Bartonia (*four yellowish, greenish, or purple petals; minute scales instead of leaves, these borne singly in one species, paired in the other*).

III. *A genus of aquatic plants with round, floating leaves and white or yellow flowers scarcely more than an inch across:* Nymphoides.

IV. *A genus (one species) with leaf-blade divided into three segments on a long stalk; flowers an inch or more across, with long hairs on the surface of the five petals:* Menyanthes.

THE GENTIANS (GENTIANA)

The gentians are among our most admired wild flowers. They are most abundant in mountains all around the world, dwarf species being found at high, alpine altitudes; but tall species grow also at sea level and on valley floors. The flowers are commonly blue or violet, but there are species with green, bronze, and white corollas. The leaves are paired and lack stalks. The underground parts of certain species have yielded medicines. Their value is said to have been discovered by King Gentius of Illyria.

The numerous species (some of which are hard to characterize without attention to technical details) may first be clearly separated into two groups by characteristics of corolla and calyx.

*The exceptional genera have sometimes been treated as a distinct family.

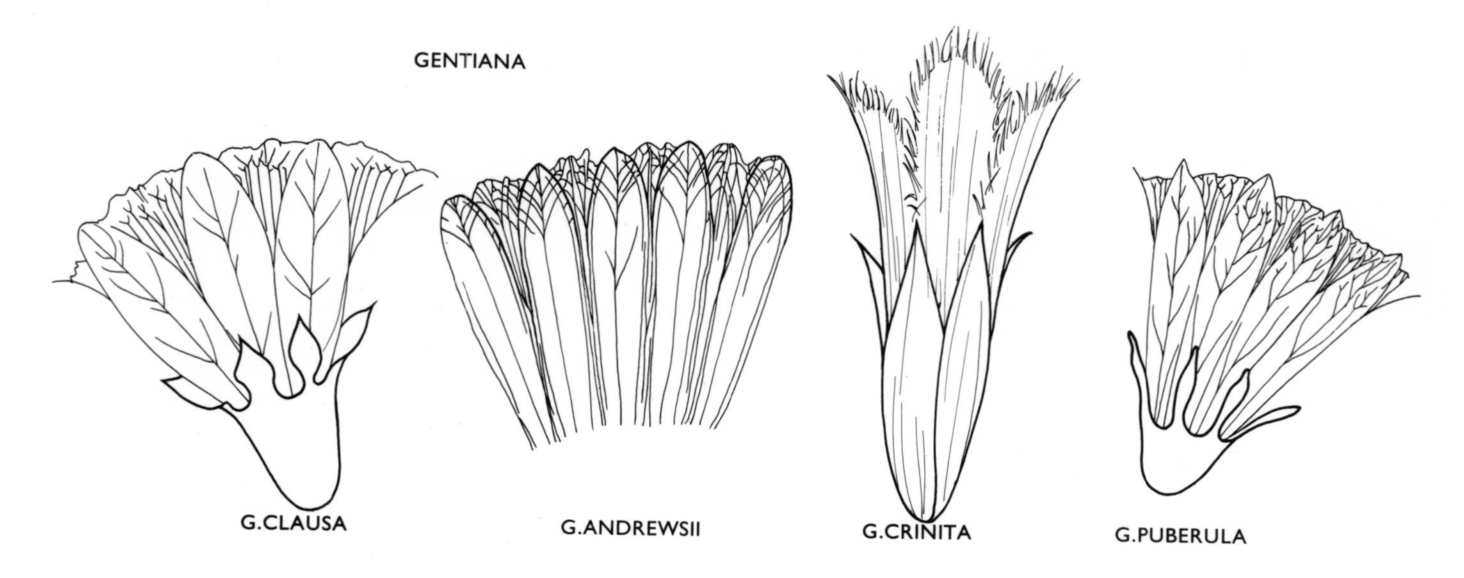

I. *Species with no extra lobes or "plaits" between the four or five teeth or lobes of the corolla; teeth of the calyx meeting at a sharp angle (see the drawings), except in G. quinquefolia.*

A. Two of these have four large and beautifully fringed lobes on the corolla.

FRINGED GENTIAN, G. CRINITA, is certainly one of our most beautiful wild flowers. The stem, from 4 to 40 inches tall, bears long-stalked flowers at the tips of the stem and its branches. The corolla is often 2 inches long; its lobes are at first twisted together, later flaring outward. The teeth of the calyx are unequal, two very narrow and two wider.

August to November: in meadows and damp woodland from Maine to Manitoba and southward to Georgia, Ohio, Indiana, and northern Iowa. *Plate 92*.

NARROW-LEAVED FRINGED GENTIAN, G. PROCERA, is quite similar to *G. crinita* but has very narrow leaves and the lobes of the corolla are toothed rather than fringed.

August to October: in bogs and swamps and on wet banks from New York to Alaska and southward to Ohio, Illinois, Iowa, and North Dakota. *Plate 92*.

B. The remaining species of group I have generally five teeth at the end of the corolla.

FELWORT, G. AMARELLA, grows from 2 to 20 inches tall. The small flowers, about an inch long, are crowded in the axils of leaves, on short stalks. The corolla is pale violet. It has a short fringe just inside the teeth.

July to September: in damp places from Labrador to Alaska and southward to Maine, Vermont, Minnesota, South Dakota, New Mexico, and California. *Plate 92*. *Amarella* is from a word meaning "bitter." The plant is called bitterwort in England. "Felwort" probably means a plant of the fell or field.

AGUEWEED, G. QUINQUEFOLIA, is from 1 to 3 feet tall. The stem is sharply angled. The flowers are pale violet or greenish-white. Each of the five teeth is tipped with a bristle.

August to November: in woods and meadows and on damp banks from Maine to western New York, southern Ontario, and Minnesota, and southward to Florida, Tennessee, and Missouri. *Plate 93*.

II. *Species with lobes or plaits between the five teeth of the corolla (see the drawings); the calyx with slender teeth arising from a cup-shaped base.*

PINE-BARREN GENTIAN, G. AUTUMNALIS, is a very slender plant from 8 inches to 2 feet tall, with very narrow leaves. The corolla spreads at the end into five pointed lobes nearly an inch long, much longer than the plaits between. The color varies greatly, usually deep blue but sometimes white or green or purple; the tube is bronze-color, spotted with brown inside.

September and October: in pine barrens and pinelands from New Jersey to Virginia and South Carolina. *Plate 93*. This is often called *G. porphyrio*.

G. PUBERULA resembles *G. autumnalis* in some ways, especially in the length of the corolla-lobes. These are broad, obtusely pointed, and spread out; the color is purple. The plaits between are much shorter and somewhat fringed.

September and October: in prairies and sandy and rocky open places from New York to North Dakota and southward to Georgia and Kansas. *Plate 93.*

CLOSED or BOTTLE GENTIAN, G. ANDREWSII, has a purple, violet, pinkish, or even white corolla whose teeth never separate; fertilization takes place within the one flower. When the flower is opened, the plaits between the teeth are found to be actually longer than the teeth, and are themselves toothed.

August to October: in meadows and prairies, even on overgrown roadsides from Quebec to Saskatchewan and southward to Georgia, Arkansas, and Nebraska. *Plate 93.*

CLOSED or BOTTLE GENTIAN, G. CLAUSA, differs from the preceding species chiefly in the plaits between the teeth of the corolla: the plaits are shorter than the teeth.

August to October: in meadows and thickets and along streams from Maine to Minnesota and southward to Maryland, the mountains of North Carolina and Tennessee, and Missouri. *Plate 93.* One must open the corolla to be sure of the species.

SOAPWORT GENTIAN, G. SAPONARIA, was named for the resemblance of its foliage to that of bouncing-Bet, *Saponaria*, in the pink family. The leaves, however, are arranged like those of the closed gentians, in pairs or fours. The flowers open very slightly, the toothed plaits being practically as long as the lobes of the corolla.

September and October: in sandy swamps and in bogs, meadows, and moist woods on the coastal plain from New York to Florida and Texas and inland in West Virginia, Indiana, Wisconsin, and Minnesota. *Plate 94.*

SAMPSON'S SNAKEROOT, G. VILLOSA, is misnamed in the Latin; *villosa* means "hairy," but the plant is smooth. It stands from 4 inches to 2 feet tall with blunt leaves widest between the middle and tip. The greenish-white or purplish corolla ends in erect teeth longer than the plaits between, which are also toothed. The teeth of the calyx are long and narrow.

August to October: in open woods from New Jersey to Indiana and southward to Florida and Louisiana. *Plate 94.* Snakeroot means a plant effective against snakebite; many species have been so designated. But who was Sampson?

NARROW-LEAVED GENTIAN, G. LINEARIS, is from 6 inches to 2 feet tall, with pairs of narrow ("linear") leaves. The blue or violet corolla ends in blunt lobes, not teeth, longer than the narrow plaits between.

July to September: in bogs, meadows, and wet roads from Labrador to Lake Superior and southward to Maryland, West Virginia, and Minnesota. *Plate 93.*

RED-STEMMED GENTIAN, G. RUBRICAULIS, is quite similar to *G. linearis*, but with broader leaves and red-tinged (*rubri-*) stem (*caulis*).

July to September: in wet meadows and woods from New Brunswick to southern Ontario and from Michigan to Minnesota and Nebraska.

YELLOW GENTIAN, G. FLAVIDA, is much like *G. rubricaulis* in foliage, but is easily distinguished by the pale yellow or greenish corolla. This has broad pointed lobes much longer than the small rounded plaits between.

August to October: in wet woods, meadows, and prairies from Pennsylvania and southern Ontario to Manitoba and southward to North Carolina and Arkansas. *Plate 94.*

Besides these, several other species may be found near the margins of our range. *G. catesbaei*, which shares the name Sampson's snakeroot with *G. villosa*, extends north on the coastal plain to Delaware; the stem is downy, the leaves lanceolate, the corolla wide open. *G. decora*, with round corolla-lobes and a small, downy calyx occurs from Virginia to Kentucky and southward.

The Cherokee gentian, *G. cherokeensis*, is found in southeastern Virginia. And the northern and western *G. tonsa* and *G. macounii* occur in Minnesota and Iowa.

CENTAURY (CENTAURIUM)

The centauries are small plants, not 2 feet tall, with pink or purple flowers not more than ½ inch across. The flowers are in mostly branched inflorescences at the summit of the stem. The leaves are in pairs, rarely more than an inch long. The tube of the corolla is long and slim; it spreads at the summit into five pointed lobes.

CENTAURY, C. UMBELLATUM, is a native of Europe. The flowers are without stalks or almost so. The petals are rose-purple. This species is established in damp places, meadows, waste land, etc. from Quebec to Michigan and southward to Georgia and Indiana.

C. PULCHELLUM, another immigrant, has flowers with stalks, in a very much branched inflorescence. The teeth of the corolla do not exceed ⅕ inch in length. The leaves are ovate or lanceolate. This centaury grows in wet fields and waste lands from New York to Illinois and Virginia.

PLATE 93

Gentiana clausa *Elbert*

Gentiana quinquefolia *Johnson*

Gentiana autumnalis *Elbert*

Gentiana autumnalis *Elbert*

Gentiana linearis *Elbert*

Gentiana puberula *Johnson*

Gentiana andrewsii *Johnson*

Two southwestern species are to be seen in Missouri: *C. texense*, with narrow leaves; *C. calycosum*, with corolla-teeth up to ⅖ inch long.

The name centaury is derived, according to Pliny, from that mythical creature the centaur, one of them having cured himself of a wound from a poisoned arrow with some plant; what this plant was we do not know. Compare *Centaurea* in the daisy family.

HALENIA

This is mainly a tropical American genus, with one species in our range.

SPURRED-GENTIAN, H. DEFLEXA, is a leafy plant from 4 inches to 3 feet tall. The paired leaves have three or five main veins running lengthwise. The flowers are in branched clusters at the tips of the stem and and its branches. The corolla is purplish-green or bronze, with four petals. The spurs, hollow tubes formed on the petals, distinguish this from other genera.

July to September: typically in moist woods, but also (a variety) near the sea on open slopes and coasts from Labrador to British Columbia and southward to New York, northern Illinois, South Dakota and Montana; also in Mexico. *Plate 178.*

OBOLARIA

There is only one species of *Obolaria;* we have it.

PENNYWORT, O. VIRGINICA, is a smooth plant not more than 6 inches tall with paired round leaves only ½ inch long. The flowers are usually in threes in the axils, with one on the tip of the stem. There are only two sepals, which resemble the leaves. The four petals are white.

March to May: in moist woods from New Jersey to Illinois and southward to Florida and Texas. *Plate 94.* An *obolos* was a Greek coin, likened to a penny; the reference is to the round leaves.

MARSH-PINKS AND THEIR KIN (SABATIA)

Most species of *Sabatia* have five petals and stamens, but some have as many as twelve. The petals are rose-color varying to white or yellow, often with a yellow "eye"; they seem scarcely joined. All the species described below flower from July to September, occasionally earlier or later. Most inhabit wet places near the coast.

I. *Species with typically five petals.*

MARSH-PINK or SEA-PINK, S. STELLARIS, has lovely pink flowers, with a yellow (or green) "eye" bordered with brown, on long leafless stalks. The leafy stem is from 12 to 20 inches tall, with pairs of small elliptic leaves.

In salt or brackish marshes from Massachusetts to Florida and Louisiana. *Plate 94.*

S. CAMPANULATA is very like *S. stellaris*, and forms intermediate between these two occur; they occupy in part the same range. *S. campanulata* has broader-based, narrower, and blunter leaves; the teeth of the calyx are very narrow and nearly as long as the petals. It grows in sand and peat on the coastal plain from Massachusetts to Louisiana and inland to Indiana, Kentucky, and Alabama; also in the West Indies. *Plate 94.*

ROSE-PINK or BITTER-BLOOM, S. ANGULARIS, has a sharply angled stem up to 3 feet tall, usually much branched and bushy. The leaves have lobes which "clasp" the stem. The fragrant flowers are pink with a yellow "eye," the ends of the petals being blunt and broad.

In open woodlands and fields from Connecticut to Michigan and Kansas and southward to Florida, Louisiana, and Oklahoma. *Plate 95.*

S. CAMPESTRIS resembles *S. angularis* but is not so tall (little more than a foot) with fewer branches. The yellow eye of the corolla has several prongs on each petal. The calyx bears projecting flanges running up to the gaps between the teeth.

In prairies, fields, and open woodland from Illinois to Iowa and southward to Arkansas and Texas; occasionally turning up in New England. *Plate 95.*

S. DIFFORMIS has a basal rosette of leaves which has usually disappeared by flowering time. The stems grow from 16 to 40 inches tall, unbranched except for

PLATE 94

Gentiana saponaria *Justice*

Sabatia stellaris *V. Richard*

Gentiana villosa *Johnson*

Sabatia campanulata *Elbert*

Obolaria virginica *Johnson*

Gentiana flavida *Johnson*

the flower-stalks. The leaves are lanceolate and taper to a narrow, sharp-pointed tip. The petals are white; the calyx-teeth very narrow.

In bogs and swamps and wet pine-barrens on the coastal plain from New Jersey to Florida and inland to Tennessee. *Plate 95.*

II. *Species with more than five petals. In both of these the petals are typically pink with a yellow spot at the base.*

LARGE MARSH-PINK or SEA-PINK, S. DODECANDRA, may reach a height of 2 feet. The leaves are narrow. The flowers have up to twelve (*dodeka*) pink or white petals. The lobes of the calyx tend to be widest above the middle.

In salt and brackish marshes on the coastal plain from Connecticut to Florida and Louisiana. *Plate 95.*

PLYMOUTH-GENTIAN, S. KENNEDYANA, resembles *S. dodecandra*, but differs in more commonly forming runners from the base of the stem. It is taller, up to nearly 3 feet. The leaves have tiny sharp points. The lobes of the corolla are commonly over an inch long, and the yellow spot at their base up to $\frac{1}{5}$ inch broad.

At the margins of freshwater ponds and in fresh marshes on the coastal plain in Nova Scotia, Massachusetts, and Rhode Island.

Several other species are found in southeastern Virginia; they belong to the southern flora.

SWERTIA

There are many species of *Swertia* in the West and in the Old World. We have only one in our range.

AMERICAN COLUMBO, S. CAROLINIENSIS, is a smooth plant from 3 to 8 feet tall. The leaves are mostly in fours. The flowers form an ample inflorescence at the summit. The elaborate corolla has four greenish-yellow lobes with small brownish dots; each bears a large round gland, just below the middle, surrounded by a fringe of hairs.

May and June: in meadows and woods from western New York to Wisconsin and southward to Georgia and Louisiana. *Plate 95.* Columbo (or Colombo or Calomba) was a drug erroneously thought to have come from Colombo in Ceylon. *Swertia* provided a substitute in the New World.

LOMATOGONIUM

We have one species.

MARSH FELWORT, L. ROTATUM, is only a few inches tall, with blue (or white) flowers, about an inch across, on branches from the axils. The leaves are narrow and somewhat succulent. The pistil is distinguished by lacking a style; the two stigmas lie on opposite sides of the upper part of the ovary.

July to September: in wet places near the sea from Labrador and Newfoundland to Maine; also in the Rocky Mountains, and in the Old World. *Plate 95.* For the name felwort, see under *Gentiana amarella.*

BARTONIA

Two insignificant plants compose *Bartonia* in our range. The leaves of both are represented by minute scales. The tiny flowers are on branches at the summit of the stem.

SCREW-STEM, B. PANICULATA, has leaves (scales) mostly singly borne, on a stem from 2 to 18 inches tall which often bends and sometimes twines. The whole plant is yellowish or purplish; the flowers may be white, cream-colored, or purplish, only about $\frac{1}{4}$ inch long, or less.

August to October: in wet meadows, bogs, and swamps from Nova Scotia to New England and New York, westward to Oklahoma, southward on the coastal plain to Florida and Louisiana. This species varies in color and in minor details.

B. VIRGINICA does not exceed a foot in height. It is stiff, with mostly paired leaves (scales). The plant is yellowish, including the flowers; the flowers may be $\frac{1}{5}$ inch long or less.

July to September: in dry or wet soil from Quebec to Minnesota and southward to Florida and Louisiana. *Plate 95.*

Sabatia dodecandra *Phelps*

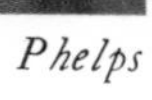

Swertia caroliniensis *Rhein*

Sabatia difformis *Rhein*

Sabatia campestris *Merkle*

Bartonia virginica *Cain*

Lomatogonium rotatum *Scribner*

Sabatia angularis *Horne*

FLOATING-HEARTS (NYMPHOIDES)

The floating-hearts have a suggestion of very small water-lilies. But the round, floating leaves have almost no stalks. The long stems that bear them (and look like leaf-stalks) also bear the clusters of small flowers. The corolla is formed of five almost separate petals. The pistil has a very short style or none.

YELLOW FLOATING-HEART, N. PELTATA, has a creeping stem bearing pairs of very unequal leaves, the larger up to 6 inches across (they are not truly "peltate" which means attached at the center of the blade rather than at the edge). The flowers are yellow, about an inch across, more or less fringed.

July to September: introduced into cultivation from Europe, and now spreading to the wild in the South and northward to New York and Missouri. *Plate 96.*

N. CORDATA has ovate leaves up to 2 inches wide. The flowers are white, from $\frac{1}{5}$ to $\frac{2}{5}$ inch across.

July to September: in quiet water from Newfoundland to Ontario and southward along the coast to Florida and Louisiana.

N. AQUATICA has round leaves up to 6 inches across. The flowers are white, from $\frac{2}{5}$ to $\frac{4}{5}$ inch across.

July to August: in quiet water on the coastal plain from New Jersey to Florida and Texas.

MENYANTHES

There is but one species.

BUCKBEAN or BOGBEAN, M. TRIFOLIATA, sends up its leaves and leafless flowering stem from a horizontal underground stem (rhizome). The leaves have stalks 2–12 inches long, expanding at the base into sheaths around the rhizome. At the summit of the stalk are the three broad segments of the blade. The flowering stem rises 4–12 inches above the surface, bearing a long cluster, a raceme, of white flowers (sometimes tinged with pink), each about an inch across. The petals, which are joined nearly half-way up, are covered by a thick "beard" of long hairs.

April to July: in shallow water and mud from Labrador to Alaska and southward to Maryland and western Virginia, Ohio, Indiana, Illinois, Missouri, Nebraska, and Wyoming; also in northern Europe and Asia. *Plate 96.*

THE DOGBANE FAMILY (APOCYNACEAE)

The *Apocynaceae* are distinguished by their milky juice, seen when a leaf or stem is broken. This "latex" may be poisonous. The flower is remarkable for having two ovaries but only one style (or none) and one stigma; two pods (follicles) are formed from each flower. The other parts of the flower are in fives, the petals joined to form a tube or bell. The leaves are generally paired, undivided and unlobed, with short stalks or none. We have three herbaceous genera, not at all difficult to characterize and identify. There are many tropical genera.

THE DOGBANES (APOCYNUM)

The dogbanes are tall, branching plants with numerous small flowers in clusters at the ends of branches. The petals form a bell or funnel; a hand magnifier will reveal little appendages inside, just below the five lobes. The style is lacking, the large stigma sitting directly on the two ovaries. The fruit is a long, thin follicle (one from each ovary); the seed is provided with a tuft of silky hair, like those of milkweeds.

The botanical name is from Greek words meaning "away with the dog" – the English being a translation of this. The original use of the name, however, was for some other plant. Our species of *Apocynum* seem to offer no threats to dogs.

SPREADING DOGBANE, A. ANDROSAEMIFOLIUM, grows from a few inches to nearly 2 feet tall, the stem often forking so that there is no one main stem. The ovate leaf-blades are on short stalks. The fragrant

Apocynum androsaemifolium *Gottscho*

Apocynum androsaemifolium *Johnson*

Nymphoides peltata *Gottscho*

Menyanthes trifoliata *Gottscho*

Vinca minor *Johnson*

Apocynum cannabinum *Johnson*

flowers are small bells many of which hang on curved stalks; they are pale pink with deeper-pink stripes inside; the five teeth curve outwards. The pods are from 3 to 8 inches long.

June to August: in dry soil in woodland and fields and on roadsides and even in waste land from Newfoundland to Alaska and southward to Georgia, Tennessee, Missouri, Texas, and Mexico. *Plate 96.*

INDIAN HEMP, A. CANNABINUM, has an erect stem from 3 to 5 feet tall, with branches from the axils. The leaf-blades are ovate or lanceolate, those on the main stem with short stalks, those on the branches often without stalks. The flowers are greenish-white and more or less erect on straight stalks. The pods are from 5 to 8 inches long.

June to September: in open ground and thickets from Quebec to British Columbia and southward throughout the United States; less common in New England. *Plate 96.* This is a very variable species, divided by botanists into several varieties. It also interbreeds with *A. androsaemifolium*, yielding a confusing array of plants that cannot be classified in either species. Some of these have been treated as a species under the name *A. medium.* There are also smaller plants with stalkless leaves and somewhat intermediate flowers known as *A. sibiricum* – probably a form of *A. cannabinum.*

True hemp is *Cannabis*, in another family.

PERIWINKLE (VINCA)

The species of *Vinca* are European, brought to America for use in gardens, and spreading to the wild in places. They are trailing or creeping plants, sometimes forming thick mats in the woods. The creeping stems bear dark green, short-stalked leaves in pairs, from the axils of which arise single flowers on stalks. The flowers are generally violet-blue, sometimes white, the petals forming a funnel-shaped tube flaring at the end into the five broad lobes all in one plane.

The common species in our range is *V. minor*, often called myrtle (though it is no relation to *Myrtus*, the true myrtle). *Plate 96. V. major* is larger, with erect stems up to 3 feet tall and flowers 2 inches across. It is found from Virginia southward.

The Roman naturalist Pliny named this or some similar plant *vincapervinca*, bonds through bonds – a tangle; whence Italian *pervinca*, French *pervenche*, and English periwinkle! One name in southern England is old-woman's-eye.

These plants are said to induce love between husband and wife, the leaves being eaten (but one recipe calls for the addition of powdered earthworms). They also had many supposed medicinal uses. They form excellent ground-covers.

AMSONIA

This western and southern genus contributes one species to the midwestern states of our range.

BLUE-STAR, A. TABERNAEMONTANA, has a cluster of stems from 1 to 3 feet tall with lanceolate or elliptic leaf-blades on short stalks. The leaves are borne singly but so close as to often appear paired. The flowers are in a much-branched cluster at the summit. The petals are pale blue, narrow and pointed, separate nearly to the base, ½ inch or more across.

April to June: in woods and on banks from New Jersey to Tennessee, Indiana, Illinois, and Kansas and southward to Georgia, Louisiana, and Texas. *Plate 97.* The tongue-twisting botanical name commemorates the German herbalist Jakob Theodor Müller, who called himself Tabernaemontanus from a mountain near which he was born (in the sixteenth century).

Two other species occur in Missouri: *A. illustris*, with shining ("lustrous") leaves; and *A. ciliata*, with very numerous, very narrow leaves.

TRACHELOSPERMUM

We have one species of this Asiatic genus.

CLIMBING-DOGBANE, T. DIFFORME, is a slightly woody vine with twining stem which climbs high on bushes and trees. The leaves are lanceolate, or wider towards their tip. The small yellow flowers are in close clusters at the ends of branches from the axils. They form small stars about ½ inch across on a tube about ⅛ inch long. The fruits are very narrow and up to 10 inches long.

June and July: in swamps and wet woods on the coastal plain from Delaware to Florida and Texas and inland to southern Indiana, southern Illinois, Missouri, and Oklahoma.

THE MILKWEED FAMILY (ASCLEPIADACEAE)

The milkweed family derives its name from the thick white sap that oozes from any cut or broken surface of the plants (but one of our milkweeds lacks this). The flowers have an extraordinary structure, the tips of the five stamens being more or less joined with each other and with the broad stigma. Two styles from the two ovaries support this one stigma. The pollen of each stamen forms two waxy masses of microscopic grains; one mass of one stamen is connected with the neighboring mass of the adjacent stamen (see under *Asclepias* for further details). The fruit is a pod (follicle), two from each flower but not formed by many of the flowers in a cluster. The seeds are tipped with long silky hairs, which were used in war time as a substitute for Kapok.

THE MILKWEEDS (ASCLEPIAS)

The milkweeds have a deeply cleft corolla, the five lobes turned backwards down the stalk and concealing the sepals. Just above the junction of the lobes and on a short tube rise five cups, forming the crown or *corona*. From within each cup rises a curved horn, the point directed towards the stigma. This apparatus, the corona, is generally the most conspicuous part of the flower. The flowers are generally sweetly fragrant.

Adjoining masses of pollen are attached to a cleft triangular gland. Insects alighting on the stigma may get their feet in this; if they are large enough they can pull it out and fly away with two masses of pollen dangling – to rub off grains on the next stigma visited. Many kinds of insects cannot exert enough pull and die in the trap.

I. *Species with leaves mostly borne singly; no milky sap.*

BUTTERFLY-WEED or PLEURISY-ROOT, A. TUBEROSA, is distinguished by bright orange flowers. Opinion is divided over the origin of the first name, some attributing it to the bright flowers, others pointing to the crowds of butterflies sometimes seen on it. The deep, tough rhizome ("root") was reputed to furnish a cure for pleurisy. The leaves are roughly hairy, long, narrow, and pointed, in some forms with an indented base.

June to September: in dry open soil from New Hampshire to Minnesota and Colorado and southward to Florida, Texas, and Arizona. *Plate 97*. A very variable species.

II. *Species with at least some leaves in circles.*

FOUR-LEAVED MILKWEED, A. QUADRIFOLIA, has ovate or lanceolate leaves in pairs and fours, on a stem not usually reaching 3 feet in height. The flowers are in several clusters near the summit. The petals are pale pink, with white corona.

May to July: in dry woods from New Hampshire to Minnesota and southward to North Carolina, Alabama, Arkansas, and Kansas. *Plate 97*.

A. VERTICILLATA has a generally unbranched stem from 1 to 4 feet tall. The leaves are very narrow with the margins rolled underneath, from three to six in a circle. The flowers are small, greenish-white.

June to September: in dry, open soil from Massachusetts to southern Ontario, Michigan, and Saskatchewan and southward to Florida, Texas, and Mexico. *Plate 97*.

III. *Species with all leaves in pairs. This group includes most of the milkweeds, some of which are not easy to characterize without technicality.*

A. The following species have short stalks to the leaves (they are inconspicuous and must be sought carefully).

1. Three of these species have rather narrow, lanceolate or ovate leaves tapering upward from about the middle or below; the veins leave the midrib at an acute angle, slanting towards the margin (compare 2).

SWAMP MILKWEED, A. INCARNATA, has a stem from 1 to 5 feet tall crowned with flattish clusters of rose-crimson flowers. The pods are slender, tapering to both ends.

June to August: in swamps and on shores from Quebec to Manitoba and Wyoming and southward to Florida, Louisiana, and New Mexico. *Plate 97*. A variable species.

A. PERENNIS has the narrowest leaves of this group, up to 6 inches long and not more than 2 inches broad. At the summit are a few clusters of small white flowers. The ripe slender pods hang down from curved stalks.

June to September: from Indiana to Missouri and southward to Florida and Texas.

A. EXALTATA has ovate or lanceolate leaves on a stem from 2 to 5 feet tall. The flowers are in clusters at the summit and from the upper axils. The petals are greenish with a white corona. The pods are slender, tapering to both ends; they stand erect on stalks that are bent down.

June and July: in woods from Maine to Minnesota and southward to Georgia, Kentucky, and Iowa. *Plate 97.*

2. Five species of this group (A) have stalked leaves with broader, thicker blades; the veins leave the midrib almost at right angles, curving only slightly as they approach the margin. (See also *A. sullivantii* under B.)

In the first of these the pods are ribbed and warty; in the others they are smooth.

COMMON MILKWEED, A. SYRIACA, was thought by Linnaeus to have come from Syria; hence the misnomer, which the rules of nomenclature do not permit us to change. This is the common tall and robust plant, up to 7 feet tall, seen everywhere along roadsides and in fields. The leaves are large, oblong, and blunt, coated with a grayish down on the under surface. The flowers are in several clusters from the axils of the leaves; the color is a dull crimson-purple or greenish. The thick warty follicles are mentioned above and shown in a photograph; they are sufficient to identify the species; they stand erect on stalks which are bent down.

June to August: on roadsides, in fields and thickets from New Brunswick to Saskatchewan and southward to Georgia, Tennessee, and Kansas. *Plate 98.* The young shoots are sometimes gathered and cooked for "greens."

PURPLE MILKWEED, A. PURPURASCENS, grows to 3 feet tall. The leaves taper to the point; their under surface is very finely downy. The petals are dark purple. The cups of the corona are considerably taller than those of *A. syriaca*, hiding the horns. The pods are finely woolly but not warty.

May to July: in woods and thickets from southern New Hampshire to southern Ontario, Minnesota, and South Dakota, and southward to North Carolina, Tennessee, Arkansas, and Oklahoma. *Plate 99.*

A. RUBRA is a southern species with purplish-red flowers. The cups of the corona are tall and slim and the horn rather straight. The plant grows up to 4 feet tall. The leaves taper to the points. The pods are smooth.

June and July: in wet pinelands and bogs on the coastal plain from Long Island to Florida and Texas. *Plate 98.*

A. VARIEGATA is a smooth plant from 1 to 3 feet tall, with oblong, blunt leaves. The flowers are white with purplish centers. The cups of the corona are round and not toothed. The pods are slender, tapering both ways.

May to July: in woods from Connecticut to Missouri and Oklahoma and southward to Florida and Texas. *Plate 98.*

A. OVALIFOLIA is a northwestern species. The plants are small, not exceeding 2 feet in height. The leaves are oval or ovate, rather short, downy on the under surface. The petals are greenish-white with a yellowish corona. The pods are tapering, downy.

June and July: in prairies and oak woodlands from Manitoba to Alberta and southward to Illinois, Iowa, and Nebraska. *Plate 98.*

Another western species, *A. speciosa*, occurs in Minnesota. It may be known by the dense white wool on its young growth, and by the purplish lobes of the corolla which are up to $\frac{1}{2}$ inch long.

B. The remaining species with paired leaves have no stalks to the leaves.

A. AMPLEXICAULIS is readily recognizable by its leaves, which are "amplexicaul" – embracing the stem by the base of their blades – and wavy-margined. They are oblong and blunt. The flowers are greenish-purple or pinkish with pink corona, the horns protruding above the cups. The pods are slender and practically smooth.

May to July: in dry open places from New Hampshire to Minnesota and Nebraska and southward to Florida and Texas. *Plate 98.*

A. SULLIVANTII is very smooth, from 2 to 5 feet tall, with oblong blunt leaves, more or less indented at the base (some may have short stalks). The general aspect is much like that of *A. syriaca*, but the pods are slender and smooth and the flowers vary from purplish to nearly white. The cups of the corona are tall like those of *A. purpurascens*, concealing the horns.

June and July: in prairies and low grounds from southern Ontario to Minnesota and southward to Kentucky, Kansas, and Oklahoma.

A. LANCEOLATA is a coastal species, smooth, up to 5 feet tall, with very narrow leaves (not more than an inch wide). The flowers are showy, with a tall orange-yellow or scarlet corona on red petals. The pods are smooth, standing erect on stalks that are bent down.

June to August: in wet pinelands and swamps on the coastal plain from New Jersey to Florida and Texas. *Plate 99.*

A. MEADII is a rare plant, reported only from Indiana, Illinois, Wisconsin, Iowa, and Kansas. The leaves "clasp" the stem by projecting lobes on either side; they taper to the tip. The plant has a whitish bloom. There is a single cluster of greenish-yellow flowers. It grows on prairies, flowering in June.

PLATE 97

Asclepias incarnata *Johnson*

Amsonia tabernaemontana *Roche*

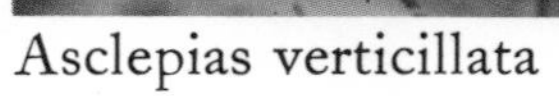
Asclepias verticillata *Johnson*

Asclepias exaltata *Gottscho*

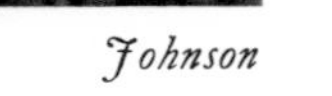
Asclepias quadrifolia *Johnson*

Asclepias tuberosa *Johnson*

THE GREEN-MILKWEEDS (ACERATES)

The plants in this genus resemble the milkweeds (some of which also have greenish flowers) and by some botanists are placed in *Asclepias*. The distinction is that the five cups that form the corona (see under *Asclepias*) have no horns. The flowers are small and greenish.

A. VIRIDIFLORA is from 1 to 3 feet tall. The leaves are much like those of common milkweed, thick and blunt, with nearly parallel sides, and veins running out nearly at right angles to the midrib. The flowers are in dense clusters.

June to August: in dry woodland from Massachusetts to Ontario, Manitoba, and Montana and southward to Florida, Kentucky, Louisiana, Texas, and New Mexico. *Plate 99*.

A. HIRTELLA is western and southwestern, from 16 to 40 inches tall. The leaves are narrow, roughish; their side veins leave the midrib at a slight angle and curve towards the midrib. The plant is found in prairies from Michigan westward and southward; flowering from June to August.

A. LANUGINOSA is a low plant, from 6 to 12 inches tall, and hairy (*lanugo* is "wool"). The leaves are something like small replicas of those of *A. viridiflora*. This is a rare western plant, growing from Wisconsin to Minnesota and Nebraska (and Wyoming?) and southward to Illinois and Kansas; it flowers in May and June. *Plate 99*.

A. LONGIFOLIA is a plant of the southeastern coastal plain. The leaves are very narrow – from 3 to 6 inches long but not more than $\frac{2}{5}$ inch wide. The stem is from 1 to 2 feet tall. It occurs in moist pinelands from Delaware to Florida and Mississippi, flowering in July and August.

ASCLEPIODORA

This genus is largely Mexican. One species extends into our southern regions.

SPIDER-MILKWEED, A. VIRIDIS, differs from *Asclepias* and *Acerates* in the shape of the cups that make up the corona (see under *Asclepias* for this). They extend out horizontally, then curve upward. There are no horns, but a thin plate divides the cavity of each cup lengthwise. The petals also spread horizontally, instead of being bent sharply downward. The pods stand erect on short stalks that are bent down.

May and June, and often later: in prairies and dry woods from Indiana to Missouri and southward to Florida and Texas. *Plate 99*. The name of the genus means, of course, "with the fragrance of the milkweeds." *Viridis* means "green." Spider-milkweed presumably because of the spreading petals. The species is also known as antelope-horn.

CYNANCHUM

This is a large genus of tropical and near-tropical plants, of which one species strays into northeastern North America, where it is sometimes cultivated.

BLACK SWALLOW-WORT, C. NIGRUM, has a twining stem which climbs to a height up to 7 feet. The leaf-blades are lanceolate or ovate, on short stalks. The small, purplish-brown flowers (not black, *nigrum*, as the name would have us think) are in short-stalked clusters in the axils. The lobes of the corolla spread out starlike. The corona forms an inconspicuous, fleshy cap around the stamens.

June to September: on roadsides and in woods; a native of southern Europe.

AMPELAMUS

We have one species.

BLUE-VINE or SAND-VINE, A. ALBIDUS, is a twining vine up to 7 feet tall. The leaf-blades are somewhat triangular with an indented base. The flowers are white (*albidus*) in clusters in the axils. Each cup of the corona is deeply cleft in two.

July and August: a weed in low woods and fields from Pennsylvania to Kansas and southward to Georgia and Texas. *Plate 99*. The English name is a translation of the Latin: sand-vine. From the vine alone this could be mistaken for something in the morning-glory family; but the deep and *wide* indentation at the base of the leaf is distinctive.

PLATE 98

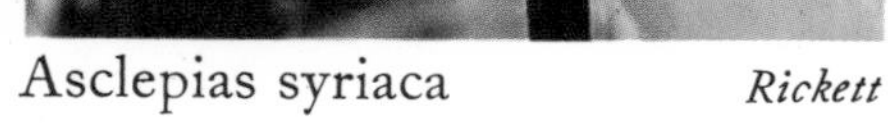

Asclepias syriaca *Rickett*

Asclepias syriaca *Johnson*

Asclepias variegata *Allen*

Asclepias rubra *Gottscho*

Asclepias amplexicaulis *V. Richard*

Asclepias ovalifolia *Johnson*

THE ANGLE-PODS (GONOLOBUS)

The angle-pods are twining vines with heart-shaped leaves. In twining stem and heart-shaped leaves they resemble *Ampelamus*, but the corona is quite different, forming a shallow, inconspicuous cup around the stamens. The five narrow petals extend out like the points of a star. The pods (follicles) of some species are angular in cross section. The genus is mainly tropical. Our species are southern.

G. OBLIQUUS is our most widespread species. The leaves are ovate or even round, with a wide indentation at the base. The flowers are brownish-purple.

June and July: in woods and thickets from Pennsylvania to Missouri and southward to North Carolina and Tennessee. The vine may be easily mistaken for a morning-glory (*Ipomoea*) if flowers are lacking; but compare the leaves carefully. The indentation at the base is wider and round in *Gonolobus*.

G. CAROLINENSIS has brownish-purple flowers like those of *G. obliquus*, but the corona has five sharp teeth, and the petals do not taper gradually.

May and June: in moist woods from Tennessee and Mississippi and northward along the coast to Delaware and in the interior to Missouri.

G. DECIPIENS is distinguished from *G. obliquus* only by the size and shape of the lobes of the corolla: they are from $\frac{1}{2}$ to $\frac{4}{5}$ inches long and widest above the middle. This species occurs across the southern states and extends northward to Indiana and Missouri, flowering in May and June. *G. gonocarpos* has a very short-spreading corona, forming a small disk scalloped around the edge. The lobes of the corolla are brownish-purple, narrowly lanceolate. It grows in wet woods in the South and extends northward to southeastern Virginia, southern Indiana, and southern Missouri.

THE PHLOX FAMILY (POLEMONIACEAE)

The flowers of the phlox family have a bell-shaped corolla or a tube that flares into five lobes. The stamens are attached to the inside of the corolla. The family is most easily recognized by the three branches of the style. The three-chambered ovary develops into a pod (capsule).

The genus *Phlox* has mostly paired leaves, undivided and unlobed. *Collomia* also has undivided and unlobed leaves, but they are borne singly. *Ipomopsis* and *Polemonium* have pinnately divided leaves borne singly, the first with narrow tubular flowers flaring into five teeth, the second with bell-shaped flowers.

PHLOX

This is the genus represented in our gardens by all the varieties of phlox and the moss-pink. The flowers are in close clusters at the ends of the stem and branches. The calyx is deeply cleft into narrow teeth. The corolla is a narrow tube flaring at the end into five equal lobes. The stamens adhere to the inside of the tube.

The species may be separated into two groups by their manner of growth.

I. *Species with creeping woody stems and erect branches covered with narrow leaves, forming mats or tufts.*

MOSS-PINK, PHLOX SUBULATA, has needle-like or awl-like ("subulate") leaves nearly an inch long on branches that rise to heights of from 2 to 8 inches. The flowers are numerous, rose or rose-purple or sometimes white, forming sheets of color.

April to July: in sandy or rocky soil from Maine to Michigan and southward to North Carolina and Tennessee. *Plate 100*. This is widely cultivated and plants found in New England have escaped from cultivation. The garden forms have flowers of many colors from white to lavender, lilac, and blue besides the pink and purple shades; but rather garish colors predominate in the gardens and roadside banks throughout the eastern states. The name moss-pink underlines the resemblance between the phlox and the pink families. It is superficial, however. In the pink family the petals are separate and the ovary has but one chamber.

P. BIFIDA forms tufts up to 8 inches tall or even more. The leaves are narrow and stiff. It is readily distinguished from *P. subulata* by the lobes of the corolla, which are deeply cleft ("bifid"). The color is usually a pale purple, with two dots at the base of each lobe.

PLATE 99

Asclepias lanceolata *Elbert*

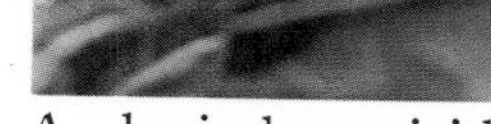

Asclepiodora viridis *Johnson*

Asclepias purpurascens *Johnson*

Ampelamus albidus *D. Richards*

Acerates lanuginosa *Johnson*

Acerates viridiflora *Johnson*

April to June: in dry soil and on rock ledges from southern Michigan to Iowa and southward to Tennessee, Arkansas, and Oklahoma. *Plate 100.*

II. *Species with erect or spreading stems bearing mostly lanceolate or ovate leaves.*

These may be separated into two groups by observing with a hand magnifier the stamens and style.

A. In one group of species the stamens and style do not reach the opening of the tube of the corolla.

WILD-SWEET-WILLIAM, P. DIVARICATA, is a common and beautiful plant of midwestern fields and woodland. The stem is erect or nearly so, though the base may lie on the ground, curving upward; the plants are from 1 to nearly 3 feet tall. The calyx and flower-stalks are beset with glands. Runners are formed at the base.

April to June: in fields and woods from Quebec to Minnesota and Nebraska and southward to Florida and Texas. *Plate 100.* Often cultivated. The English name is unfortunate, since sweet-William of gardens is in the pink family. (Compare moss pink.)

P. PILOSA is from 8 to 30 inches tall. The leaves are mostly very narrow and sharp-pointed. The stem is often downy. The flowers are redder than those of *P. divaricata.*

April to July: in prairies and woodland from Connecticut to Manitoba and southward to Florida and Texas. *Plate 100.* A variable species; some forms have ovate leaves.

B. Species whose stigma is carried by a longer style beyond the mouth of the corolla-tube.

CREEPING PHLOX, P. STOLONIFERA, has runners with several pairs of small leaves. The erect or spreading, hairy flowering stems are from a few inches to a foot tall. The inflorescence also is hairy. The flowers are violet or purple.

April to June: in bottomlands and moist woods from Pennsylvania and Ohio southward to Georgia. *Plate 100.*

WILD-SWEET-WILLIAM, P. MACULATA, is named in the Latin for the reddish spots (*maculae*) usually present on the stems. There is an underground stem, but no runners. The flowering stem is from 1 to 3 feet tall. The flowers are reddish-purple.

May to September: in meadows and bottomlands from Quebec to Minnesota and southward to North Carolina and Georgia, Mississippi, and Arkansas.

MOUNTAIN PHLOX, P. OVATA, grows from 6 inches to 3 feet tall. The leafy stems tend to lie on the ground, the flowering branches with their tips turning upward. The leaves are elliptic or widest near their tips, the lower ones with distinct stalks. The flowers are pink or red-purple.

May to July: in open woods and meadows from Pennsylvania to Indiana and southward in the mountains to Georgia and northern Alabama.

P. GLABERRIMA sends up from a crown one or a few erect, smooth flowering branches to a height of from 10 inches to 5 feet. The leaves have narrow, sharp-pointed blades. The flowers are red-purple or violet.

May to July: in wet woods and thickets from Ohio to Wisconsin and southward to Florida and Texas. *Plate 100.*

P. CAROLINA resembles both *P. ovata* and *P. glaberrima*, its leaves being either broad or narrow. The flowers are pink or purplish-red, in a round cluster.

May to July: in open woods and thickets from Maryland to Illinois and southward to Florida and Mississippi.

PERENNIAL PHLOX, P. PANICULATA, may be distinguished (with the following species) by the distinct leaf-veins which branch from the midrib and curve forward near the margin to join with the vein above. It is a tall plant, up to 7 feet, with from fourteen to twenty-five pairs of elliptic or ovate leaves, or even more (the upper leaves may be single). The flowers vary greatly in color, but are typically purplish-pink.

July to October: in open woods, thickets, bottomlands, etc. from New York to Iowa and Kansas and southward to Georgia, northern Mississippi, and Arkansas. This is often cultivated (hence its many flower-colors), and often escapes in other parts of the country.

BROADLEAF PHLOX, P. AMPLIFOLIA, has the same sort of veins in the leaves as the preceding species. It is somewhat lower, up to 5 feet tall, with not more than fifteen pairs of leaves. The leaves are wider. The inflorescence is beset with numerous glandular hairs.

June to September: on wooded slopes and banks from western Virginia to Indiana and Missouri and southward to North Carolina and Alabama.

P. drummondii, from Texas, is cultivated and is sometimes found wild. *P. buckleyi*, with evergreen rosettes of narrow leaves and flowering stems a foot tall bearing purple flowers is a rare species in western Virginia and West Virginia. *P. amoena*, with dense inflorescence and finely downy stem, is found in Kentucky.

PLATE 100

Phlox stolonifera *Lee*

Phlox divaricata *Johnson*

Phlox subulata *Johnson*

Phlox bifida *Rickett*

Phlox pilosa *Johnson*

Phlox glaberrima *Johnson*

COLLOMIA

We have one species of this western genus.

C. LINEARIS has narrow leaves borne singly. The flowers resemble those of *Phlox* but are smaller, the flat part about $\frac{1}{6}$ inch across, from lilac to white.

May to August: in dry open places from Quebec to British Columbia and southward to Wisconsin, Nebraska, New Mexico, and California, and occasionally farther eastward. *Plate 101*.

IPOMOPSIS

One very wide-ranging species reaches our area.

STANDING-CYPRESS, SPANISH-LARKSPUR, or SCARLET-GILIA, I. RUBRA, has leaves pinnately divided into very narrow, hairlike segments. The flowers form a narrow inflorescence at the summit. The red petals (yellow inside) form a long tube with five flaring teeth.

May to September: a native of the southern United States from North Carolina to Florida, Oklahoma, and Texas, often cultivated and escaped in our range from New England to Michigan and Missouri and southward. In gardens it is found sometimes under the name *Gilia coronopifolia* – an "illegitimate" name. The first two English names above seem singularly inappropriate – but perhaps I am missing a point. By many botanists the genus is merged with the larger, western genus *Gilia*.

POLEMONIUM

This genus is distinguished by its pinnately divided leaves, the segments being lanceolate or elliptic, not hairlike. The leaves are borne singly. The flowers are blue and bell-shaped, hanging. They form a cluster at the end of an arching stem.

GREEK-VALERIAN, or JACOB'S-LADDER, P. REPTANS, is a spring-flowering species with a stem from 8 to 16 inches long. The flower-cluster is rather loose; the stamens do not project beyond the corolla.

April to June: in moist woods from New York to Minnesota and southward to Georgia, Alabama, Mississippi, and Oklahoma. *Plate 101*. This is typically a midwestern and southern species, often cultivated in eastern gardens. A related form with reddish flowers and blunter leaf-segments has received the name *P. longii;* it was found in Pennsylvania.

P. VAN-BRUNTIAE has a longer stem than *P. reptans*, up to 3 feet or more. The inflorescence is not so loose. The corolla is purplish, and the stamens extend noticeably beyond its margins.

June and July: in swampy woods and bogs from Vermont and northern New York to Maryland and West Virginia, chiefly in mountains. *Plate 101*.

The summer-flowering *P. caeruleum*, from the Old World, escapes from cultivation. The western *P. occidentale*, with tightly clustered flowers, is found in bogs in northern Minnesota.

PLATE 101

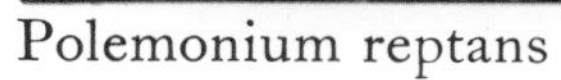
Polemonium reptans
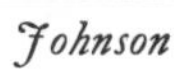
Johnson

Hydrophyllum appendiculatum *Johnson*

Hydrophyllum virginianum *Johnson*

Polemonium van-bruntiae *Rhein*

Phacelia dubia *Elbert*

Collomia linearis *Dobbs*

GROUP XI

SEPALS five, petals five; petals joined, radially symmetric. Stamens as many as the petals; leaves borne singly on the stem. Exceptions: *Echium* has a bilaterally symmetric flower. Some species of *Lithospermum* have paired leaves.

I. *Flowers in a false raceme – arranged along one side of a stem which is curved or coiled at the tip (except* Ellisia *in the waterleaf family).*

A. Plants with mostly lobed, cleft, or divided leaves: waterleaf family (except *Hydrolea*).

B. Plants with leaves not lobed, cleft, or divided: forget-me-not family.

II. *Flowers growing singly or in small clusters from the axils of leaves or midway between leaves, or at the tip of the stem, or in a true raceme or spike.*

A. Flowers with mostly white or yellow petals (blue in *Nicandra* and a few small species of the forget-me-not family); ovary superior: potato family; some small-flowered genera of the forget-me-not family.

B. Flowers with red, white, or blue petals and twining or trailing stem; ovary superior: morning-glory family.

C. Flowers with mostly blue or violet petals (or white in a few species); ovary inferior: bluebell family.

THE WATERLEAF FAMILY (HYDROPHYLLACEAE)

The name of the family apparently refers to a species that has extremely watery or juicy leaves. Some have taken it as derived from the pale markings – "watermarks" – on the leaves of some species; but this is hardly likely. The family is typically North American and has numerous species in the West. It may be distinguished from other families with radial corollas of five joined petals and leaves borne singly by the ovary with a style cleft into two at the summit (or in one genus two styles). The leaves of most of our species are pinnately cleft or divided.

The genera may be distinguished as follows:

I. *Plants with leaves lobed, cleft, or divided.*

A. Of these, some have flowers clustered in an inflorescence: *Phacelia* (unbranched inflorescence); *Hydrophyllum* (forking inflorescence).

B. Others have flowers growing singly from the axils: *Ellisia* (leaves oblong in outline, deeply cleft pinnately).

II. *Plants with leaves not lobed, cleft, or divided:* Hydrolea.

PHACELIA

Phacelia is one of the big western genera, with very numerous species: in the Northeast we have only five. They are small plants with pinnately cleft or divided leaves and pretty blue or white flowers in false racemes (these are at first coiled, straightening as they mature). It is curious that most of these attractive little plants seem to have never acquired English names (other than scorpion-weed, which refers to the coiled inflorescence, and which smells like the invention of some botanist).

Our species are not hard to distinguish, by a combination of characteristics of flowers and leaves, and with due regard to their geographic distribution.

MIAMI-MIST, P. PURSHII, has beautifully fringed blue petals which distinguish it from all others of our range (except the southern white-flowered *P. fimbriata*, which comes into Virginia). It grows some 6 to 20 inches tall. The lower leaves are pinnately divided, the upper pinnately cleft into narrow lobes.

April to June: in woods and fields from Pennsylvania to Minnesota and southward to Alabama and Oklahoma. *Plate 102*.

P. DUBIA grows from 4 to 16 inches tall. The lower leaves are pinnately divided, the upper divided or deeply cleft into rather broad segments or lobes which tend to be wider towards their tips. The stalks of the stamens are conspicuously hairy.

March to June: in woods and thickets and open places from New York to Ohio and Missouri and southward to Florida and Texas. *Plate 101*. The western plants may be a distinct species (*P. gilioides*), distinguished by the white-hairiness of stems and leaves.

P. RANUNCULACEA is a weak plant not more than a foot tall. The leaves are long-stalked, the blades mostly pinnately divided into several segments which are often cleft – especially those at the end. The flowers are few. The corolla is blue, small, with rather short lobes. The stalks of the stamens are smooth, hairless.

April and May: in moist woods in Maryland and Virginia and from Indiana to Arkansas.

P. BIPINNATIFIDA may reach 2 feet in height. The formidable name indicates that the leaves are *twice* pinnately divided or cleft; i.e. that the divisions or lobes of the blade are themselves pinnately divided or cleft. The flowers are bright blue, about $\frac{1}{2}$ inch long. The stamens project beyond the corolla, their stalks being long and hairy.

April to June: in moist woods from Virginia to southern Iowa and southward to Georgia, Alabama, and Arkansas. *Plate 102*.

P. FRANKLINII is the only truly northern species. The stem is from 8 inches to 2 feet tall and hairy. The leaves are pinnately divided or cleft into numerous narrow segments or lobes, those near the end often being themselves lobed; they are hairy like the stem. The flowers are blue or white. The stamens project only slightly.

June to August: in dry sandy soil, burned land, etc. from Ontario to Yukon and southward to Michigan, Wyoming, and Idaho.

The southern *P. hirsuta*, with densely hairy stem and divided leaves, is found in southern Missouri.

WATERLEAF (HYDROPHYLLUM)

The plants called waterleaf are rather tall (up to 3 feet, or more), hairy plants with broad leaves which are palmately or pinnately lobed, cleft, or divided. The flowers are white or pale bluish-purple, in clusters (cymes) which are at first dense and coiled in several directions, later unrolling and revealing themselves as several false racemes. The stamens characteristically project beyond the corolla, which is a small funnel cleft into five lobes which do not flare outward.

JOHN'S-CABBAGE, H. VIRGINIANUM, has pinnately divided or cleft leaves, the segments or lobes, usually five, sharply toothed and tapering to sharp points. The calyx is bristly. The flower-clusters are carried on long stalks well above the leaves.

May to August: in moist woods and along streams from Quebec to Manitoba and southward to Virginia, Tennessee, Arkansas, and Kansas. *Plate 101*. Some plants of the mountains have deep violet flowers.

H. MACROPHYLLUM also has leaves pinnately divided or cleft into seven or more segments or lobes, which are very coarsely toothed. The whole plant is roughly hairy.

May and June: in moist woods from Virginia to Ohio and Illinois and southward to Georgia and Alabama.

H. APPENDICULATUM has palmately lobed leaves. Leaves and stem are hairy, and the calyx also. The clusters of pale lavender flowers are carried well above the leaves on long stalks.

May and June: in moist woods from Ontario to Minnesota and southward to Pennsylvania, Tennessee, Missouri, and Kansas. *Plate 101*. The botanical name refers to small teeth *between* the calyx-lobes.

H. CANADENSE has palmately lobed leaves. The plant is smooth or nearly so. The clusters of nearly white flowers are on short stalks and do not rise above the leaves.

May to July: in damp woods from Vermont to Ontario and Michigan and southward to Georgia, Alabama, and Missouri.

ELLISIA

There is only one species.

E. NYCTELEA is a little weed, not much more than a foot tall, with leaves pinnately cleft into from seven to thirteen narrow lobes, which may be toothed. The flowers hang singly on stalks from the leaf-axils, each a white bell not more than $\frac{1}{3}$ inch long.

April to July: in woods, along streams, and in poor lawns from Michigan to Alberta and southward to Indiana, Oklahoma, Colorado, and New Mexico; from New Jersey and Pennsylvania to North Carolina; occasionally farther eastward.

HYDROLEA

This southern and southwestern genus barely enters our range. It is our only genus of this family with undivided and unlobed leaves. In *H. uniflora* they are lanceolate, on a stem that lies on the ground at the base. The blue flowers are in dense clusters in the axils. The range is from Indiana to Missouri and southward to Mississippi and Texas; the time of flowering from June to September.

THE FORGET-ME-NOT FAMILY (BORAGINACEAE)

The forget-me-nots and their relatives are characterized by that peculiar type of flower-cluster called in this book a false raceme (technically a *cincinnus*). In this the flowers are arranged along a stem as in a raceme, the lower ones opening first and often forming fruit while those at the tip are still unopened buds. This inflorescence differs from a true raceme in that the flowers are *all on one side* of the stem, extending alternately to right and left as you face this flower-bearing side. If any bracts are present, they are *opposite* the flower stalks, not beneath them. The stem is usually coiled at first (it is sometimes called *scorpioid*, scorpion-like), straightening from the base as the fruits mature.

The parts of the flowers of *Boraginaceae* are in fives except the ovary of the pistil, which has four chambers and is generally deeply four-lobed with the style rising from their midst. Each lobe becomes a small nutlike body containing one seed. Many of the species are rough with stiff hairs.

Several species of this family are found in gardens: heliotrope, forget-me-not, alkanet, Virginia-bluebells; and several others of our native species deserve to be cultivated.

Genera of Boraginaceae

I. *Plants with bilaterally symmetric flowers, blue with red stamens:* Echium.

II. *Plants with radially symmetric flowers.*
These may be grouped by the shape of the corolla.

A. Corolla shaped like a funnel without distinct lobes at the margin: *Mertensia.*

B. Corolla forming a tube which flares into five round lobes (compare C).

These may be partly separated by the color of their flowers (but note that white flowers are found in both the first two groups, blue in both the first and third).

1. Flowers blue or white (perhaps with a yellow center), leaves without stalks: *Myosotis* (flowers usually in two false racemes); *Lycopsis* (flowers in a short leafy cluster; plant bristly); *Lappula* (flowers blue, $\frac{1}{8}$ inch across or less; plant rough-hairy; fruit a small bur); *Hackelia* (flowers white, $\frac{1}{12}$ inch across; plants rough or downy; fruit a small bur).

2. Flowers yellow, orange, or white: *Lithospermum* (flowers crowded at the summit or single in the axils; plants roughish); *Amsinckia* (small yellow flowers in coiled clusters; plants bristly).

3. Flowers red or blue; lower leaves with stalks: *Cynoglossum.*

PLATE 102

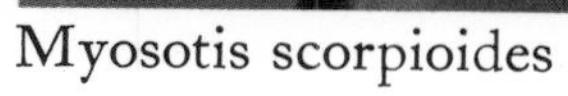
Myosotis scorpioides 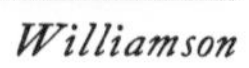 *Williamson*

Phacelia purshii *Johnson*

Echium vulgare *Rickett*

Phacelia bipinnatifida *Uttal*

Echium vulgare *Horne*

Mertensia virginica *Mayer*

C. Corolla divided at the margin into five more or less sharp teeth: *Onosmodium* (flowers white or yellow; styles projecting; mostly very rough plants); *Heliotropium* (flowers minute, blue, in a long, coiled false raceme, or white and borne singly); *Symphytum* (flowers dull blue or yellow or whitish; plants harsh).

ECHIUM

Except for certain rare waifs, the species below is our only representative of this Old-World genus.

VIPER'S-BUGLOSS, BLUEWEED, or BLUE-DEVIL, E. VULGARE, is a rather beautiful if obnoxious weed. The bristly stem grows from 1 to 3 feet tall. In the axils of the upper leaves are numerous short, curved false racemes. The corolla is unique among our *Boraginaceae* in being clearly bilateral in its symmetry, the upper lobes much longer than the lower. The stamens project from the corolla, their red stalks making a striking contrast with its bright blue.

June to October: on roadsides, in fields and waste land from Quebec to Ontario and southward; a stray from Europe. *Plate 102*. The name of the genus is from the Greek word for viper, and alluded to a fancied resemblance of the corolla or perhaps of the nuts to a viper's head. In consequence of this and of the doctrine of signatures (see *Hepatica*), the plant was supposed to cure snake-bite. Various other virtues were ascribed to this weed, including that of curing melancholy! Bugloss is not "bug-loss" but "bu-gloss," from two Greek words meaning "ox-tongue"; this name originally belonged to some other plant.

THE LUNGWORTS (MERTENSIA)

The lungworts or Virginia-bluebells are among our most beautiful wild flowers, often cultivated in gardens. The blue flowers (pink before they open) hang in a cluster (a false raceme, or several) from the arching tip of the stem. The petals form a funnel with scarcely any lobing of the margin. The plants are smooth or nearly so.

The English name was given to a related plant in the genus *Pulmonaria*. This has white-spotted leaves and so was thought to be divinely indicated as a remedy for diseased lungs! The name was transferred to our genus doubtless because of likeness of the flowers.

VIRGINIA-BLUEBELLS or VIRGINIA-COWSLIP, M. VIRGINICA, grows from 8 inches to 2 feet or more tall. The flowers are from $\frac{3}{4}$ to 1 inch long.

March to June: in woods and wet meadows from New York to Ontario and Minnesota and southward to South Carolina, Alabama, Arkansas, and Kansas. *Plate 102*. Our names are unfortunate, since so many unrelated plants are called bluebells (see, for instance, *Campanula*), and the cowslip is an European primrose with yellow flowers.

A wooded slope or wet field covered with this species is a lovely sight. It is also successful in informal gardens.

SEA-LUNGWORT or OYSTERLEAF, M. MARITIMA, differs from *M. virginica* in the size of its flowers: they are only about $\frac{1}{3}$ inch long. The plant has a whitish bloom and is very smooth. The stems spread out to a length of 3 feet.

June to September: on beaches along the coast from Greenland to Massachusetts; and from Alaska to British Columbia; also in Europe.

LANGUID-LADIES, M. PANICULATA, is erect, more or less hairy or downy, from 8 to 30 inches tall. The flowers are about $\frac{1}{4}$ inch long.

June to September: in woods and thickets and on shores from Quebec to Alaska and southward to Michigan, Iowa, Montana, and Washington.

FORGET-ME-NOTS (MYOSOTIS)

There is an old German tale of a knight who, drowning in a river, caught a sprig of *Myosotis* (which grows in wet places) and tossed it to his lover on the bank, crying (in old German, of course), "Forget me not." The story was introduced to England by Coleridge in a poem published in 1802 (though the name had previously been used for a different plant), and forget-me-not is now the accepted name for all the species of this genus. It was the symbol of love, and is still known in some parts of England as love-me and in France as *aimez-moi*.

The plants are small, often bending over, the single stem ending usually in a forked inflorescence (two false racemes), each branch coiled at first and

PLATE 103

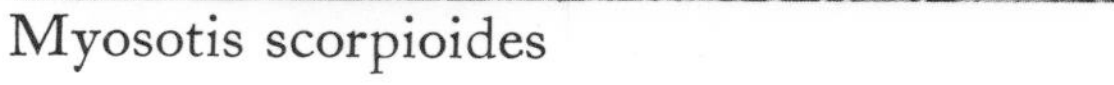

Myosotis scorpioides *Rickett*

Lithospermum canescens *Johnson*

Myosotis laxa *Becker*

Hackelia americana *Elbert*

Lappula redowskii *Elbert*

Lithospermum croceum *Johnson*

Lithospermum carolinense *Johnson*

uncoiling at the base as fruits ripen. The flowers have five blue or white lobes, often with a yellow center. At the "throat" of the corolla – the mouth of the tube – there are five small scales.

We have several native species, but some from the Old World are just as common. They are not easy to distinguish, their identity depending largely on the characteristics of the minute hairs and other traits that are seen only under a magnifier.

TRUE FORGET-ME-NOT, M. SCORPIOIDES, has stems up to 2 feet long and blue flowers up to $\frac{1}{3}$ inch across, with a yellow center. The five scales at the "throat" are lobed so that they seem to be ten. The calyx bears straight hairs which lie against the surface. The stem is angular.

May to October: in wet places especially along small streams from Newfoundland to Ontario and southward to Georgia, Tennessee, and Louisiana, and in the Pacific states; an immigrant from the Old World. *Plates 102, 103.* The Latin name means "scorpion-like," referring to the branches of the inflorescence, coiled like the tail of that creature.

M. LAXA is similar to *M. scorpioides;* the flowers are smaller, up to $\frac{1}{4}$ inch across; there are usually some leaves in the inflorescence; the corolla has a tube only just equal to the calyx.

May to September: in shallow water and wet ground from Newfoundland to Ontario and Minnesota and southward to Georgia and Tennessee; and from British Columbia to California. *Plate 103.* A native species.

M. VERNA has more erect stems which may reach 16 inches in height. The flowers are white, only $\frac{1}{12}$ inch across, or even less.

April to July: in relatively dry soil in woodland and open ground from Maine to Minnesota and Kansas and southward to Florida and Texas. A native of North America.

M. MACROSPERMA is a southern species with white flowers like those of *M. verna*, to be distinguished chiefly by the small bristles with a hooked tip on the calyx.

March to June: in moist woods and bottomlands from Maryland to Indiana and Missouri and southward to Florida and Texas.

Other species from the Old World are *M. arvensis*, with the lobes of the blue corolla forming a cup (Newfoundland to Minnesota and New Jersey); *M. sylvatica*, characterized by blue, pink, or white corolla with flat lobes up to $\frac{1}{3}$ inch across (Quebec to Ontario, Michigan, and New York); *M. stricta*, with blue, funnel-shaped corolla barely projecting from the calyx (in dry places scattered from Quebec to Michigan, New Jersey, and Indiana); *M. versicolor*, with pale yellow corolla, changing to blue and violet, and very short flower-stalks (in fields, etc. from Quebec to Massachusetts, New York, and Virginia, and in the Pacific states).

LYCOPSIS

One species of *Lycopsis* has become a weed in the northeastern United States.

BUGLOSS, L. ARVENSE, is a rough-hairy plant from 4 inches to 2 feet tall. The leaves are narrow, without stalks, and blunt. The flowers are in dense clusters at the summit of the stem. The petals are blue, forming a short tube, no longer than the calyx, crowned with five round lobes about $\frac{1}{5}$ inch across.

June to September: in dry fields and waste land from Newfoundland to Ontario and southward to Virginia, Ohio, and Nebraska; commonest in the northeastern part of this range. For the name bugloss see *Echium*.

THE STICKSEEDS (LAPPULA)

The stickseeds are really "stickfruits." The four-lobed ovary develops into four small nutlike or seed-like fruits each containing a seed; they are covered with barbed prickles which adhere to the coats of animals and the clothing of man. The plants are rough-hairy with branching stems, each branch ending in a long inflorescence of small blue or white flowers.

BEGGAR'S-LICE, L. ECHINATA, is a plant from 8 to 30 inches tall, with very narrow leaves up to 2 inches long. There are several inflorescences. The flowers are blue and not more than $\frac{1}{8}$ inch across.

May to September: an immigrant weed from the Old World, now established on roadsides and in waste land from Newfoundland to Alaska and southward to New Jersey, Kentucky, Missouri, and Texas.

L. REDOWSKII is a native of the West which has spread eastward to Michigan and Missouri and is occasionally found in New England. It differs from *L. echinata* only in details of the barbed hairs on the fruits: these are in only one row along the edge, instead of in two. *Plate 103.*

STICKSEEDS (HACKELIA)

The species of *Hackelia* resemble those of *Lappula*, differing only in a few technical details of the fruits. The flowers in our species are white or pale blue and minute – about $\frac{1}{12}$ inch across.

H. VIRGINIANA grows up to 4 feet tall, with numerous branches; each branch bears towards its end the small flowers or the "stickseeds" which they form. It is found in woods and thickets from Quebec to Minnesota and South Dakota and southward to Georgia, Louisiana, and Oklahoma, flowering from June to September.

H. AMERICANA closely resembles the preceding species. The nutlets bear prickles only at their edges instead of all over their backs. The flowers are more likely to be opposite bracts. The plants grow in woods from Quebec to British Columbia and southward to Vermont, Michigan, Iowa, Kansas, Montana, and Washington. *Plate 103*.

PUCCOONS AND GROMWELLS (LITHOSPERMUM)

The genus *Lithospermum* comprises a number of plants of diverse aspects but all hairy with rather narrow leaves. The flowers are yellow, orange, or white, in clusters at the tip of the stem or single in the axils. The corolla forms a tube with five lobes at the margin.

I. *Species with yellow or orange flowers; puccoons. Puccoon is an Indian name given to many plants that yielded a coloring matter (compare bloodroot).*

YELLOW PUCCOON or INDIAN PAINT, L. CANESCENS, is sometimes called hoary puccoon from its covering of short, fine, white hairs. The flowers are orange-yellow, up to $\frac{3}{5}$ inch across, in a flattish cluster at the summit of the stem. The plant is from 4 to 18 inches tall.

April to June: in sandy soil from Ontario to Saskatchewan and southward to Georgia, Alabama, and Texas. *Plate 103*.

PUCCOON, L. CAROLINIENSE, is similar to *L. canescens*, but with harsher and longer hairs. It grows from 8 to 40 inches tall and has numerous leaves. The corolla may be up to an inch across.

April to July: mostly in sandy soil from Ontario to Saskatchewan and southward to Florida, Texas, and Mexico. *Plate 103*. The more northern plants, including all in our range, are by some authors treated as a distinct species, *L. croceum* (*Plate 103*). The differences are slight – minute characteristics of the veins of the petals and the shape of the hairs.

L. INCISUM is easily distinguished by the lobes of the corolla: they are toothed or fringed at the edge. The color is a rather pale yellow. The plants are rough, from 4 to 20 inches tall.

April to July: in prairies and other open places from southern Ontario to British Columbia and southward to Indiana, Missouri, Texas, and northern Mexico. *Plate 104*.

II. *Species with whitish flowers. These are the inconspicuous weeds called gromwells – an old French name of doubtful meaning. Because of the hard little seeds –* Lithospermum *means "stone-seed" – they were used in medieval medicine for treatment of "the stone."*

CORN GROMWELL or BASTARD ALKANET, L. ARVENSE, is a hoary little plant up to 2 feet tall, or taller. The numerous leaves are very narrow. The flowers are small, only $\frac{1}{8}$ inch across, borne singly in the axils.

April to July: an European plant now established in sandy fields and on roadsides practically throughout southern Canada and the United States. True alkanet is *Anchusa*, which may escape from cultivation.

GROMWELL, L. OFFICINALE, is very like *L. arvense*. The leaves are more numerous and crowded. The tiny white corolla scarcely emerges from the calyx.

May to August: introduced from Europe and now found from Quebec to Ontario and southward to New Jersey and Illinois.

L. LATIFOLIUM is a native American gromwell, easily distinguished from the two immigrant gromwells by its broader leaves, with side veins branching from the midrib. The flowers are nearly $\frac{1}{4}$ inch across.

May and June: in woods and thickets from New York to Minnesota and southward to West Virginia, Tennessee, Arkansas, and Kansas. The southern *L. tuberosum*, a rough plant with a rosette of basal leaves as well as leaves on the stem, and yellowish-white flowers in clusters at the stem-tips, extends northward into Kentucky.

FIDDLENECKS AND TARWEEDS (AMSINCKIA)

In waste places from New England to Ohio and Illinois one may occasionally find specimens of several species of this western genus. They are characterized by small yellow flowers in coiled inflorescences, and rough-hairy, narrow leaves. The commonest species in our range are *A. barbata* and *A. spectabilis*, the former with a "beard" (*barba*) of soft white hairs on the lobes of the calyx.

COMFREYS (CYNOGLOSSUM)

The comfreys are coarse plants with large, usually oblong, rough leaves and small dull reddish or blue flowers in several false racemes. There are five small scales at the "throat" of the corolla as in the forget-me-nots.

HOUND'S-TONGUE, C. OFFICINALE, is a hairy plant from 2 to 3 feet tall or taller. The lower leaves have stalks; the upper are lanceolate blades without stalks. The flowers are dull reddish-purple, about ⅓ inch across. The whole plant has a "mousey" odor. The small nuts are beset with hooked bristles which cause them to stick to the clothing of men and other animals.

May to August: in fields and waste places and on roadsides from Quebec to British Columbia and southward to South Carolina, Alabama, Arkansas, and California; a weed from Europe. *Plate 104*. The name of the genus is from the Greek words for "dog" and "tongue," referring to the roughish, tongue-shaped leaves.

WILD COMFREY, C. VIRGINIANUM, is a native plant somewhat (but not closely) resembling the European comfrey (*Symphytum*). The leaves "clasp" the stem; i.e. they are indented at the base, a lobe extending on either side of the stem at the point of attachment. The plant is clothed with bristly hairs. The flowers are pale blue, violet, or even white, less than ½ inch across.

April to June: in open woodland from Connecticut to Missouri and southward to Florida and Texas. *Plate 104*.

NORTHERN WILD COMFREY, C. BOREALE, is slimmer than *C. virginianum*, with the lower leaves stalked or at least not "clasping." The corolla is blue and not more than ⅓ inch across.

May and June: in moist woods and thickets from Quebec to Ontario and Minnesota and southward to Connecticut, Indiana, and Iowa; also in British Columbia.

THE FALSE GROMWELLS (ONOSMODIUM)

The false gromwells resemble the gromwells (*Lithospermum*) somewhat in foliage, but they are rougher and hairier, and their flowers are in distinctly coiled inflorescences at the stem-tips. The styles project noticeably from the flowers. The leaves commonly have several lengthwise veins.

O. VIRGINIANUM has yellow flowers. The plant is from 1 to 2 feet tall or taller; stem and leaves are clothed with harsh hairs.

May to July: in dry woods and barren places from Massachusetts and New York to Florida and thence to Louisiana.

O. MOLLE has white or greenish flowers with triangular lobes. The plant is from 16 to 24 inches tall and clothed with fine grayish hairs (*molle* means "soft").

May to July: on dry hillsides from Illinois to Kentucky and Tennessee.

O. HISPIDISSIMUM is from 2 to 4 feet tall, clothed with stiff hairs (it is "hispid"). The flowers are white.

June and July: in dry prairies, fields, and woodland from New York to Ontario, Minnesota, and Nebraska and southward to North Carolina, Tennessee, and Texas. *Plate 104*.

O. OCCIDENTALE, from 20 to 40 inches tall, very densely hairy, has white or greenish flowers about ½ inch long.

May to July: in dry prairies and open woodland from Minnesota to Alberta and southward to Illinois, Oklahoma, Texas, and New Mexico.

PLATE 104

Solanum americanum *Murray*

Lithospermum incisum *Johnson*

Cynoglossum virginianum *Johnson*

Cynoglossum officinale *D. Richards*

Symphytum officinale *Phelps*

Onosmodium hispidissimum *Johnson*

THE HELIOTROPES (HELIOTROPIUM)

The heliotropes of the garden are old-fashioned and no longer the favorites they were. They are members of a large genus mostly native in the warmer parts of the earth. They are unique among *Boraginaceae* in certain technical details of the ovary (it is not deeply four-lobed), but (except one species) have the characteristic coiled inflorescence of small blue or white flowers – in some species a very long inflorescence. The garden varieties are descended from two Peruvian species. They were valued for their fragrance, which was perhaps responsible for their American name – cherry-pie. The botanical name means "turning to the sun"; an ancient belief about these plants.

We have one species native in North America and several that have wandered northward as weeds.

H. TENELLUM is our native heliotrope, and is the species that lacks the coiled inflorescence; one would not, without close examination, place it in this family. The plant stands only from 6 inches to a foot tall, or a little taller, much branched, with very narrow leaves whitened with close hairs. The tiny white flowers are borne singly in the axils of the leaves and at the tips of the branches.

June to August: in dry, open woodland and prairies from Kentucky to Iowa and southward to Alabama and Texas.

SEASIDE HELIOTROPE, H. CURASSAVICUM, is a tropical American plant, smooth, with rather thick leaves, and white or bluish flowers in a forked, coiled inflorescence. It is found along the seashore and in salt marshes as far north as Delaware and perhaps farther.

TURNSOLE, H. INDICUM, is, as its name indicates, an Asian species, or perhaps originally Brazilian, now a weed in all warm regions and extending northward to Virginia, Kentucky, and Missouri, and occasionally farther. It is remarkable for the very long coiled inflorescence (up to 6 inches). The stem grows 2 or 3 feet tall. The leaf-blades are ovate, stalked.

H. europaeum, a native of southern Europe, and *H. amplexicaule*, from South America, are occasional strays into our range. The former is hoary with white hairs; the leaf-blades are elliptic, stalked. The latter has lanceolate, hairy leaves on spreading stems.

SYMPHYTUM

Several species of this Old-World genus are found in our range. They are coarse plants with large, rough leaves and small whitish, dull blue, reddish, or yellowish flowers hanging in false racemes.

COMFREY, S. OFFICINALE, is the commonest of these. It stands from 20 to 40 inches tall. The leaves are lanceolate or ovate, the lower ones up to 8 inches long, the upper ones continuous with projecting flanges or "wings" on the stem (so that they seem to run down onto the stem). The flowers run the whole gamut of dull colors from white and yellow to red or blue.

May to September: on roadsides and in fields and waste places from Newfoundland to Ontario and southward to Georgia, Tennessee, and Louisiana. *Plate 104.*

Other species occasionally found are *S. asperum*, whose leaves do not "run down" onto the stem, and *S. tuberosum*, similar to *S. asperum* but less hairy.

THE POTATO FAMILY (SOLANACEAE)

The potato family is notable for the number of poisons provided by its species: atropine (belladonna), hyoscyamine, scopolamine, nicotine, and others. Several of these alkaloids are used in medicine.

A number of species are familiar in gardens, either for ornament or for the table: petunia, ornamental tobaccos, *Salpiglossis*, *Nierembergia*, *Schizanthus*, potatoes, tomatoes, the green and red peppers, and eggplant.

The family is characterized by parts in fives and an ovary generally with two chambers which becomes a berry or pod (capsule). The leaves are mostly rather coarsely and irregularly toothed; in some species pinnately lobed or divided. The manner in which the flowers are borne is peculiar in most of our species; this is taken up in the descriptions of the several genera below.

The genera may be recognized by the following characteristics:

I. *Genera whose stamens clearly project beyond a flaring corolla; the pollen is discharged through holes in the tips of the stamens:* Solanum.

II. *Genera whose stamens project scarcely or not at all beyond a funnel-shaped corolla; the pollen is discharged through lengthwise splits in the stamens. We may further divide this group by the type of fruit formed.*

A. Fruit a berry: *Physalis* (flowers yellow, borne singly); *Chamaesaracha* (flowers white with yellow center, in clusters); *Nicandra* (flowers blue).

B. Fruit a pod (capsule): *Datura* (flowers white; pods prickly); *Hyoscyamus* (flowers dull purple and yellow; pods smooth).

NIGHTSHADES, HORSE-NETTLES, AND OTHERS (SOLANUM)

The genus *Solanum* is vast, comprising probably about two thousand species, mostly tropical. A number of species are native in the West, and some of these reach the northeastern states. Some others have come from the Old World and are now common everywhere. And there are a few natives also common in our range. The potato is *Solanum tuberosum*. The Jerusalem-cherry is *S. pseudo-capsicum*.

The flowers are in short clusters (cymes) which in most of our species arise from the stem between the points at which leaves are attached: a most unusual pattern. Sometimes they are paired with leaves, on the opposite side of the stem. The fruit is a berry with many seeds in its two chambers.

I. *We may first distinguish species without prickles.* (*See also* S. elaeagnifolium *in group II.*)

BITTERSWEET or NIGHTSHADE, S. DULCAMARA, is a twining plant generally recognizable by the shape of the leaves: some at least of the leaves have two distinct lobes or segments at the base of the ovate blade. The corolla is purple, with a cone of yellow stamens in the center. The berries are red.

May to September: an European plant found in thickets from Nova Scotia to Minnesota and southward to North Carolina and Missouri; also in the West. *Plate 105. Dulcamara* combines two Latin words meaning "sweet" and "bitter"; the berries are said to taste first sweet then bitter. Experiments to test this are not recommended. Leaves and unripe berries contain the poisonous alkaloid solanine, and, while the ripe berries are apparently harmless, they may be injurious if taken too freely. This bittersweet should not be confused with the woody vine of that name, in another family. In England the plant is also known as snakeberry, snakeflower, snake's-food, and snake's-meat; and – being good against witchcraft – as witch-flower.

BLACK NIGHTSHADE, S. AMERICANUM, is a slender, branched plant up to 3 feet tall, with undivided and unlobed leaves. The flowers are white or tinged with purple, in small clusters. The berries are black.

June to November: in dry woods and thickets and waste or cultivated land from Maine to North Dakota and southward to Florida and Texas. *Plate 104.* This is the American counterpart of the European black nightshade, *S. nigrum;* and the two have generally been confused. True *S. nigrum* is apparently quite rare in North America. Both have often been called deadly nightshade, and the leaves and unripe berries do indeed contain enough solanine to cause the death of a small animal. This disappears from the berries as they ripen, and they are eaten, cooked or raw, by many persons without harm. However, some persons are apparently poisoned by them, so that caution is wise. The name deadly nightshade is applied in England to a much more poisonous plant, *Atropa belladonna*, which is not found wild in this country.

II. *Other species commonly encountered are beset with prickles on leaves and stems.*

HORSE-NETTLE, S. CAROLINENSE, is a common weed. The stem is up to 3 feet tall or taller. The leaves are coarsely toothed or even lobed, with prickles on the lower side along the midrib. The flowers are white or pale violet. The berries are yellow.

May to October: in waste land, often in sandy places, from New England to Washington and southward to Florida and Texas; commoner southward. *Plate 105.*

BUFFALO-BUR, S. ROSTRATUM, is a western plant that has invaded the East. It is about 2 feet tall. The stem and leaves are beset both with yellowish prickles and with starlike hairs. The leaves are pinnately deeply cleft or divided with rounded lobes or segments which may themselves be lobed. The flowers are bright yellow. One stamen, the lowest, is longer than the other four and stands out from them at an angle. The berry is surrounded by the close-fitting and very prickly calyx; this forms the "bur."

July to October: in prairies, fields, etc. from North Dakota to Wyoming and southward to Mexico; and straying eastward through our range. *Plate 105.* It is not uncommon at least in Missouri.

WHITE HORSE-NETTLE or SILVERLEAF NIGHTSHADE, S. ELAEAGNIFOLIUM, is another westerner that has penetrated our range. It stands up to 3 feet tall or taller. The leaves are not lobed or divided; they may have wavy margins. The stem and leaves are covered with silvery hairs which are branched to form stars; there are usually prickles also. The flowers are violet, the corolla not deeply cleft into lobes, consequently rather more plate-like.

July to September: in prairies, waste places, and dry woodland from Ohio to Kansas, Texas, and Arizona and southward to Florida. *S. torreyi*, a rather smoother plant of the Southwest, is found in Missouri. Several other western and southern species are found rather rarely as weeds in our range.

THE GROUND-CHERRIES (PHYSALIS)

The ground-cherries are small plants, mostly about 2 feet tall, with branched stems and flowers borne singly. Although the lower leaves on the stem are single, the upper ones, where flowers are borne, are generally paired; a flower-stalk arises between the two opposite leaf-stalks. Sometimes a branch arises at the same level, in the axil of one of the leaves; then the flower seems to grow in a fork of the stem. The petals (yellow in our species) form a hanging bell, the edge of which is scarcely divided into lobes. The calyx is at first small but enlarges greatly so as to completely enclose the berry (the "ground-cherry"); it forms a sort of pointed bladder (whence the botanical name of the genus, which is Greek for "bladder") with five angles, often indented at the top.

The berries are edible though rather tasteless. They have been much used by country people for preserves. The plants bloom in late summer in most of our range, earlier farther south. Our species all grow in prairies, open woodland, waste land, etc.

The shape of the leaves, their hairiness, and other characteristics by which we generally try to distinguish species are very variable. The species have accordingly been quite confused. Recent study indicates that the best criteria are such minutiae as the length and color of the pollen-bearing heads of the stamens and the presence or absence of dark spots in the corolla. On this basis several supposed species are reduced to being varieties of more easily distinguished species.

I. *Species with five dark spots inside and deep down in the corolla.*

P. VIRGINIANA is perhaps our most widespread species. It may be smooth or hairy but not sticky. The stamens have yellow heads from $\frac{1}{12}$ to $\frac{1}{6}$ inch long.

April to July: from Connecticut to Minnesota and Oregon and southward to South Carolina, Alabama, Arkansas, Arizona, and California. *Plate 176.*

P. HETEROPHYLLA is another common species. Its stem and leaves usually bear short, sticky (glandular) hairs mixed with longer, soft hairs. The stamens have yellow heads from $\frac{1}{8}$ to $\frac{1}{5}$ inch long.

April to September: from Quebec to Minnesota and Colorado and southward to Florida and Texas. *Plate 176.*

P. PUBESCENS has stem and leaves covered densely or sparingly with long sticky hairs. The stamens have blue heads not longer than $\frac{1}{12}$ inch.

May to November: from Maine to Washington and southward to North Carolina, Alabama, Texas, and California.

P. SUBGLABRATA is smooth or almost so. It grows from 1 to 5 feet tall. The leaves have mostly ovate blades with scarcely any teeth, on stalks 2 inches or more long. The corolla is greenish with a purple center, about an inch across.

June to September: in meadows, fields, and waste places from Vermont to Ontario, Michigan, and Iowa and southward to Georgia and Texas. *Plate 105.*

II. *Species which lack dark spots in the corolla.*

P. ANGULATA has (as the name perhaps suggests) more sharply-toothed and long-toothed leaves than most species. It is smooth or nearly so throughout. The heads of the stamens do not exceed $\frac{1}{10}$ inch, and are bluish.

May to September: from Connecticut to Kansas and southward to Florida and Texas.

P. MISSOURIENSIS is hairy and often sticky. The heads of the stamens are minute, not exceeding $\frac{1}{20}$ inch in length, and bluish.

June to October: in rocky woods and barrens from Missouri and Kansas to Arkansas and Oklahoma (and perhaps Texas).

P. PUMILA has hairs standing straight out on the stem, thus making it look bristly. The stamens have yellow heads from $\frac{1}{10}$ to $\frac{1}{8}$ inch long.

Several other species – southerners – may be found in our range, notably *P. viscosa*, with yellow stamen-heads $\frac{1}{8}$ inch long, and starlike branched hairs. *P. alkekengi*, whose calyx is bright red at fruiting time, is cultivated as Chinese-lantern-plant (from Europe), and is sometimes found growing wild.

PLATE 105

Physalis subglabrata *McDowell*

Solanum dulcamara *Johnson*

Chamaesaracha grandiflora *Rhein*

Solanum rostratum *Rickett*

Datura stramonium *D. Richards*

Solanum carolinense *Gottscho*

CHAMAESARACHA

We have a single species of this genus; others inhabit Asia and South America.

WHITE-FLOWERED-GROUND-CHERRY, C. GRANDIFLORA, somewhat resembles *Physalis* and is by some botanists placed in that genus. The stem grows from 6 to 36 inches tall, bearing long sticky hairs. The leaves usually lack teeth. Several flowers are borne at the same level. The corolla is white with a yellow center, up to 2 inches across.

June to August: in sandy or rocky places from Quebec to Saskatchewan and southward to Vermont, Michigan, Wisconsin, and Minnesota. *Plate 105*. There are several similar South American species.

NICANDRA

There is one species of *Nicandra*, a native of Peru and escaped from cultivation in North America.

APPLE-OF-PERU, N. PHYSALODES, resembles *Physalis*, but the corolla is blue. The berry is enclosed in a bladderlike calyx, as in *Physalis* (*physalodes* means "resembling *Physalis*"). The plant grows up to 5 feet tall. The leaves are stalked, the blades coarsely and irregularly toothed.

July to September: in waste grounds, usually near old gardens and abandoned dwellings from Nova Scotia to Missouri and southward to Florida. *Plate 180*.

DATURA

One species of this genus is a common weed of our range. Others occasionally escape from cultivation and are found growing wild. All are natives of the Southwest or of Asia.

JIMSONWEED or THORN-APPLE, D. STRAMONIUM, is a smooth plant up to 5 feet tall with coarsely toothed leaves. The flowers are borne singly in forks of the stem, accompanied in that position by a single leaf. The corolla forms a flaring funnel up to 4 inches long, the lower half enclosed in a tubular, angled calyx. The fruit is a spiny capsule containing many black seeds.

July to October: in waste ground and fields throughout our range; more abundant southward. From Asia. *Plate 105*. The stem may be green or purple, the flowers white or pale violet. The plant contains several dangerously poisonous alkaloids; the preparation known as stramonium, containing several of these, has been used medicinally.

An old negro couple, becoming too poor to buy coffee, brewed the seeds of jimsonweed, from their back yard, as a substitute. They were found a day or two later, quite dead.

The name jimsonweed is said to be a corruption of Jamestown weed, the plant having appeared around the dwellings of that early settlement.

HYOSCYAMUS

One species of this poisonous genus is a native of Europe that has been introduced to cultivation in North America and is sometimes found growing wild.

HENBANE, H. NIGER, grows 3 feet or more tall. It is a clammy, hairy, ill-smelling plant with numerous coarsely toothed leaves. The flowers also are numerous, in a false raceme at the summit, each flower opposite a leaflike bract. The flowers are large, an inch or more long, the dull yellow corolla, with purple veins, sheathed at the base by a vase-shaped calyx. The seedpod is later enclosed in this calyx.

June to August: on roadsides and in waste places and fields from Quebec and Ontario southward to New York and westward to South Dakota. *Plate 178*. This plant yields the well-known alkaloid hyoscyamine.

THE MORNING-GLORY FAMILY (CONVOLVULACEAE)

In temperate countries the *Convolvulaceae* are mostly twining or trailing plants (there are erect woody species in the tropics). The petals are joined to form a funnel which in some species flares at the end into a disk; there may be five lobes or teeth, or none. The corolla of most species is twisted or rolled up in the bud – and often twists again as it withers. The sepals are almost separate; some species have bracts just below the calyx. The ovary has two or three chambers, in each of which two seeds are formed.

The genera of our range may be separated as follows:

Style not cleft; two stigmas: *Convolvulus.*

Style not cleft; one stigma with two or three lobes: *Ipomoea.*

Style not cleft; one stigma, not lobed; red flowers: *Quamoclit.*

Style cleft into two branches; two stigmas; small white flowers: *Stylisma.*

Besides these, the parasitic plants called dodder (*Cuscuta*) are usually included in this family. They have orange stems which twine in a tangled mass over other plants, no leaves, and small white flowers.

BINDWEEDS (CONVOLVULUS)

The bindweeds are common twining or trailing weeds with handsome pink or white funnel-shaped flowers. The margin of the corolla is only very shallowly lobed or notched, with five small points marking the tips of the five petals. There are two bracts below the calyx. The leaves are more or less arrow-shaped or heart-shaped, with pointed or blunt lobes extending down from the base of the blade. The single style is tipped with two stigmas.

HEDGE BINDWEED, C. SEPIUM, has long-stalked leaves with blades varying from heart-shaped to arrow-shaped. The corolla is 2 or 3 inches long, white or pink. The two broad bracts may be mistaken for sepals; but the calyx is within them.

May to September: in thickets and waste places from Newfoundland to British Columbia and southward to Florida, Alabama, Texas, New Mexico, and Oregon. *Plate 106.* This is a variable species: some forms are native in North America, others introduced from Europe. A form that grows from Pennsylvania to North Dakota and southward has been amusingly called variety *fraterniflorus:* the flowers are often fraternally in pairs! In England this weed has acquired many names, being variously attributed to the devil (devil's-garter, devil's-guts), to father (daddy's-white-shirt), and to grandmother (granny's-nightcap). And in Somerset it was called morning-glory before our American *Ipomoea* took over that name. The name hedge-bindweed has more meaning in England, where it does clamber over the tall hedges and cover them with its bloom. Some modern botanists place this in a genus distinct from the following species, named *Calystegia.*

FIELD BINDWEED, C. ARVENSIS, is a smaller plant than *C. sepium* but a more troublesome pest (one English name is hellweed). It has smaller, shorter-stalked leaves with blunter tips and a triangular outline. The stems twine or trail and often form tangled mats that cover the ground or the supports on which it climbs. The corolla is less than an inch long and generally white but sometimes tinged with pink. There are two bracts as in other species of *Convolvulus*, but they are narrow and not close to the calyx – sometimes ½ inch below.

May to September: a native of Europe, established in fields and waste land and on roadsides throughout the United States. *Plate 106.*

LOW BINDWEED, C. SPITHAMAEUS, grows more or less erect, from 3 to 20 inches tall. The leaves have short stalks and oblong blades either indented or tapering at the base. The few flowers grow from the lower axils, the corolla white and from 2 to 3 inches long. The calyx is concealed by two bracts much as in *C. sepium.*

May to July: in sandy and rocky places from Maine and Quebec to Minnesota and southward to Maryland, the mountains of Georgia, Tennessee, and Iowa. *Plate 106.* The plants are generally downy; some with woolly leaves have been treated as a separate species, *C. purshianus.*

MORNING-GLORIES (IPOMOEA)

The corolla of a morning-glory flares into more or less of a flat disk at the end of the trumpet. The style bears only one stigma, but this may have two or three lobes. There are no bracts immediately below the calyx. The leaf-blades are generally heart-shaped with rounded lobes at the base (or cleft into three in one species).

This is mainly a tropical genus. Many species (and hybrids) are cultivated; the most popular kinds grown in northern gardens are descended from a Mexican species. The sweet potato is an *Ipomoea*. Others, as noted below, are beautiful but pernicious weeds.

COMMON MORNING-GLORY, I. PURPUREA, has blue, purple, red, white, or variegated flowers, generally two or more on each long stalk from an axil. The corolla is about 2 inches long or longer. The stigma has three lobes and the ovary three chambers.

July to October: on roadsides and in waste land and cultivated fields in many parts of the United States. *Plate 180*. This is a tropical species, originally brought into cultivation in this country and now established as a troublesome weed. A familiar sight in the south-central states is a tangle of morning-glory covering the corn-stalks, the handsome flowers appearing among the leaves of the corn.

I. HEDERACEA receives its name from the resemblance of its leaves to those of some forms of ivy (*Hedera*); the blades are usually cleft into three; at the base there is a deep indentation. The flowers are blue (or white), becoming redder as they age. The corolla is less than 2 inches long. The stigma is three-lobed and the ovary three-chambered.

July to October: like common morning-glory, a tropical species now established as a weed in waste and cultivated land, from New England to Minnesota and southward to Florida and Arizona.

MAN-OF-THE-EARTH, I. PANDURATA, is chiefly remarkable for its extraordinary tuberous root; this may be several feet long and 20 pounds in weight, and is edible. Such a root would be treasure trove to a hungry Indian family! The rest of the plant resembles common morning-glory, but may be recognized by the large white flowers (up to 3 inches long; they may have a purple center) and the two-lobed stigma.

June to September: in woods and thickets from Connecticut to southern Ontario, Michigan, and Kansas and southward to Florida and Texas. *Plate 106*.

I. LACUNOSA is a small plant, a diminutive morning-glory, with flowers that are generally white (sometimes purplish) and less than 2 inches long. The stigma is two-lobed.

August to October: in moist thickets and fields and on roadsides from New Jersey to Kansas and southward to Florida and Texas. *Plate 106*.

CYPRESS-VINES (QUAMOCLIT)

The cypress-vines are distinguished from the morning-glories chiefly by their bright scarlet flowers, which flare into a small, nearly flat, five-lobed disk. (Some botanists, however, place them in *Ipomoea*.) Both stamens and style project from the corolla. The stigma is single and unlobed. Both species are natives of tropical America. They have been cultivated and have spread to the wild.

RED MORNING-GLORY or QUAMOCLIT, Q. COCCINEA, has a slender twining stem with generally heart-shaped leaf-blades; the leaf-blades are often angled and sometimes cleft into three or more lobes. There are one or two flowers on each flower-stalk.

July to October: in thickets and waste places and on roadsides from Massachusetts to Michigan and Missouri and southward to Florida and Arizona. *Plate 106*. Quamoclit is a Mexican name.

CYPRESS-VINE, Q. PENNATA, is distinguished by its leaves, pinnately divided into narrow segments.

August to October: occasionally in waste places from Virginia to Missouri and southward to Florida and Texas.

STYLISMA

One species is found within our range.

S. PICKERINGII is a slender, branching, trailing plant with very narrow leaves generally an inch or two long. Flowering branches grow from the axils of leaves, each bearing two narrow bracts and, above these, one or several flowers. The corolla is white, about ½ inch long, funnel-shaped. The style is divided into two branches.

July to September: in dry prairies and sandy barrens from New Jersey to Florida and from Illinois to Iowa and Texas. *Plate 106*. Former name, *Breweria*.

PLATE 106

Quamoclit pennata *Johnson*

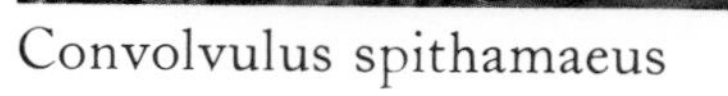

Convolvulus spithamaeus *Ward*

Convolvulus sepium *Johnson*

Ipomoea lacunosa *Elbert*

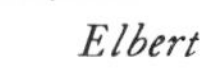

Stylisma pickeringii *Allen*

Ipomoea pandurata *Uttal*

Convolvulus arvensis *Johnson*

DODDER (CUSCUTA)

There are some fifteen species of dodder in our range (ten times as many in the world), to be distinguished only by minute characteristics of the tiny flowers. All are leafless plants with yellow or orange stems and clusters of almost spherical, generally white flowers. The seeds germinate in the ground, but the slender twining stems soon attach themselves to other plants, and connection with the ground is lost. The dodder sends suckers into the stem of its "host"; these absorb the food and other substances by which it lives. Tangles of the repulsive yellow or orange stems may eventually cover green plants.

The calyx generally has five divisions, and the corolla five teeth; but in some species the number four predominates. There are in most species two styles. In many species there are bracts below the calyx.

Few amateurs will want to undertake the formidable task of identifying the species; the ambitious must tackle the technical manuals.

Two of the most widespread species are *C. polygonorum* and *C. gronovii*. *Polygonorum* means "of the polygonums" or smartweeds; this dodder is found on smartweeds and also on other kinds of plants. It lacks bracts beneath the flowers; the sepals, petals, and stamens are mostly four; the teeth of the corolla extend straight up. *C. gronovii* is named for the Dutch botanist Jan Frederik Gronovius, teacher of Linnaeus. This also lacks bracts; the flower-parts are mostly in fives; the teeth of the corolla extend out at right angles. *Plate 180*.

THE BLUEBELL FAMILY (CAMPANULACEAE)

Many plants, naturally, are called bluebells: many plants have bell-shaped flowers colored blue. But the name has been attached by botanists to one such group of plants, the "bluebells of Scotland" (which grow in many other lands, including America). It was for these plants that Leonhard Fuchs in the sixteenth century coined the name *Campanula*, from the Latin *campana*, a "bell." And on these plants the bluebell family is built.

The *Campanulaceae* have five petals joined into a bell or spreading into a five-lobed disk, generally blue or white. Our species have three stigmas atop the single style. The ovary is inferior. The fruit is a capsule which usually sheds its many seeds through pores on the sides.

BLUEBELLS (CAMPANULA)

The true bluebells are generally distinguished by the bell-shaped corolla; but one species has no such corolla but five spreading petals joined only at the base. This species, like some of the others, has flowers in a tall raceme.

HAREBELL, BLUEBELLS-OF-SCOTLAND, C. ROTUNDIFOLIA, is named, curiously, for the round (*rotundi-*) leaves (*-folia*) at the base of the stem which are seldom seen. A few other leaves with roundish or ovate blades on stalks succeed the first round-bladed ones, but most of the leaves are very narrow. The stem is slender and much branched, the blue bells hanging from the tips of branches.

June to September: on rock ledges, and beaches, in meadows and woods throughout Canada and southward to Pennsylvania, West Virginia, Illinois, Nebraska, Texas, and California; and in the Old World. *Plate 107*. There is great variation in stature, branching, length of leaves, and other characteristics; a white-flowered form is known.

The "Scotland" in the name should not conceal the worldwide distribution of the species. The English, attached to their own exquisite bluebells which are in the lily family, are careful to call this species harebell. The hare is a witch animal, and this plant has magic in it. It is also called old-man's-bell (the "old man" is the devil), fairy-bells, fairies'-thimble, and witches'-bells.

MARSH BLUEBELLS, C. APARINOIDES, belies its name by having flowers that are almost white. The stem is slender and weak, up to 2 feet tall but usually supported by other plants. The leaves are narrow, pointed, mostly 2 or 3 inches long. The flowers are only about $\frac{1}{2}$ inch long.

June to August: in meadows and other wet places from Quebec to Saskatchewan and southward to Georgia, Kentucky, Iowa, and Colorado. *Plate 107*.

PLATE 107

Campanula aparinoides *Elbert*

Campanula americana *Johnson*

Campanula rapunculoides *Rickett*

Campanula rotundifolia *Smith*

Campanula divaricata *D. Richards*

Jasione montana *Phelps*

Triodanis perfoliata *Johnson*

TALL BELLFLOWER, C. AMERICANA, is the species without bell-like flowers. It is a stout plant, up to 6 feet tall, with lanceolate or ovate, toothed leaves. The five spreading lobes of the corolla are light blue. The style is directed downward, with its tip bent upward. The lower flowers are in the axils of leaves, the upper in the axils of bracts, forming a raceme.

June to September: in woods and thickets and on roadside banks from southern Ontario to Minnesota and South Dakota and southward to Florida, Alabama, and Oklahoma. *Plate 107*.

SOUTHERN HAREBELL, C. DIVARICATA, is from 1 to 3 feet tall. The leaves are lanceolate or ovate, with large sharp teeth. The flowers are tiny, $\frac{1}{3}$ inch long or less, and pale blue; they are numerous, in a branched inflorescence at the top of the plant.

July to September: in dry, rocky woods from western Maryland to Kentucky and southward to Georgia and Alabama. *Plate 107*.

C. RAPUNCULOIDES is a handsome immigrant from Europe often 3 feet tall. Its blue flowers are in racemes terminating the stems. The corolla is about an inch long, and hangs lazily downward along the stem.

July to September: on roadsides and in thickets and waste land from Newfoundland to North Dakota and southward to Maryland, West Virginia, Illinois, and Missouri. *Plate 107*.

NETTLE-LEAVED BLUEBELL or THROATWORT, C. TRACHELIUM, is a plant up to 3 feet tall, often bristly, with broad ovate leaves, sharply toothed. The flowers are in loose clusters at the tip of the stem and of branches from the axils of leaves. The corolla is violet, bell-shaped with flaring lobes, an inch or more long.

August to October: on roadsides and in thickets and waste land from Quebec to Massachusetts and Ohio; an immigrant from Europe.

CLUSTERED BLUEBELL, C. GLOMERATA, is named for the arrangement of its flowers, in "glomerules" – close bunches; they have no stalks. The plant is 2 feet tall or taller, and rather hairy. The corolla is purple, about an inch long.

June and July: on roadsides and in old fields and waste land from Quebec to Minnesota and Massachusetts; introduced from Europe.

VENUS'-LOOKING-GLASS (TRIODANIS)

These plants have few or no branches. The stems bear leaves without stalks, in whose axils are the flowers, mostly single, also without stalks. The five petals spread apart, being joined only at the base; but those of many flowers never develop or separate, the flower pollinating itself within the corolla.

In most books these species are called *Specularia* (from *speculum*, "mirror"); but it has been shown that they form a genus distinct from the true *Specularia*, the Old-World Venus'-looking-glass. The name refers to the flattish, polished seeds, which suggested small mirrors.

T. PERFOLIATA is the common species, with nearly round leaves which partly encircle or "clasp" the stem. The lower axils contain flowers that do not open.

May to August: in open woods and fields from Maine to British Columbia and southward to Florida, Texas, and Mexico. *Plate 107*.

T. BIFLORA is very similar to *T. perfoliata*, with more ovate leaves whose bases do not extend around the stem. All the flowers often fail to open.

April to June: in dry open places from southern Virginia to Kentucky and Kansas and southward to Florida, Texas, and Mexico; and in Oregon and California; also in South America.

T. LEPTOCARPA has much narrower leaves than the two preceding species; they are narrowly elliptic or lanceolate, or broader near the tip.

May to August: in prairies and dry, open woods from Indiana to Minnesota and Montana and southward to Arkansas, Oklahoma, and Texas.

JASIONE

SHEEP'S-BIT, J. MONTANA, is an Old-World plant 2 feet tall or taller with narrow leaves and many minute blue flowers in a round head at the summit. It is found in waste places and fields from Massachusetts to New Jersey, flowering from June to September. *Plate 107*.

GROUP XII

SEPALS five or minute or lacking, petals from three to five; petals joined, radially symmetric. Stamens from two to four. Exceptions: *Justicia* and *Dicliptera* have bilaterally symmetric petals. *Triosteum* has five stamens. *Valeriana* has slightly bilateral petals.

I. *Flowers with five petals and two or four stamens; ovary superior:* acanthus family.

II. *Flowers with three or four petals and stamens of the same number; ovary inferior:* bluet family.

III. *Flowers with five petals and three stamens; ovary inferior:* valerian family.

IV. *Flowers with five petals and four (except in* Triosteum) *stamens; ovary inferior:* honeysuckle family.

V. *Flowers with five petals and three stamens; stamens and pistil in separate flowers; ovary inferior; climbing or trailing stems:* cucumber family.

THE ACANTHUS FAMILY (ACANTHACEAE)

Most species of the *Acanthaceae* are tropical. Only a few reach our range as wild flowers. A number are cultivated, including that Mediterranean *Acanthus* whose leaves inspired the capitals of columns in ancient Greece. The corolla of five joined petals in most genera forms two lips; i.e. it is bilaterally symmetric. However, in the commoner species of our range it is practically radial in symmetry. There are two or four stamens. The leaves are paired and mostly without teeth or lobes (but pinnately lobed in *Acanthus*). The fruit is a two-chambered capsule which splits into two halves.

RUELLIA

This is a large genus in the warmer parts of both hemispheres, especially the Western. The corolla is nearly or quite radial in symmetry, a trumpet with five flaring lobes. The calyx is deeply cleft into five narrow teeth. The genus is named for the sixteenth-century French herbalist Jean Ruel; none of the species seems to have acquired an English name, attractive though they are.

R. STREPENS grows from 1 to 4 feet tall; it is very nearly smooth. The flowers are violet, up to 2 inches long; the teeth of the calyx are lanceolate. The flowers are borne in the axils of leaves about the middle of the stem, singly or in clusters of two or three on a short leafy stem.

May to July: in open woods, thickets, etc. from New Jersey to Iowa and southward to South Carolina and Texas. A form is known whose flowers pollinate themselves without opening.

R. HUMILIS has a branching stem from 8 inches to 3 feet tall, usually hairy or covered with a whitish down (it is very variable in this and other respects). The flowers are pale violet, nearly 3 inches long. The teeth of the calyx are very narrow. The flowers are crowded in the upper axils, with practically no stalks.

June to August: in prairies, open woodlands, grassy slopes, etc. from Pennsylvania to Nebraska and southward to western Florida and Texas. *Plate 108.* Recent studies merge this species with the next.

R. CAROLINIENSIS is distinguished by its usually hairy stem, from 1 to 3 feet tall, and by its distinctly stalked leaves. The leaves are crowded. The flowers also are crowded, practically without stalks, in the upper axils. The corolla is lavender, from 1 to 2 inches long.

June to August: in woods and clearings from New Jersey to Indiana and southward to Florida and Texas. *Plate 108.* An extremely variable species through its wide range.

R. PEDUNCULATA has a finely downy stem from 1 to 3 feet tall. The flowers are about 2 inches long, with purple petals and threadlike calyx-teeth; they are borne singly at the ends of relatively long branches ("peduncles") from the axils of the middle leaves.

May to August: in dry woods and on rocky slopes from Illinois to Missouri and southward to Louisiana and Texas. *Plate 108. R. purshiana* of the Appalachian Mountains is very similar and may be simply a form of this species.

JUSTICIA

Of this tropical genus one species is widespread in our range. Another, *J. ovata*, just reaches our southern boundaries.

WATER-WILLOW, J. AMERICANA, receives its inappropriate English name from its narrow, willow-like leaves. It does grow in water, emerging to a height of 3 feet. The flowers are in dense heads on long branches from the axils of leaves commonly in the upper half of the stem. The flower has a bilaterally symmetric corolla, with an upper lip curving back, and a three-lobed lower lip; all pale violet or white, often with purple markings. A pretty flower especially when seen through a lens.

June to October: in shallow water and on moist shores of streams from Quebec to southern Ontario and Wisconsin and southward to Georgia and Texas. *Plate 108.*

DICLIPTERA

We have one species of this large tropical genus.

D. BRACHIATA is a smooth or nearly smooth plant from 1 to more than 2 feet tall, with ovate leaf-blades on fairly long stalks. The flowers are pink or pale purple, with a two-lipped, bilaterally symmetric corolla not much more than $\frac{1}{2}$ inch long. They are in clusters in the axils of leaves and on branches that grow from the axils, each cluster more or less enveloped by leaflike bracts. There are two stamens.

September: in moist woods and bottomlands from southern Indiana to southern Missouri and Kansas and southward to Florida and Texas.

THE BLUET FAMILY (RUBIACEAE)

The *Rubiaceae* are an enormous tropical family, of which we have only a few outlying members. To call it the bluet family will seem absurd to the botanist who knows the family as a whole; but the bluets (*Houstonia*) may serve to represent the family in our part of the world.

In our species the flower-parts – sepals, petals, stamens – are mostly in fours (some have them in threes). The leaves are paired or in circles. The ovary is inferior.

Coffee and quinine are derived from trees of this family.

BLUETS (HOUSTONIA)

Bluets are small plants with a single flower or a cluster of flowers at the tip of each stem. The leaves on the stem are paired, and in many species there is a rosette of leaves at the base. There are four lobes to the

PLATE 108

Houstonia nigricans *Rickett*

Ruellia humilis *Johnson*

Ruellia caroliniensis *Uttal*

Houstonia caerulea *Gottscho*

Ruellia pedunculata *Uttal*

Justicia americana *Johnson*

corolla. The style bears two narrow stigmas. These are pretty little plants with blue or white flowers, some appearing in early spring.

The numerous species may easily be placed in two groups.

I. *Species whose corolla is a narrow tube crowned by four lobes spreading sharply at right angles. The flowers are single at the tips of branches.*

QUAKER-LADIES or INNOCENCE, H. CAERULEA, grows in tufts. From the basal rosettes arise leafy stems from 2 to 8 inches tall, branching, each branch tipped with one flower. The petals are pale blue or lavender with a yellow center, about $\frac{1}{2}$ inch across. White-flowered plants also occur.

April to June: in meadows and fields and in woods from Nova Scotia to Ontario and Wisconsin and southward to Georgia, Alabama, and Missouri. *Plate 108.* This small plant rejoices in at least a dozen names in English.

STAR-VIOLET, H. MINIMA, is tiny, scarcely over 4 inches tall, branching, with a purple flower $\frac{1}{4}$ or $\frac{1}{3}$ inch broad ending each branch. The calyx nearly equals the tube of the corolla.

March and April: in fields and open woodland from Illinois to Iowa and Kansas and southward to Arkansas and Texas. Small as they are, these plants will sometimes color a field blue.

STAR-VIOLET, H. PUSILLA, is similar to *H. minima*, differing chiefly in having sepals much shorter than the tube of the corolla.

March and April: in fields and open woods from Virginia to Nebraska and southward to Florida and Texas.

H. SERPYLLIFOLIA is a creeping plant, the stem covered with small round leaves. The flower-stalks arise from the axils of leaves and stand erect. The flowers resemble those of *H. caerulea*.

April to July: on damp slopes and along streams in the mountains from Pennsylvania and West Virginia to Georgia and Tennessee. The name comes from the resemblance of the leaves to those of thyme – *Thymus serpyllum*.

II. *Species whose corolla ends in four teeth flaring gradually, not sharply bent outward; the flowers are clustered.*

H. PURPUREA has ovate or lanceolate leaves each with three or five main veins extending from the base. The stem is from 4 to 20 inches tall. The numerous small flowers are pale violet or white.

May and June: in open woods and clearings from Delaware and Pennsylvania to Iowa and southward to Georgia, Louisiana, and Oklahoma. *Plate 109.* See also the following paragraph.

H. LANCEOLATA, H. LONGIFOLIA, and H. TENUIFOLIA are three species upon which botanists are unable to agree and which the amateur cannot hope to distinguish with certainty. They have very narrow leaves, with one main vein, except those at the base, which may be elliptic; this basal rosette is often conspicuous, since they are mostly plants of open, rocky or sandy places. The flowers are generally white or nearly so. The calyx of *H. lanceolata* has comparatively long teeth, like those of *H. purpurea* (about $\frac{1}{8}$ inch); the others have very short and very narrow teeth (about $\frac{1}{12}$ inch). They have been variously treated as varieties of *H. purpurea* or of *H. longifolia*, and plants have been labeled as hybrids between these species! If one wants a name, perhaps *H. longifolia* is as good as any. The photographs on *Plate 109* are as labeled by the photographers and are probably as correct as is possible.

H. CANADENSIS may be distinguished by its leaves, which, though narrow, have a curved outline, and usually bear a row of hairs at the margin. The corolla is purple.

May to August: on rocky slopes from Maine to Minnesota and southward to western Pennsylvania, Tennessee, and Arkansas. *Plate 109.*

H. NIGRICANS is named for its propensity to turn black when it is dried. It is narrow-leaved, like the preceding group of species; in fact, the leaves of many plants are almost hairlike. The plants are frequently very bushy.

June to October: in dry soil from Michigan to Nebraska and southward to Florida, Texas, and Mexico. *Plates 108, 109.*

HEDYOTIS

This is a vast genus of the tropics; by some botanists it is made even larger by putting *Houstonia* in it. Two species enter our range: *H. uniflora*, a rather straggling small plant, with elliptic or ovate leaves and small white flowers singly in their axils; the corolla has a short tube and four lobes shorter than the sepals (from Long Island and southern Missouri southward); and *H. boscii*, with very narrow leaves (in southeastern Virginia and southeastern Missouri). Both flower from July to October.

PLATE 109

Mitchella repens *Johnson*

Houstonia purpurea *V. Richard*

Diodia teres *Allen*

D. Richards

Houstonia canadensis

Houstonia lanceolata *Rhein*

Houstonia nigricans *Johnson*

Houstonia longifolia *Johnson*

MITCHELLA

One species of *Mitchella* is found in North America.

PARTRIDGE-BERRY or RUNNING-BOX, M. REPENS, is a creeping (*repens*) plant, with small evergreen leaves in pairs, the shining roundish blades on stalks. The fragrant white flowers are in pairs at the ends of the branches. The petals are hairy on the inside. The style bears four narrow stigmas. The fruit develops curiously from the lower parts of both flowers, which fuse into a single scarlet berry. At the end of this one can see the remains of both sets of sepals.

May to July: in woods from Newfoundland to Minnesota and southward to Florida and Texas. *Plate 109*. Sometimes the two flowers of a pair are united. The dry, seedy berries are edible but probably more appreciated by partridges than by man. The leaves are sometimes variegated.

BUTTONWEEDS (DIODIA)

These are small plants with weak stems and short stalkless leaves in pairs. The two opposed leaves are connected, around the stem, by small membranes from which long bristles project. The flowers are small, without stalks, from one to three in each axil.

D. TERES has stiff, narrow leaves and white, pink, or purplish flowers about $\frac{1}{4}$ inch long. There are four petals and four sepals; the style is single, with one stigma.

June to October: in dry ground, often a weed in sandy roads (*Diodia* is from a Greek word for "thoroughfare") from Rhode Island to Michigan and Iowa and southward to Florida and Texas. *Plate 109*.

D. VIRGINIANA has elliptic leaves and white or pink flowers nearly $\frac{1}{2}$ inch long. There are four petals but only two sepals; the style is branched, each branch ending in a narrow stigma.

THE BEDSTRAWS AND CLEAVERS (GALIUM)

The genus *Galium* includes some species that make a handsome show of very many very small white or yellow flowers; and others that are weedy, with fewer, white or greenish flowers. There is practically no calyx. There may be three or four petals, joined only at the base; the stamens are of the same number. Two styles rise from the inferior ovary, which forms a pair of small globular dry fruits on one stalk.

It is scarcely necessary here to describe all the weedy species of *Galium*, but only a few of the commoner ones, with the more attractive species.

I. *Species with very numerous white or yellow flowers closely clustered: bedstraws.*

YELLOW or LADY'S BEDSTRAW, G. VERUM, is a stiff plant 2 or 3 feet tall, with very narrow leaves, six or eight in a circle. The upper part of the plant is covered with the little yellow, fragrant flowers.

June to September: introduced from Europe and now quite common in fields and on roadsides from Newfoundland to Ontario and North Dakota and southward to Virginia, Indiana, and Kansas. *Plate 110*. The plants have had many uses for many centuries. The leaves curdle milk and have been used in Europe, throughout the history of that land, in the preparation of cheese. One English name of the plant is cheese-rennet. The name *Galium* seems to be derived from the Greek for "milk." A red dye is obtained from the roots. A beverage may be distilled from the leaves and has been used in medicine. And its astringency brought it into use for stopping the flow of blood. The foliage is sweet-smelling when dried, and was on this account much used as a bed – the medieval mattress. It may also have decreased the flea population in such beds – another English name is fleaweed. And finally it was believed to make childbirth easier, whence the common English name. At some time in the Middle Ages this was changed into Our-Lady's-bedstraw, becoming associated with various legends about the birth of Christ; and in Latin it became *Galium verum*, the "true galium" of those stories.

NORTHERN or WHITE BEDSTRAW, G. SEPTENTRIONALE, has leaves in fours on a stem from 1 to 3 feet tall or taller. It is variable in hairiness and the fruit may or may not be bristly. The massed little white flowers are very fragrant.

June to August: on roadsides and in fields and meadows from Quebec to Alaska and southward to Delaware, West Virginia, Kentucky, and Missouri; and in the West. *Plate 110*. This is frequently regarded as identical with *G. boreale* of northern Europe and Asia. It shares several properties with yellow bedstraw, particularly that of furnishing sweet-smelling straw for a bed, and providing a red dye from its roots.

PLATE 110

Galium mollugo *Murray*

Galium septentrionale *Johnson*

Galium verum *Gottscho*

Galium asprellum *Gottscho*

Valeriana pauciflora *Johnson*

Galium aparine *Rickett*

Galium aparine *Rickett*

G. MOLLUGO has leaves on the main stem in circles of eight; on the branches they may be in sixes. They tend to be wider near the tip. The flowers are small, numerous, white.

May to August: an European plant naturalized from Newfoundland to Ontario and southward to Virginia and Tennessee. *Plate 110.*

II. *Species with white or greenish flowers in more open inflorescences or scattered among the leaves.*

A. Of these, several have bristly fruits, like minute burs.

CLEAVERS or GOOSEGRASS, G. APARINE, may be mentioned as a weed which occurs throughout North America and Europe and Asia. It is a straggler, a leaner on other plants, sometimes 4 feet tall. Stem and narrow leaves (mostly in eights) are furnished with minute backward-pointing hooks, by which it clings or cleaves to the hair or clothing of passing animals. The minute white flowers are on branches from the axils. The bristly fruit also cleaves. *Plate 110.* The seeds, roasted, have been used as a substitute for coffee. The plants are eaten by geese, and, chopped up, used in feeding geese.

WILD-LICORICE, G. CIRCAEZANS, is a species with bristly fruits and rather broad, elliptic, three-veined leaves in fours. The greenish flowers and burs are without stalks, attached on the sides of forking, often zigzag branches that come from the axils.

June and July: in woods and thickets from Quebec to Minnesota and Nebraska and southward to Florida and Texas.

G. LANCEOLATUM resembles the preceding and is also called wild-licorice. The leaves are more lanceolate. The flowers are yellowish turning purple.

June and July: in dry woods from Maine and Quebec to Minnesota and southward to the mountains of North Carolina and Tennessee.

SWEET-SCENTED BEDSTRAW, G. TRIFLORUM, has narrowly elliptic or lanceolate leaves mostly in sixes, the upper ones with a minute spine at the tip. The greenish flowers and burs are stalked, in threes, on the end of branches from the axils.

May to September: in woods and thickets from Newfoundland to Alaska and southward to Florida, Tennessee, Texas, and Mexico. The foliage is sweet-scented when dried.

G. PILOSUM has elliptic leaves in fours, each about an inch long and with only one conspicuous vein. The plants are generally softly hairy ("pilose"). The greenish or purplish flowers are stalked, on forking branches from the axils.

June to August: in dry woods from New Hampshire to southern Ontario and Michigan and southward to Florida and Texas.

B. Species with smooth fruits.

ROUGH BEDSTRAW, G. ASPRELLUM, like cleavers has hooked prickles on stem and leaves. The leaves are in fours, fives, or sixes, elliptic or lanceolate. The white flowers are in many clusters at the ends of the numerous branches and from the upper axils.

July to September: in moist thickets and woods from Newfoundland to Ontario and southward to North Carolina, Ohio, and Nebraska. *Plate 110.*

G. CONCINNUM is a slender, erect plant up to 2 feet tall, smooth or nearly so, with narrow leaves in sixes. The whitish flowers are numerous on many forked branches from the axils.

June and July: in woods and thickets from New Jersey to Minnesota and southward to Virginia, Kentucky, Arkansas, and Kansas.

G. TINCTORIUM generally clambers by means of the prickles on the angles of the stem. The leaves are narrow, mostly in fives and sixes. The whitish flowers, generally in threes, are at the ends of forking branches. There are mostly three petals.

June to September: in wet places practically throughout Canada and the United States. The name indicates that – like several other species in this family, for instance madder and woad – it has been used for dyeing. A similar widespread species is *G. trifidum*, with leaves chiefly in fours.

Two other small weeds invade the southern parts of our range: *Spermacoce glabra*, with bristles between paired leaves as in *Diodia*, and numerous flowers crowded in small round masses in the axils; and *Richardia scabra*, sometimes called Mexican clover, very hairy, the flowers packed into heads under which are two or four leaves broader and shorter than the rest; the flower parts are mostly in sixes. Field madder, *Sherardia arvensis*, from Europe, is found as a weed in cultivated and waste land through much of our range and southward. It has a weak stem which bears circles of narrow leaves like some species of *Galium*. The flowers are minute ($\frac{1}{8}$ inch long), pink or blue, in heads surrounded by narrow bracts joined together at the base.

THE VALERIAN FAMILY (VALERIANACEAE)

The valerian family is in our range a small and unimportant group, and is nowhere very impressive. Our species have no proper sepals, five petals, three stamens, and one pistil with an inferior ovary. The bristly or feathery calyx of some species is remarkable.

THE VALERIANS (VALERIANA)

The valerians are abundant in the West. We have four native species and one that has escaped from gardens. The flowers are small, white or pink, and clustered at the tip of the stem. They have feathery white or pink bristles in place of a calyx; these are at first rolled up, unrolling as the fruit develops. The leaves on the stem are pinnately divided (those at the base may be undivided).

V. PAUCIFLORA has a stem up to about 3 feet tall. The leaves at the bottom have ovate blades, indented at the base, on stalks. The leaves on the stem have a broad, ovate blade with one or several pairs of smaller segments below. The corolla is pink, and large for this group: about ½ inch long.

May and June: in moist woods and bottomlands from Pennsylvania to Illinois and southward to Virginia and Tennessee.

V. CILIATA is from 1 to 4 feet tall. The leaves at the base have short stalks bearing blades with several parallel veins, with hairs along the edge ("eyelashes," *cilia*). The leaves on the stem are pinnately cleft or divided into narrow lobes or segments. The flower cluster is long, with several branches. The flowers have only stamens or only pistil, on separate plants. The flowers are only ⅛ inch long or less.

May and June: in wet prairies and swamps from southern Ontario to Minnesota and southward to Ohio, Illinois, and Iowa. The root of the closely related *V. edulis* of the West was valued as food by the Indians, and probably this species is similarly useful.

V. ULIGINOSA grows up to 3 feet tall. The basal leaves are stalked, their blades either cleft or not. The leaves on the stem are pinnately deeply cleft or divided into lanceolate lobes or segments, the end one the largest. The white or pink flowers are up to ⅓ inch long.

May to July: in swamps and wet woods from Quebec to Michigan and southward to New York and Ohio.

GARDEN-HELIOTROPE, V. OFFICINALIS, is a stout plant from 2 to 5 feet tall. All the leaves are divided into from eleven to twenty-one narrow segments. The flowers are pinkish, in a rather broad, dense cluster at the summit of the stem.

May to August: commonly cultivated and escaped from cultivation into fields, thickets, and roadsides from Quebec to Minnesota and southward to New Jersey and Ohio. *Plate 176*. The flowers have what L. H. Bailey describes as a spicy fragrance, which perhaps accounts for the misleading name. The species is also known as cat's valerian (is it an acceptable substitute for catnip?) and St.-George's-herb. The root yields a valuable drug.

CORN-SALAD AND LAMB'S-LETTUCE (VALERIANELLA)

The species of *Valerianella* differ from the valerians in not having bristles in place of the calyx: they have no visible calyx at all. The leaves, also, are undivided and unlobed. The flowers are white or nearly white, very small, crowded at the tips of stems and branches.

There are at least seven species in our range: rather insignificant little plants distinguished chiefly by characteristics of the minute fruits and the bracts around the flower-clusters. The four most widely occurring species are mentioned here; two are illustrated.

CORN-SALAD or LAMB'S-LETTUCE, V. OLITORIA, is aptly named, having been cultivated for many years in Europe for salads. It has angular stems up to 20 inches tall. The leaves are oblong, blunt, without stalks, toothed in the lower half. The bracts around the flower-clusters generally bear marginal hairs. The flowers are bluish.

April to June, and October: in waste places, fields, and gardens from New England to Indiana and southward to South Carolina and Tennessee. *Plate 111*. Introduced from Europe.

V. RADIATA has leaves similar to those of *V. olitoria*, the bracts with fewer marginal hairs or none and tapering more to the tips. The flowers are white.

April and May: in damp woods and meadows from New Jersey to Missouri and southward to Florida and Texas. *Plate 111*.

V. CHENOPODIFOLIA has leaves, especially the upper ones, that taper towards the tip. The bracts also taper and generally lack marginal hairs.

May and June: in meadows and fields from New York to southern Ontario and Wisconsin and southward to Pennsylvania and Indiana.

V. UMBILICATA bears its flowers more loosely; the bracts generally lack marginal hairs.

May and June: in meadows and on roadsides from New York to Illinois and southward to North Carolina and Tennessee; not abundant.

THE HONEYSUCKLE FAMILY (CAPRIFOLIACEAE)

Most species of *Caprifoliaceae* are woody, shrubs or vines (*Viburnum*, honeysuckles, elderberry). We are concerned here only with two small genera. These are herbaceous plants with flower-parts in fours or fives and an inferior ovary. The leaves are in pairs.

LINNAEA

We have one species of this genus.

TWINFLOWER, L. BOREALIS, is a delicate small plant with a creeping stem from which erect branches arise. Each of these bears several leaves and a pair of fragrant pink flowers hanging from its curved tip. The leaves have roundish blades on short stalks, the edges of some notched or toothed. The flowers are like funnels upside-down, only ½ inch long, with five small teeth which curve upward. There are four stamens.

June to August: in woods, bogs, and peaty places from Greenland to Alaska and southward to Maryland, West Virginia, Indiana, South Dakota, Utah, and northern California. *Plate 111*. Our plants are a variety of the species that grows across northern Europe and Asia. It was named after Linnaeus, the father of modern botany, who liked to have his portrait painted holding or wearing a sprig of twinflower. The plants are pretty and fragrant inhabitants of cool northern woods.

TINKERWEEDS (TRIOSTEUM)

The tinkerweeds are not very attractive plants which have nevertheless acquired a surprising number of common names; feverwort, doubtless from popular medicinal use; wild coffee, the seeds having been used by the "Pennsylvania Dutch" as a substitute for coffee; horse-gentians – the "horse," as in other names, meaning "coarse" or even "false"; but why gentians? The origin of "tinkerweed" I do not know.

They are coarse, strong plants with rather narrow, paired leaves – making a surprising contrast with our only other wild flower in this family. The flowers have no stalks and are seated in the axils. They have dull yellow or reddish petals joined to make a funnel-shaped corolla with five erect lobes; this is not strictly radially symmetric, since it is slightly swollen on one side near the base. There are five stamens. The fruit is a yellow or red berry-like stone-fruit crowned by the five sepals (the ovary being inferior).

There are three fairly common species, easily distinguished.

T. PERFOLIATUM is named for its leaves: the two leaves of each pair meet and are joined around the stem so that the stem seems to grow through the leaves (*per*, "through"; *folia*, "leaves"). The stem is downy with gland-tipped hairs. The flowers are greenish, yellowish, or dull purple; the fruits dull orange.

May to July: in rocky open woods and clearings from Massachusetts to southern Ontario, Minnesota, and Nebraska and southward to Georgia and Kansas. *Plate 111*.

T. AURANTIACUM differs in that its opposite leaves are not joined but taper to their bases; there is generally a ridge connecting them across the stem. The hairs of the stem are not glandular. The flowers are red-purple; the fruit orange-red.

May to July: in woods and thickets from Quebec to Minnesota and southward to Georgia, Kentucky, and Kansas. *Plate 111*. This is a variable species.

PLATE 111

Triosteum aurantiacum *Johnson*

Sicyos angulatus *Johnson*

Linnaea borealis *Rhein*

Triosteum perfoliatum *Elbert*

Valerianella radiata  *Johnson*

Valerianella olitoria *Johnson*

Some plants have their lower leaves joined, and some have glandular hairs, as if they were intermediate between this and the preceding species (some botanists put them all in one species). There are also hairy and smooth forms. This species is said to blossom a little later than the preceding; crossing may or may not occur.

T. ANGUSTIFOLIUM has stems covered with stiff, glandless hairs. The leaves taper to narrow bases. The flowers, mostly only one in an axil, are greenish or yellowish; the fruits orange-red.

April to June: in woods and thickets from Connecticut to Missouri and southward to North Carolina, Alabama, and Louisiana.

THE CUCUMBER FAMILY (CUCURBITACEAE)

This, the family that contributes cucumbers, squashes, pumpkin, and melons to our tables, is represented in the wild through our range only by a few weeds. The flowers are of one sex, either staminate or pistillate. They have mostly five sepals and five petals, but only three stamens; these may be joined. The plants are vines, our species with tendrils.

ECHINOCYSTIS

The genus *Echinocystis* has flowers with six petals. Staminate and pistillate flowers are on the same plant. The stamens are joined.

BALSAM-APPLE or WILD-CUCUMBER, E. LOBATA, is a high-climbing vine with palmately lobed leaves. The staminate flowers are on a stem that rises from the axil of a leaf; each has six white petals, separate nearly to the base, about $\frac{1}{3}$ inch long. A pistillate flower hangs at the base of one of these flowering branches. Its ovary develops into an oval fruit covered with soft prickles and containing four seeds.

June to October: forming tangled masses with other weeds in waste land and thickets from New Brunswick to Saskatchewan and southward to Florida and Texas. *Plate 112*.

MELOTHRIA

We have a single species of *Melothria* along the southern borders of our range.

CREEPING-CUCUMBER, M. PENDULA, has very small flowers borne much as in *Echinocystis;* each has five petals. The fruit is smooth and contains numerous seeds.

June to September: in woods and thickets from Virginia to Missouri and southward to Florida, Texas, and Mexico.

SICYOS

The species of *Sicyos* have staminate and pistillate flowers on the same plant. The petals are five. The stamens are joined. The style bears three stigmas.

BUR-CUCUMBER, S. ANGULATUS, is a high-climbing vine with broad, palmately lobed leaves. The staminate flowers have stalks, the pistillate almost none; both are in small dense clusters. The petals are white or greenish. The fruits are covered with prickles. Each contains a single seed.

July to September: on river-banks and in damp soil from Maine to Minnesota and southward to Florida, Texas, and Arizona. *Plate 111*.

CUCURBITA

This is the genus that includes the squashes and pumpkins, natives of South America no longer found wild. The wild species are natives of the Southwest, one found to some extent in our range, chiefly along railroad tracks.

FETID WILD-PUMPKIN or MISSOURI GOURD, C. FOETIDISSIMA, is a trailing plant with stems that may be several yards long, and large, rather triangular, rough leaves. The flowers have a yellow funnel-shaped corolla up to 4 inches long. Staminate and pistillate flowers grow on the same plant. The three stamens are twisted at the ends and united. The style bears three stigmas, each divided into two. The fruit is roundish, smooth, many-seeded.

May to July: in dry soil from Indiana to Nebraska, Texas, Mexico, and California.

GROUP XIII

SEPALS five, petals from three to five; petals joined and bilaterally symmetric. Stamens two, four, five, or eight.
Exceptions: in some species of the mint family sepals are joined in such a way that there are apparently only four. Some species of the mint and snapdragon families have almost radially symmetric petals.

Echium in the forget-me-not family (XI) may seem to belong here. The petals of *Valeriana* and *Triosteum* (XII) also are very slightly bilateral in symmetry.

I. *Stamens two or four (five in* Verbascum).

A. Stems mostly square and leaves mostly paired: vervain family (corolla spreading into four unequal, rather flat lobes at the end of a tube); mint family (corolla evidently two-lipped, or ending in four unequal small lobes).

B. Stems mostly round; leaves various: snapdragon family.

C. Plant sticky; flower large; fruit hard, with two curved horns: unicorn-plant family.

D. Plants without green color (parasites): broomrape family.

E. Calyx, after corolla falls, turned down against the stem: lopseed family.

F. Plants insect-catching; either with small, sticky, butter-yellow leaves at the base of the stem, or growing in water or wet mud and having petals prolonged into a spur below: bladderwort family.

II. *Stamens five; corolla open on top:* lobelia family.

III. *Stamens eight; lowest petal fringed:* milkwort family.

THE VERVAIN FAMILY (VERBENACEAE)

The vervain family is here represented chiefly by the genus *Verbena*, with the mainly southern and western *Lippia*. The corolla is bilaterally symmetric, with five petals. The stamens are four in our species, in two pairs of different length. The ovary is slightly four-lobed but not deeply as in the mint and forget-me-not families.

THE VERVAINS (VERBENA)

The vervains form a large genus chiefly native in temperate and tropical America. The flowers are in spikes which in some species are very short at flowering time so that the flower-cluster appears flat; in others the spikes are long and slender, with only a few flowers open at one time. At first glance the corolla, which is a tube crowned with five flaring lobes, may seem radial in symmetry; but a second look will show that the lobes on one side – the lower or outer – are much larger than the others. The leaves of all our species are sharply toothed or lobed or cleft, often rather jaggedly.

The English name seems to be an Anglo-Saxon version of the Latin *Verbena*, a name used by the naturalist Pliny. In classical Rome a verbena was a sacred herb or branch used in various ceremonies, carried in processions, and so forth; many kinds of plants were used. How this name became limited to one genus I do not know.

We may here group the species of *Verbena* by the shape of the inflorescence at flowering time.

I. *Species with flowers (when they first appear) in a flattish cluster.*

ROSE VERVAIN, V. CANADENSIS, is a usually hairy, spreading plant. The leaves are ovate, variously toothed, lobed, or cleft. The flowers are handsome, from rose to lavender or even blue, or sometimes variegated, changing color as they age, the lobes spreading to form a disk $\frac{1}{2}$ inch broad or broader. The flowers stand next to bracts about as long as the calyx.

April to October: in sandy and rocky open places from southern Pennsylvania to Colorado and southward to Florida and Texas. *Plate 112*. This makes a good garden plant. It must be remembered that when this species was named *canadensis* by Linnaeus, "Canada" meant much more of North America than it does now; sometimes everything west of the Mississippi.

V. BRACTEATA resembles *V. canadensis* in general stature. Its leaves are apt to be pinnately lobed or cleft; its inflorescence is longer, less flat; and the bracts by the flowers are much longer than the calyx. The corolla is much smaller, the flat part only about $\frac{1}{8}$ inch across.

April to October: in sandy prairies, fields, and waste places from Virginia to Florida and from southern Ontario to British Columbia and southward to Texas and Mexico. *Plate 112*.

II. *Species with flowers, when they first open, in a slender spike. Among these we may distinguish first two common species with rather thick, dense spikes.*

A. Species with fairly thick spikes.

BLUE VERVAIN, V. HASTATA, sometimes called simpler's-joy from some medicinal use ("simples" are drugs), is from 1 to 5 feet tall, with several erect spikes branching out near the summit. The leaves have generally lanceolate blades, distinctly stalked. The spikes are pointed, with a circle of bright violet-blue flowers somewhere along their length; each flower about $\frac{1}{6}$ inch across.

June to September: in meadows, fields, prairies, and swamps from Nova Scotia to British Columbia and southward to Florida, Tennessee, Missouri, Texas, and California. *Plate 112*. In spite of having only a few small flowers open at one time, the color is so intense that the plants can make a handsome show.

HOARY VERVAIN, V. STRICTA, grows from 1 to 4 feet tall. Its stem and leaves are whitened with dense hairs. The leaves are ovate or elliptic, practically without stalks, rather thick, coarsely toothed. There may be one or several spikes at and near the summit; they are blunt at the end, somewhere along their length encircled by bright violet-blue flowers about $\frac{1}{3}$ inch across.

June to September: in fields, prairies, and rocky open places, and on roadsides from southern Ontario to Montana and southward to Tennessee, Arkansas, and New Mexico; also from New England to Delaware and West Virginia, where the species has been introduced. *Plate 112*.

B. Species with slender spikes.

V. SIMPLEX is from 4 inches to 2 feet tall, and smooth or nearly so. The leaves are generally narrow and lanceolate, tapering to the base and not stalked. The flowers are lavender or purple, about $\frac{1}{4}$ inch across.

May to September: in dry fields, in rocky open places, and on roadsides from Quebec to Minnesota and Nebraska and southward to Florida, Louisiana, and Oklahoma. *Plate 112*.

V. RIPARIA grows up to 5 feet tall, sparsely hairy or smooth, much branched, with ovate, stalked leaf-blades which are pinnately lobed or cleft. The several spikes are very slim and long, the flowers being separated on the lower part. The corolla is light blue, spreading to a width of about $\frac{1}{8}$ inch or a little more.

June to September: on river-banks (this is the meaning of *riparia*) and in thickets from New Jersey to North Carolina and West Virginia; not common. Compare the following species.

V. OFFICINALIS is an European species, reaching a height of 2 feet, nearly or quite smooth. The lower leaves are pinnately cleft, the upper lobed or merely toothed. The flower-spikes are slim, with the flowers separated as they mature. The corolla is violet or purple, the flat part only about $\frac{1}{6}$ inch across.

June to October: in waste places and fields from Massachusetts to Florida and Louisiana; also in the West. The differences between this species and the preceding are very small and technical. *V. officinalis* was formerly esteemed for medicinal virtues.

V. URTICIFOLIA is a straggling, unattractive, weedy species up to 5 feet tall with ovate, coarsely toothed leaves with short stalks, the blades often blotched with mildew. The spikes are almost thread-like, the flowers minute and separated. The corolla is white, only about $\frac{1}{12}$ inch across.

PLATE 112

Echinocystis lobata *Johnson*

Verbena stricta *Rickett*

Verbena hastata *Rickett*

Verbena simplex *Rickett*

Verbena bracteata *Johnson*

Verbena canadensis *Love*

June to September: at the edges of woods and in fields and waste places from Quebec to South Dakota and southward to Florida and Texas. *Plate 113.* An unattractive weed, included here only to complete the roster of common species.

Several other species are found along the borders of our range: *V. brasiliensis* in Virginia, with many short, dense spikes; *V. scabra* in southeastern Virginia, with lanceolate stalked leaf-blades and long, slim spikes; and *V. bipinnatifida* in Missouri, with an inflorescence like that of *V. canadensis* but leaves deeply pinnately cleft or divided and the lobes or segments themselves pinnately cleft or lobed.

LIPPIA

A few species of this tropical genus are found in the northeastern states, only one at all widespread.

FOG-FRUIT, L. LANCEOLATA, is a sprawling little plant with erect flowering branches. The leaves are lanceolate, toothed, tapering to each end, generally without distinct stalks. The flowers are in small dense heads, with many pale blue, pink, or white flowers mixed with broad, sharp bracts.

May to October: in bottom lands, roadside ditches, even along railroad tracks, from southern Ontario to Minnesota and Nebraska and southward to Florida, Mexico, and California. *Plate 113.*

THE MINT FAMILY (LABIATAE)

The *Labiatae* take their English name from the genus *Mentha*, mint, familiar as sources of flavors (peppermint, spearmint) and medicines (menthol). The substances so used are "essential oils"; similar materials are found in many, perhaps most of the genera of this family. Various species are cultivated for their essential oils, which are of considerable importance in our civilization. Besides the mints themselves, lavender, rosemary, sage, thyme, marjoram, savory, basil, and others contribute to the pleasures of life. Many of the wild species have characteristic odors.

The botanical name of the family refers to the general form of the corolla, which has two "lips" (*labia*), upper and lower as the flower extends horizontally; the corolla is thus bilaterally symmetric. Five petals contribute to the two lips; they are all joined at the base, forming a tube or funnel, and commonly flare into a two-lobed upper lip and a three-lobed lower lip; but there is a great variety, and in some genera no lips are evident, the petals forming an almost radially symmetric corolla. There are typically four stamens, but in a number of genera one pair is abortive and only two stamens bear pollen. The pistil has a four-lobed ovary which forms four small nuts; the style ends in a forked stigma which imitates a snake's tongue.

The family may often be recognized by the characteristic square stem bearing paired leaves (but these features occur also in other families).

Because the flowers of most species are small, and because the general aspect of many species – leaves and stems – is much the same, it is hard to separate the genera of *Labiatae* without recourse to minute details of stamens and other technical characteristics. The guide to genera which follows depends on easily visible characteristics, such as type of inflorescence, as far as possible. But it is necessary to introduce certain features of corolla and calyx, and one must expect to make good use of the hand magnifier. If the flowers are so minute that a "10×" lens is not enough to reveal things, one might try *Lycopus*, *Hedeoma*, and *Satureia*.

Using the hand lens, we first separate genera into two groups by the number of stamens per flower.

I. *Genera with two stamens* (*two others may be present but undeveloped, sterile*).

A. Some of these have small flowers almost radially symmetric: *Cunila* (low, branching plant with leaves almost plain-edged): *Lycopus* (taller, less branched, the leaves sharply toothed or even lobed).

B. The other genera with two stamens have flowers distinctly bilaterally symmetric with upper and lower "lips" which differ from each other: *Monarda* (corolla with very long, narrow upper

PLATE 113

Verbena urticifolia *Rickett*

Lamium album *Rickett*

Cunila origanoides *Uttal*

Lippia lanceolata *Johnson*

Lycopus americanus *Donahue*

Lycopus virginicus *Rickett*

Lycopus amplectens *Murray*

lip; flowers in dense heads at the summit of the stem, with or without clusters in the axils below); *Blephilia* (calyx with unequal teeth; flowers in dense head or spike at summit of stem, with clusters in axils of upper leaves); *Salvia* (flowers in a long spike interrupted by short lengths of bare stem; leaves mostly basal); *Collinsonia* (flowers yellow with fringed lower lip; inflorescence much branched at the summit of the leafy stem); *Hedeoma* (flowers minute; calyx two-lipped with three teeth above and two longer teeth below).

II. *Genera with four stamens.* (Collinsonia – *see above – occasionally matures four stamens*).

A. In this group also we can recognize genera whose flowers are almost radially symmetric: *Mentha* (the odor of mint identifies it; flowers in dense clusters at the tip of the stem or in axils of leaves or both); *Isanthus* (flowers in axils of narrow leaves); *Elsholtzia* (flowers in narrow spikes).

B. The remaining genera with four stamens have bilaterally symmetric flowers with distinct upper and lower lips. These can be separated to some extent by their inflorescences, but in this feature there is much variation even in one species, so that it is difficult to be precise.

1. Genera with flowers massed at the summit of the stem and branches in dense heads (first three genera), spikes (next ten genera), or racemes (last three genera). There may be similar clusters in the axils of leaves just below the topmost one. In heads, spikes, and racemes the flowers may be mixed with bracts but not with ordinary leaves; spikes may be interrupted by short lengths of stem without flowers, and such inflorescences may easily be confused with those of group 2 below. Genera with flowers in heads, spikes, or racemes are: *Pycnanthemum* (minty odor; flowers on numerous branches); *Satureia* (one species; leaves scalloped); *Origanum* (bracts and flowers crimson; leaves stalked, without teeth); *Prunella* (bracts conspicuous, round with a sharp point, often colored); *Teucrium* (corolla split along upper side); *Galeopsis* (teeth of calyx end in spines); *Nepeta* (leaves coarsely toothed, stalked, the base of the blade indented); *Thymus* (creeping plant with small, roundish leaves); *Dracocephalum* (spike very dense, with conspicuous, sharp-pointed bracts); *Stachys* (spike more or less interrupted; leaves mostly toothed or notched); *Ajuga* (creeping stems with erect branches; spike more or less interrupted); *Agastache* (one species with small greenish flowers in a dense continuous spike; another with purplish flowers in an interrupted spike); *Hyssopus* (flowers small; leaves narrow); *Physostegia* (flowers large with concave upper lip; leaves narrow, sharply pointed and toothed); *Meehania* (leaves with long stalks, the blades indented at the base); *Perilla* (leaves long-stalked, the blades tapering to the stalk); *Scutellaria* (some species; calyx with a hump on the upper side).

2. Genera with four stamens and bilaterally symmetric flowers growing singly or in clusters from the axils of ordinary leaves: *Scutellaria* (some species with flowers single in the axils; calyx with a hump on the upper side); *Glecoma* (creeping stems with erect branches; leaf-blades round, notched); *Lamium* (calyx teeth mostly ending in spines; upper lip of corolla forms a hood, hairy on top); *Ballota* (leaf-blades ovate, toothed, indented at base, on rather long stalks); *Marrubium* (teeth of calyx ending in spines; flowers in dense clusters); *Leonurus* (teeth of calyx ending in spines; flowers in dense clusters; leaf-blades tapering to the stalks, in one species lobed); *Melissa* (two lower teeth of calyx ending in spines, three upper teeth triangular; leaves coarsely toothed); *Synandra* (flowers few and large, single in the axils; calyx with four teeth); *Conradina* (somewhat shrubby, with narrow, blunt leaves); *Satureia* (delicate species with flowers singly or few in the axils and narrow leaves, and a coarser species with denser, stalked flower-clusters); *Trichostema* (flowers in long-stalked clusters; stamens very long, projecting, curved downward).

CUNILA

One species is common through the southern half of our range.

DITTANY, C. ORIGANOIDES, has a wiry, branching stem about a foot tall. The leaves are ovate, toothed, and practically without stalks. The small purplish or white flowers are in little tufts in the axils of leaves and at the tips of branches. The five lobes of the corolla are almost equal.

July to November: in dry open woods and prairies from southern New York to Oklahoma and southward to Florida and Texas. *Plate 113*. The name dittany is a corruption of *Dictamnus*, the name of the gas-plant, an European plant often cultivated in America. Our plant should properly be distinguished as American dittany. The plant has long been familiar in popular medicine, for fever and snake-bite. The leaves have a pleasant minty odor and a tea may be brewed from them. The plant is interesting also as being an "ice-plant": under certain conditions long threads of ice appear from cracks in the stem. The botanical name means "like *Origanum*."

BUGLEWEEDS AND WATER-HOREHOUNDS (LYCOPUS)

The bugleweeds have the general aspect of mints but lack the minty odor. They are mostly 2 or 3 feet tall. The leaves are toothed or sometimes lobed or cleft pinnately, and mostly have short stalks or none. The flowers, generally white, are tiny, in a dense mass that encircles the stem at each pair of leaves. The calyx has four or five teeth or lobes. The corolla is two-lipped.

The botanical name is from two Greek words signifying "wolf's-foot"; this and the English names apply to the European plant, whose leaves were apparently thought to resemble a canine paw, and which grows in water; it was at one time named *Marrubium*, now the name of the horehound. "Bugle" is not the martial and musical instrument, but a descendant of a medieval Latin word for some sort of pin worn in ladies' hair.

The species are best distinguished by the calyx, small as it is, taken together with the leaves.

I. *Species whose calyx-teeth are not tipped with a spine; they may be blunt or pointed.*

L. VIRGINICUS has blunt calyx-lobes. The leaves are ovate, tapering to both ends, with more or less distinct stalks.

July to October: in damp soil from Maine to Minnesota and Nebraska and southward to Georgia and Texas. *Plate 113.*

L. UNIFLORUS has sharp but not spine-tipped calyx-teeth. The leaves are lanceolate, gradually tapered at both ends.

June to September: in moist ground from Newfoundland to Alaska and southward to Maryland, the mountains of North Carolina, Ohio, Illinois, Oklahoma, Montana, and California. The name is based on an error: it has more than one flower. Edible white tubers are formed below ground; these are said to have a pleasant flavor, boiled or raw.

II. *Species with spine-tipped calyx-teeth. These species are hard to distinguish without the use of the technical manuals. The first two are apt to have their leaves, especially the lower ones, pinnately lobed or cleft; but in this characteristic they vary greatly. The leaves of the first three are likely to have stalks.*

L. AMERICANUS is smooth or almost so. The leaves are generally lanceolate. The calyx-teeth end in long spines.

June to September: in low, moist ground practically throughout the United States and southern Canada. *Plate 113.*

L. EUROPAEUS is hairy, with leaves tending to be ovate. The calyx-teeth are narrow and spine-tipped. The corolla is spotted with purple.

August to October: the European plant now found on roadsides and in waste places, not very common, from Massachusetts to southern Ontario and southward to Virginia, Alabama, and Louisiana.

L. RUBELLUS is a tall species, up to 4 feet, with leaves varying from elliptic to lanceolate or ovate and toothed; stem and leaves are generally smooth, sometimes downy.

July to October: in damp woods and thickets from New England to Michigan and Missouri and southward to Florida and Texas.

L. AMPLECTENS is smooth or hairy. The leaves vary from elliptic to lanceolate or ovate, with a few broad teeth on each side. The calyx-teeth are narrow and spine-tipped.

August and September: in wet sand and peat mostly on the coastal plain from Massachusetts to Florida and Mississippi; also in Indiana. *Plate 113.* This species has edible tubers like those of *L. uniflorus.*

L. ASPER has a hairy stem and rough (*asper*) leaves. The leaves are lanceolate, sharply toothed. The calyx-teeth are triangular, tapering to sharp points.

July and August: in marshes from Michigan to Alaska and southward to Illinois, Missouri, and California; occasionally found farther to the east. This species, like *L. uniflorus*, has tubers, whether tasty or not I do not know; Dr Steyermark says they are eaten by muskrats.

THE HORSE-MINTS (MONARDA)

"Horse" signifies "coarse," and these are certainly large plants, coarser than the true mints (*Mentha*); but they are handsome plants, some cultivated and more worthy of it. Their flowers are long and narrow and very markedly two-lipped; the upper lip continues the tube, the lower turns downward and is broader. The style and the two stamens lie close under the upper lip. The flowers are grouped in dense heads at the summit of the stems, with or without similar clusters in the axils below. The clusters are surrounded by broad bracts which may be colored white, pink, or purplish.

We may group the species by the shape of the corolla and by the presence or absence of flower-clusters in the axils.

I. *Species with flower-clusters* usually *only at the summit of the stems, rarely with similar clusters encircling the stem in the axils. The long upper lip curves slightly either upward or downward, or may be straight.*

Among these we may distinguish two groups by color.

A. Species of group I with lilac, lavender, pinkish, or white petals.

WILD BERGAMOT, M. FISTULOSA, is the common and handsome species of eastern roadsides and woods. It stands from 1 to 5 feet tall. The leaves are stalked, with lanceolate, toothed blades. Stem and leaves may be smooth or softly hairy. The flowers are lilac or pink. The upper lip of the corolla bears a tuft of long hairs at its end.

June to September: in open woods and on roadsides from Quebec to British Columbia and southward to Virginia, the mountains of Georgia, Mississippi, Texas, Arizona, and Mexico. *Plate 114*. In most parts of the United States this is the most abundant species, forming dense colonies which spread a pleasant minty fragrance for many yards around.

M. CLINOPODIA is rather like *M. fistulosa*. It does not generally exceed 3 feet in height. The bracts around the flower-cluster are more or less white. The corolla is whitish or pink, the lower lip with purplish spots. The calyx has very short teeth.

June to September: in woods from Connecticut to Illinois and southward to North Carolina and Alabama; also elsewhere in New England where it has escaped from cultivation. *Plate 114*.

M. BRADBURIANA grows from 1 to 2 feet tall, with a few hairs, or smooth. The leaves may have short stalks or none. The calyx-teeth are like small spines. The corolla is pale rose or white dotted with purple.

May and June: in woods and thickets from Indiana to Iowa and Kansas and southward to Alabama and Texas. *Plate 114*.

M. RUSSELIANA is similar to *M. bradburiana* (and some botanists treat them both as one species). The chief difference is in the length of the corolla; it is shorter in *M. russeliana*.

May and June: in woods from Kentucky and Missouri to Arkansas and Texas.

B. Species of group I with scarlet or dark purplish-crimson flowers.

BEE-BALM or OSWEGO-TEA, M. DIDYMA, has scarlet flowers and leaves of a bronze hue. The bracts also are red; the inner ones are bristle-like.

June to September: in woods and thickets from New York to Michigan and southward in the mountains to Georgia and Tennessee; widely cultivated and escaped into the wild from Quebec to New Jersey. *Plate 114*. The red color is attractive to birds, and nectar is often sipped from the long tubes of the flowers by humming-birds.

M. MEDIA is much like *M. didyma* but with flowers of a dark rose-purple color. The bracts, also purplish, bear bristles on their margins.

July to September: in woods and thickets from New York to Ontario and southward to North Carolina and Tennessee; like *M. didyma* widely cultivated and found wild in New England.

II. *Species with several flower-clusters below that at the summit. The upper lip of the flowers is arched, the tip pointing downward.*

DOTTED MONARDA or HORSE-MINT, M. PUNCTATA, has a stem from 1 to 3 feet tall, hairy or nearly smooth, with stalked leaves whose blades are lanceolate. The bracts around the flowers are colored lilac or whitish. The flowers are pale yellow with purple spots.

In open sandy places from Vermont to Minnesota and southward to Florida, Texas, Arizona, and Mexico. *Plate 114*. The western plants are generally hairy, the southeastern ones very finely downy. A handsome species worthy of cultivation. The western lemon-mint, *M. citriodora*, strays into our range. It is like *M. punctata*, but the bracts are densely downy and so shaped as to form a cup around the flowers; they taper to very long, sharp tips. The flowers are white or pink, with light purple spots.

WOOD-MINTS (BLEPHILIA)

These plants have something the aspect of the horse-mints, but their flower-clusters are in several of the upper axils, encircling the stem, as well as at the summit. The bracts around these clusters have a characteristic pattern of veins. The flowers are readily distinguished from those of *Monarda* by the calyx: that of *Blephilia* is two-lipped, bilaterally symmetric like the petals, with three spine-tipped teeth on the upper side and two shorter, blunter teeth on the lower side. The calyx of *Monarda* has five equal teeth. The

PLATE 114

Hedeoma pulegioides *Rickett*

Monarda punctata *Johnson*

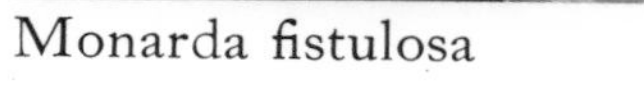

Monarda fistulosa *Johnson*

Monarda clinopodia *D. Richards*

Blephilia ciliata *D. Richards*

Monarda didyma *Scribner*

Monarda bradburiana *Johnson*

corolla is much like that of *Monarda* but shorter. They lack the minty smell. The name *Blephilia* is from the Greek word for eyelashes, suggested by the row of long hairs on the edge of the bracts.

There are two species.

B. CILIATA is downy. The leaves are practically without stalks but taper towards the base as well as towards the tip. The plants are from 1 to 3 feet tall. The leaves are whitish underneath.

May to August: in woods from Massachusetts to Iowa and southward to Georgia, Mississippi, and Texas. *Plate 114.*

B. HIRSUTA is hairy ("hirsute"). The leaves are long-stalked with blades round or even indented at the base. The plants are often taller than *B. ciliata.*

May to September: in moist shady places from Quebec to Minnesota and southward to Georgia, Tennessee, Arkansas, and Texas.

SALVIA

The genus *Salvia* probably surpasses in number of species any other genus in the country, but fortunately for us most of them are tropical and South American. The sage used in cooking is *S. officinalis. S. splendens* is the commonly cultivated scarlet salvia. Several other species are cultivated. The genus is distinguished by an inflorescence generally called an "interrupted spike"; i.e. a number of clusters, encircling the stem, in the axils of bracts, all occupying the upper part of the stem. The lower lip of the corolla is long and conspicuous. The calyx also is two-lipped, with three short teeth above and two longer teeth below.

Ten species have been listed for our area, but only four are native; the others are cultivated European species occasionally found growing wild. Of the native kinds, one, blue salvia, *S. azurea*, is western, entering our area only in Missouri.

CANCER-WEED, S. LYRATA, has a rosette of leaves at the base of the flowering stem. Some of these leaves usually have the outline called by botanists – curiously – "lyrate" (i.e. shaped like a lyre, which they are not). The stem rises from 1 to 2 feet, with circles of blue or violet flowers at intervals. The flowers are about an inch long.

April to June: in sandy woodland from Connecticut to Missouri and southward to Florida and Texas. *Plate 115.*

S. URTICIFOLIA has leaves on the stem, with ovate, sharply toothed blades on short, flat stalks (the name means "nettle-leaved"; *Urtica* is a nettle). The corolla is variegated blue and white, about ½ inch long.

May to July: in woods from Pennsylvania and Ohio to Florida and Louisiana.

S. REFLEXA has narrow, lanceolate leaves, with or without teeth, on its 2-foot stem. The flowers are blue, the corolla only ⅓ inch long or less, scarcely projecting from the calyx.

May to October: in dry prairies and on open rocky slopes from Ohio to Michigan, Wisconsin, and Montana and southward to Arkansas, Texas, and Mexico; occasionally found farther to the east.

COLLINSONIA

This is a small North American genus. Besides the single species described below, the southern *C. verticillata*, differing in having regularly four stamens, is found in Virginia.

HORSE-BALM, RICHWEED, or STONEROOT, C. CANADENSIS, is a somewhat bushy plant from 2 to 5 feet tall, with large, ovate, toothed leaves. The flowers are in a much-branched inflorescence which crowns the stem. The corolla is yellow, rather more than ½ inch long, with the lower lip longer than the upper and delicately fringed. The flowers and leaves have a pleasant lemony odor.

July to September: in wet woods from Massachusetts and Vermont to Ontario and Wisconsin and southward to Florida and Arkansas. *Plate 115.* Like many plants of this family the essential oil responsible for the odor has been used medicinally, the leaves being brewed to make a tea. So "horse-balm" does not mean a balm for horses, but a coarse sort of plant comparable to balm (*Melissa*) in its uses. "Stoneroot" does not mean that it has a hard root, but that its root (or, more accurately, its rhizome) was used medicinally in treating the affliction known as "stone." This species regularly has two stamens.

PLATE 115

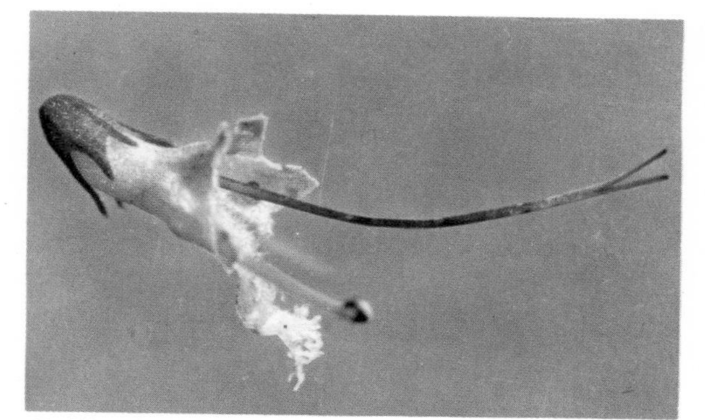

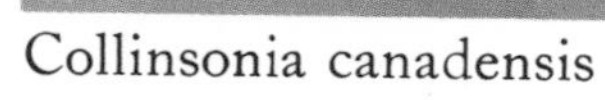

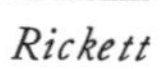

Collinsonia canadensis *Rickett*

Mentha arvensis *Elbert*

Salvia lyrata *Johnson*

Pycnanthemum incanum *D. Richards*

Collinsonia canadensis *Rickett*

Mentha piperita *Donahue*

Mentha spicata *Rickett*

THE AMERICAN PENNYROYALS (HEDEOMA)

Hedeoma is an American genus with two species in the northeastern United States. They have clusters of small blue flowers in the axils. Both calyx and corolla are two-lipped. The plants (especially those of the first species) have a minty odor resembling that of true pennyroyal, the European *Mentha pulegium*.

H. PULEGIOIDES is the strongly aromatic kind, often perfuming the air throughout a field or woodlot. The plants stand only about a foot tall. The leaves are lanceolate or ovate. The calyx is characteristic, with three short, broad teeth above and two long, narrow, curved teeth below.

July to September: in old fields and dry woods from Quebec to Minnesota and South Dakota and southward to Florida, Tennessee, Arkansas, and Kansas. *Plate 114*. The second half of its name means "like *Pulegium*," the old name of true pennyroyal which was derived from the Latin for "flea." The aroma was supposed to keep fleas away – as well it might; I have suspected a certain immunity from mosquitoes in a small tent in the Ozarks to be due to the presence of *Hedeoma*.

H. HISPIDA is less strongly scented. It is smaller, rarely over 8 inches tall. The leaves are very narrow, often with a few stiff hairs. The calyx is, as in *H. pulegioides*, two-lipped; but all five teeth are narrow.

May to August: in sandy soil and dry banks and prairies from New York and Ontario to Alberta and southward to Mississippi and Texas; occasionally found farther to the east.

THE MINTS (MENTHA)

In a sense all the *Labiatae* are "mints," and certainly many of them have aromatic essential oils with a more or less minty odor. But the true mints, used commercially and familiar to everyone for flavoring, are species of *Mentha*. The genus is characterized by almost radially symmetric flowers – it might be said that the true mints are not typical of the mint family! The corolla is four-lobed, the upper lobe usually broader than the others and often notched. There are four stamens.

The species are the despair of botanists, having long been cultivated in the Old World and having been repeatedly crossed. Many of the resulting races have been brought to this country and have escaped from cultivation. Most of these, being only occasionally found and distinguishable by very minor characteristics, are not here described. We have one native species (found also in the Old World), which itself is extremely variable. The odors of the different species are to some extent distinguishable.

FIELD MINT, M. ARVENSIS, is our native species. The flowers are all in clusters in the axils of the paired leaves and encircling the stem. There are often none in the upper axils. The leaves are stalked, with blades varying from ovate to lanceolate. The stem may be hairy or downy or even smooth.

July to September: in woods, wet or dry, and on slopes and shores, from Newfoundland to Alaska and southward to Virginia, Kentucky, Missouri, Nebraska, New Mexico, and California. *Plate 115*.

PEPPERMINT, M. PIPERITA, is a smooth plant, up to 3 feet tall. The flowers are in a dense spike at the summit, with a few clusters below, in the axils of bracts, separated by short lengths of stem.

June to October: along streams and in wet meadows and swamps practically throughout the United States. *Plate 115*. This is thought to be of hybrid origin, one of the parents having been plants of the following species. Different races of peppermint are widely cultivated for the oil used in flavoring sweets and for the menthol used medicinally (the latter is also obtained from a mint grown in the Orient).

SPEARMINT, M. SPICATA, grows up to 3 feet tall, usually smooth. The leaves are lanceolate, toothed, nearly or quite without stalks. The flowers are in long slim spikes, with the lower clusters separate; the bracts are narrow.

June to October: in wet places throughout; like peppermint, widely cultivated for flavoring and escaped. *Plate 115*.

Other kinds that may be recognized are *M. aquatica*, with flowers in a round or oval head at the summit of the stem; *M. cardiaca*, with flowers in clusters in the axils of bracts, extending to the summit – an "interrupted spike"; *M. longifolia*, like *M. spicata* but with narrower leaves.

ISANTHUS

There is a single species of *Isanthus*.

FALSE PENNYROYAL, I. BRACHIATUS, is a small, clammy plant up to 8 inches tall, with narrow, lanceolate leaves; these have short stalks or none, few or no teeth, and one or three lengthwise veins. The flowers are quite small ($\frac{1}{5}$ inch long), pale blue, and, like those of the mints, almost radially symmetric.

July to September: in dry soil from Quebec to Minnesota and Nebraska and southward to Florida, Tennessee, Texas, and Arizona. *Plate 116.*

For other pennyroyals see page 372.

ELSHOLTZIA

Our only species of this mainly Asian genus is an immigrant from that continent.

E. CILIATA is a smooth plant from 1 to 2 feet tall. The leaves have ovate blades on stalks. The flowers are in small dense spikes which grow at the tip of the stem and in the axils of the leaves. The flowers are tiny (about $\frac{1}{6}$ inch long), with a pale blue, almost radially symmetric corolla shaped like a funnel.

July to October: on roadsides and in old fields from Quebec to Massachusetts and Vermont.

THE MOUNTAIN-MINTS (PYCNANTHEMUM)

The botanical name signifies "densely flowered"; the flowers are crowded in heads at the tips of the numerous branches and often also in the axils of leaves. The plants are rather stiff, with a strong minty odor. The flowers are white or purplish, strongly two-lipped, and mixed with conspicuous bracts. All flower from June or July to September. The English name is not appropriate; a few species grow in mountains, but most are common in the lowlands. The name basil has also been applied to them, misleadingly. The herb basil is a species of another genus, *Ocimum*.

Most species of *Pycnanthemum* belong to the eastern United States. They are not easy to distinguish, and probably hybridize in nature. We here separate those with narrow leaves from those with broader leaves.

I. *Species with leaves at least five times as long as wide, lanceolate or narrower with parallel sides.*

P. VIRGINIANUM is from 1 to 3 feet tall; the stem is hairy *on the angles*. The leaves are lanceolate, smooth, sometimes with a few small teeth but commonly with plain margins; they vary greatly in width.

In wet meadows and thickets and by water from Maine to North Dakota and southward to Georgia and Oklahoma. *Plate 116.*

P. PILOSUM grows up to 5 feet tall. The stem is hairy. The leaves are wide for this group, some nearly an inch wide, but more usually less than $\frac{1}{2}$ inch. They mostly lack teeth. They are hairy on the lower side; and the bracts around the flower-heads are hairy on the *upper* side.

In dry woodland from southwestern Ontario to Kansas and southward to Tennessee and Oklahoma; also, as strays, in the Atlantic states from Massachusetts to Pennsylvania.

P. TORREI somewhat resembles *P. pilosum*, without the hairs on bracts and leaves. The leaves are lanceolate, with a few small teeth or none, their length from five to ten times their breadth.

In dry woods and thickets from Connecticut to Kansas and Arkansas and southward to Virginia, the mountains of Georgia, and West Virginia. The southern plants may have a hairy stem.

P. TENUIFOLIUM is a much branched plant up to 3 feet tall, with very narrow (*tenui-*) leaves without teeth. The whole plant is smooth or nearly so.

In dry open ground and thickets from New England to southern Ontario and Minnesota and southward to Georgia and Texas. *Plate 116.* (Often miscalled *P. flexuosum*, which is the name of a southern species.)

P. VERTICILLATUM has a minutely downy stem bearing leaves much like those of *P. virginianum* but more uniformly toothed.

In dry or moist thickets and peat from southwestern Quebec to Michigan and southward to Virginia and the mountains of North Carolina.

II. *Species with ovate leaves, their length not five times their width (but see* P. virginianum, P. torrei *under group I).*

P. INCANUM is distinguished by the covering of white hairs (*incanum* means "hoary") on the stem, and on the upper side of the leaves and bracts. The blades are ovate, stalked. The lower surface also is usually hairy.

In dry woods and thickets from southern New Hampshire to Illinois and southward to Florida and Alabama. *Plate 115.* In Missouri we find the very similar *P. albescens* instead of this species; from West Virginia to Illinois and southward the more hairy *P. pycnanthemoides* intrudes.

P. MUTICUM has ovate leaf-blades with round or even indented base, on short stalks, and toothed. The stem and bracts are hoary with minute white hairs. The leaves are smooth or nearly so.

In woodland from Maine to Michigan and southward to Florida and Texas.

P. SETOSUM may be known by the rather loose disposition of its flower-heads and the stiff sharp points of its bracts. The stem grows nearly 3 feet tall, and is minutely downy. The leaf-blades are elliptic or ovate, on very short stalks.

In sandy woods and marshes on the coastal plain from New Jersey to Florida.

THE SAVORIES AND CALAMINTS (SATUREIA)

The plants of *Satureia* are of such various aspects that it seems unlikely that they belong in the same genus (and in fact they have sometimes been classed in two). They all have in common a tubular calyx with sharp, narrow teeth, and the two-lipped corolla illustrated in *Plate 116.* Various cultivated culinary herbs are species of *Satureia:* summer savory, (winter) savory, basil-thyme, and mother-of-thyme; any of these may be found growing wild. In fact the name savory is a corruption of the botanical Latin *Satureia.*

WILD-BASIL, S. VULGARIS, has creeping stems from which hairy flowering branches rise from 1 to 2 feet. The leaves also are hairy, the blades elliptic or ovate, usually without teeth, stalked. The pale purple or pink flowers are in dense heads at the tips of branches, with a pair of leaves immediately beneath.

June to September: in woods and on open slopes from Newfoundland to Manitoba and southward to Delaware, the mountains of North Carolina, Tennessee, and Wisconsin; and in the West. *Plate 116.* Through most of this range the plants belong to a variety native to America; the hairier plants are of the European variety, escaped from cultivation. The English name is unfortunate, since true basil is a species of *Ocimum.*

S. GLABELLA represents the other extreme of the genus. It is a smooth plant (except for some hairs where leaves join the stem) about 2 feet tall with narrowly elliptic leaves, and small clusters of flowers in the axils. The corolla is less than ½ inch long.

May to August: in dry woods (on limestone) and barren open ground from Kentucky and Tennessee to Missouri and Arkansas. *Plate 116.* A similar form, sometimes treated as a distinct species, sometimes as a variety of *S. glabella*, with still narrower leaves and no hairs on the stem, grows in similar situations from Ontario to Minnesota and southward to western New York, Illinois, Arkansas, and Texas.

ORIGANUM

We have one species of this Old-World genus.

WILD-MARJORAM, O. VULGARE, is a hairy plant from 1 to 3 feet tall, with ovate leaf-blades tapering to fairly long stalks, and generally without teeth. The stem and branches are crowned by dense heads of flowers, both the flowers and the bracts beneath them colored bright crimson.

June to October: on roadside banks and in old fields and open woods from Quebec and southern Ontario to North Carolina; a striking decoration of roadsides in Connecticut, New York, and elsewhere. *Plate 116.* The culinary herb marjoram is *Majorana hortensis.* Wild-marjoram was much used as a medicinal tea by the colonists in America. The herb oregano is obtained from several species in the mint and vervain families – but not from origanum.

PRUNELLA

These mostly Old-World plants are characterized by a dense head or cylindric spike of flowers (without stalks) at the tip of each stem and branch. The calyx is two-lipped, the upper lip as if cut-off, with three small points, the lower lip of two narrow teeth. The corolla also is conspicuously two-lipped, the upper lip forming

PLATE 116

Isanthus brachiatus *D. Richards*

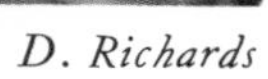

Pycnanthemum tenuifolium *Johnson*

Pycnanthemum virginianum *D. Richards*

Prunella vulgaris *Gottscho*

Satureia vulgaris *Rickett*

Origanum vulgare *Elbert*

Satureia glabella *Johnson*

a long concave hood, the lower lip bent sharply down. The flowers are mixed with broad bracts toothed at the end.

SELF-HEAL, P. VULGARIS, is a common weed, usually with many widely spreading branches, often reaching a height of 2 feet. It is extremely variable in the form of the leaves (lanceolate, ovate, with sharp or blunt points, tapering or round at the base), in hairiness, and in color of flowers and bracts.

May to October: in fields and gardens and on roadsides throughout the United States and southern Canada. *Plate 116*. Some of the many varieties are native to this country; others have been introduced from the Old World. The name (originally *Brunella*) is derived from the German for quinsy, which it was supposed to cure. It was also much esteemed for healing wounds.

Another Old-World species, *P. laciniata*, with pinnately cleft leaves, has been found in a few places.

TEUCRIUM

One species occurs commonly in our range.

AMERICAN GERMANDER or WOOD-SAGE, T. CANADENSE, has a stem from 8 inches to 3 feet tall, bearing lanceolate, toothed leaves. It is readily recognized by the disposition and structure of the flowers. They are in a tall spike. The upper lip is cleft, with a small horn rising on either side; the stamens and later the style project through this cleft. We have no other species of mints with a corolla like this.

June to September: in woods and thickets and on shores throughout the United States and southern Canada. *Plate 117*. The species varies greatly in the amount and kind of hairiness and the shape of the leaves. Other kinds of germander grow in Europe.

HEMP-NETTLES (GALEOPSIS)

The genus *Galeopsis* is characterized by the calyx, which has five triangular teeth ending in sharp spines. The corolla has a hoodlike upper lip and a three-lobed lower lip, at the base of which, inside, a hand magnifier will reveal two projections. The flowers are in dense masses in the axils of leaves. Both the species described below are introductions from the Old World.

G. TETRAHIT is recognized by the white bristles on the outside of the upper lip of the corolla. The stem also is bristly. The leaves are ovate.

June to September: in waste places and fields from Newfoundland to Alaska and southward to North Carolina, Ohio, Michigan, and South Dakota. *Plate 117*. This species has the curious distinction of a hybrid origin; i.e. it resulted from a cross between two other European species. That this is so has been shown by actually synthesizing *G. tetrahit* from a cross of the two parental species.

G. LADANUM is downy rather than bristly. The leaves are lanceolate or narrower, often without teeth. The flowers are red or pink, often yellow-spotted.

June to September: in waste land chiefly on the coastal plain from Quebec to New England, and more rarely inland to Michigan and Indiana.

NEPETA

One species of this Old-World genus has become widely established in North America.

CATNIP or CATMINT, N. CATARIA, is recognizable by its odor, beloved of cats. The stem is from 1 to 3 feet tall. The leaves are long-stalked with ovate, coarsely toothed blades. The flowers are in dense spikes at the summit of stem and branches. The upper lip of the corolla is slightly hooded, and notched. The lower, three-lobed lip hangs down.

June to October: in waste places and yards and on roadsides practically throughout the United States. *Plate 117*. It was formerly cultivated for its supposed medicinal value.

THYME (THYMUS)

Thyme is familiar to every cook and gourmet; the species used for flavoring is *T. vulgaris*. The kind found growing wild in our area has escaped from gardens. Many other species grow in the Old World.

PLATE 117

Galeopsis tetrahit *Gottscho*

Stachys palustris *Scribner*

Teucrium canadense *Rickett*

Thymus serpyllum *Gottscho*

Ajuga reptans *Rickett*

Stachys tenuifolia *Rickett*

Nepeta cataria *Johnson*

WILD THYME, T. SERPYLLUM, has a creeping, branching stem bearing pairs of small leaves about $\frac{1}{2}$ inch long. There are often bunches of still smaller leaves in the axils. The flowers are in erect spikes at the ends of branches. The calyx is two-lipped, with three triangular teeth above and two longer and much narrower teeth below. The corolla has a flat, notched upper lip and a lower lip of three spreading lobes.

June to September: in woods and fields from Quebec and Ontario to North Carolina and Indiana. *Plate 117.*

DRAGONHEAD (DRACOCEPHALUM)

One species of this mainly Asiatic genus is found as an occasional weed in many parts of North America: *D. parviflorum.* This is a plant 2 feet tall or a little more, with lanceolate or ovate, toothed leaf-blades on stalks, and light blue flowers in a dense spike at the summit of the stem. The corolla scarcely emerges from the calyx.

THE HEDGE-NETTLES (STACHYS)

Many quite unrelated plants are likened to nettles (dead-nettles, horse-nettles, hemp-nettles, etc.), not because they share the "stinging" propensity of nettles but usually from the similar shape of their leaves. The hedge-nettles have leaves of various shapes, but some do have ovate, toothed blades like those of true nettles (*Urtica*). The flowers surround the stem at intervals in the axils of bracts – forming an "interrupted spike" (*stachys* means "spike"). The flower has a hooded upper lip and a longer, three-lobed lower lip which hangs down.

The species are numerous, variable, and consequently somewhat difficult to distinguish. Several are immigrants from the Old World; those that are rarely found are here omitted.

A hand magnifier will make it possible to place a plant in one of the following groups.

I. *Species with calyx downy or hairy all over (not merely along the veins.) All but one of these have hairs on the leaves, and all but the first and last have leaves with evident stalks.*

WOUNDWORT, S. PALUSTRIS, is our most wide-ranging species, extending through northern Europe and Asia and most of North America. It is extremely variable, consisting of many varieties. It may generally be recognized by the downy calyx and the stem which is downy or hairy on the flat sides as well as on the angles. The stem stands from 8 inches to nearly 7 feet tall. The leaves have very short stalks or none. The flowers are rose mottled with purple.

July to September: in roadside ditches, meadows, waste land, etc. throughout the northeastern and western parts of the United States, and most of Canada. *Plate 117.*

S. RIDDELLII and S. CLINGMANII are mainly species of the southern Appalachian Mountains (the second was named for Clingman's Dome in the Great Smokies). Both have distinctly stalked leaves, with hairs on the veins on the lower side. Both have stems with conspicuous hairs on the angles. The stem of *S. riddellii* is very finely downy on the flat sides, that of *S. clingmanii* smooth. The flowers are purplish. The hairs on the calyx are generally very small.

June and July: in wooded bottomlands chiefly in the mountains from Maryland (*S. riddellii*) and Virginia (*S. clingmanii*) to Illinois and southward to North Carolina and Tennessee.

S. ARVENSIS is an European species found only in the Atlantic states. Stem, leaves, and calyx are all beset with long, rather stiff hairs.

June to October: in waste places from Maine to Virginia. *S. sylvatica*, another European occasionally found is copiously glandular and sticky as well as hairy.

S. LATIDENS is a plant of the southern Appalachians. The stem is smooth on the sides, smooth or bristly on the angles. The leaves are ovate, coarsely toothed, indented at the base, with short stalks or none, and smooth.

June to August: in meadows and along streams from the District of Columbia and West Virginia to Georgia and Tennessee.

II. *Species with smooth calyx or with bristles only along the veins. The leaves have short stalks or none.*

S. TENUIFOLIA is a very variable, wide-ranging species to be compared with *S. palustris.* The stem may be quite hairless, or the angles may be furnished with bristles. The leaves vary from very narrow to ovate and may be smooth or bristly. The calyx also may be quite smooth or may bear conspicuous bristles on the veins.

June to September: in bottomlands, meadows, and low woods from Quebec to Minnesota and southward to South Carolina, Tennessee, Louisiana, and Texas. *Plate 117*.

S. HYSSOPIFOLIA is a small plant, from 4 to 30 inches tall, with a stem smooth except sometimes on the angles. The leaves may be very narrow, without stalk or teeth, and smooth.

June to October: in sandy places and bogs on the coastal plain from Massachusetts to South Carolina and in the southern mountains; in Michigan and Indiana; and the hairy variety from Pennsylvania to Iowa and southward to Florida and Kentucky.

AJUGA

Three species of this Old-World genus are cultivated for ornament in American gardens and are occasionally found growing wild in waste land and fields. They flower from April to July or later.

BUGLE or BUGLEWEED, A. REPTANS (*Plate 117*), has creeping leafy stems bearing short, broad leaves with wavy margins or sometimes toothed. The flowers are on upright stems, up to a foot tall, forming a sort of spike (but the lower ones in the axils of leaves). The lower lip of the blue-purple corolla is much larger than the upper, with three diverging lobes. (For the meaning of the name bugle see under *Lycopus*. *Reptans* means "creeping"; compare "reptile.") *A. genevensis* is somewhat similar but has no creeping stems and is more evidently downy. *A. chamaepitys* has yellow flowers and leaves cleft or divided into three. *A. reptans* was used in olden times to cure wounds, like *Prunella*.

THE GIANT-HYSSOPS (AGASTACHE)

Why these plants should be likened to hyssop it is not easy to see. The flowers are in dense clusters in the axils of bracts, often so close as to constitute a spike (or "interrupted spike"). This is true also in hyssop; but there the resemblance ends. Our species of *Agastache* are tall plants with stalked, toothed, ovate leaf-blades. The flowers have long protruding stamens, in two pairs in each flower; the lower pair, curiously, curves upward, the upper pair downward, so that they cross.

YELLOW GIANT-HYSSOP, A. NEPETOIDES, grows from 2 to 5 feet tall. The leaf-blades are coarsely toothed, with a round base sometimes indented. The corolla is greenish-yellow.

July to September: in thickets and open woodland from Quebec to Ontario and South Dakota and southward to Georgia, Kentucky, Arkansas, and Kansas. *Plate 118*.

PURPLE GIANT-HYSSOP, A. SCROPHULARIAEFOLIA, is very similar to the preceding species. The stem is downier or hairy. The corolla is purplish.

July to September: in woods and thickets from Vermont and Massachusetts to Wisconsin and South Dakota and southward to North Carolina, Kentucky, Missouri, and Kansas. *Plate 118*.

BLUE GIANT-HYSSOP, A. FOENICULUM, has a smooth stem up to 3 feet tall, bearing leaves that are white on the under surface with very small hairs. The corolla is blue.

June to September: in dry woodland and prairies from Ontario to Mackenzie and southward to Illinois, Iowa, Colorado, and Washington; occasionally found farther to the east, in fields. *Plate 118*. The botanical name refers to fennel (*Foeniculum vulgare*), which this plant is said to resemble in odor; it is also compared to anise.

HYSSOPUS

There is only one species.

HYSSOP, H. OFFICINALIS, well-known and long cultivated as a medicinal herb, is a plant from 1 to 2 feet tall, with narrowly lanceolate leaves without stalks and without teeth. The flowers are in dense spikes at the ends of the stem and branches, with clusters also in the axils of leaves. The corolla is purplish, about $\frac{1}{2}$ inch long.

July to October: in dry fields and on roadsides from Quebec to Montana and southward to North Carolina. *Plate 120*.

THE FALSE DRAGONHEADS (PHYSOSTEGIA)

The plants with this somewhat absurd English name are slim, mostly unbranched plants about 3 feet tall or taller, with smooth stems and leaves. The leaves are without stalks and have toothed margins. The flowers are borne in a long tapering spike at the summit of the stem. The pinkish or purplish corolla projects from the calyx; the upper lip is a hood over the four stamens; the lower lip has three diverging lobes. The English name refers perhaps to the aspect of the corolla; they cannot be "true dragonheads," since this name is reserved for another genus, *Dracocephalum* (with which *Physostegia* has been confused). Another name is obedience or obedient plants, from the amusing docility of the flowers: if you bend them to one side or up or down they will stay in that position.

I. *Species whose upper leaves are much shorter than the lower, so that the flower-spike appears to have a long stalk beneath the flowers, bearing only reduced leaves (bracts).*

P. VIRGINIANA has a stout stem up to 3 feet tall, often bearing several flowering branches. The leaves are narrowly lanceolate and furnished with sharp, incurved teeth. The corolla is about an inch long.

June to September: in swampy thickets and prairies and wet woods from Vermont to Ohio, Missouri, and Oklahoma and southward to South Carolina, Alabama, and Texas. *Plate 118*. Often cultivated and in the eastern part of its range perhaps escaped from cultivation rather than native.

P. PURPUREA (about 3 feet) differs from *P. virginiana* chiefly in the leaves, which are very narrow (often widest towards the tip) and with blunt, inconspicuous teeth and blunt tip. The corolla is about an inch long. The flowers are less crowded in the spike.

June and July: in bottomlands on the coastal plain from Virginia to Florida and Texas and northward to Arkansas and perhaps Missouri.

P. INTERMEDIA (from 1 to 5 feet tall) has narrowly lanceolate leaves, sharp-pointed, with short, blunt teeth. The corolla is about ½ inch long.

June and July: in wet meadows and prairies and swamps from Kentucky (and perhaps Illinois) to Kansas and southward to Alabama and Texas.

II. *Species whose upper leaves are of much the same length as the lower, so that the flower-spike begins above the uppermost pair of leaves and has no stalk.*

P. FORMOSIOR is a stout plant from 1 to 5 feet tall. The leaves are lanceolate, taper-pointed, with sharp teeth which curve forward. The corolla is about an inch long.

June to September: in wet woods and swampy thickets from Maine to Alberta and southward to New York, Ohio, Missouri, and Nebraska. *Plate 118*.

Besides these several southern and western species may be found along our borders: *P. nuttallii* in Minnesota and Wisconsin westward; *P. obovata* in southern Missouri.

MEEHANIA

There is but one species of *Meehania*.

M. CORDATA is a low plant, with creeping stems and erect branches 6 or 8 inches tall. The leaves are long-stalked with heart-shaped blades which have scalloped margins. The stems and leaves are hairy. The flowers are in dense spikes at the tips of the upright branches. The corolla is pale purple, an inch or more long, the upper lip forming a hood.

May to July: in woods from Pennsylvania to Ohio and perhaps to Illinois and southward to North Carolina and Tennessee. The genus is named for Thomas Meehan, a well-known writer on plants of the nineteenth century.

PERILLA

One species of this Asian genus has prospered as an escape from cultivation in America.

BEEFSTEAK-PLANT, P. FRUTESCENS, is a rank-smelling plant from 1 to 3 feet tall with long-stalked leaves; the blades are ovate, tapering at base and tip, blunt-toothed. The white flowers are in spikes at the summit of the stem and from the axils of the leaves. The corolla is small, almost equally five-lobed. The calyx is two-lipped and becomes larger after the corolla falls; all these papery calyxes make a rustling, almost a tinkling sound as one walks through a stand.

PLATE 118

Agastache foeniculum *Johnson*

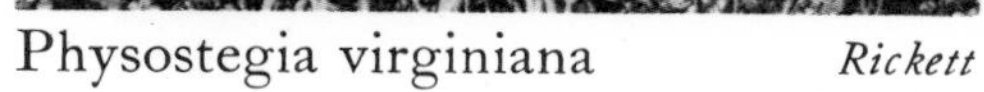

Physostegia virginiana *Rickett*

Scutellaria elliptica *Rhein*

Agastache nepetoides *D. Richards*

Scutellaria incana *Johnson*

Agastache scrophulariaefolia *D. Richards*

Physostegia formosior *Johnson*

August to October: in waste places, on roadsides, and in dry woods from Massachusetts to Kansas and southward to Florida and Texas. *Plate 120*. The plant is cultivated for its foliage, which is reddish-brown in the garden variety and gives rise to the English name. There is also a variety with cleft leaves.

THE SKULLCAPS (SCUTELLARIA)

The skullcaps form a large, worldwide genus, with many American species. We have a dozen or more in the Northeast. The flowers are distinctive. The calyx wears a hump or dish (*scutella*) on the upper side. This also has been likened to a *galerum*, a leather skullcap worn by the Romans. The corolla (blue, purple, white, or variegated) has a tube which curves upward, ending in two lips: the upper lip is hoodlike, the lower flat; the two side lobes seem connected with the upper rather than the lower lip. The plants lack the aromatic qualities of most *Labiatae*. They are bitter and were recommended for fevers by the herbalists.

The species may be easily placed in three groups according to the disposition of their flowers.

I. *Species whose flowers are in spikes or racemes at the summit of the stem, with or without other such spikes or racemes arising from the axils of the upper leaves. (Compare II and III.)*

A. Two of these have most leaf-blades heart-shaped, i.e. indented at the base (but compare *S. incana* under B).

S. OVATA (from 6 inches to 3 feet) has a hairy stem, some of the hairs often bearing glands. The leaf-blades are ovate, scalloped (notched at intervals), and are borne on long stalks. The corolla is blue with the lower lip whitish, from ½ to 1 inch long.

June and July: in rocky woods, on bluffs, and in barren places from Maryland to Ohio and Minnesota and southward to South Carolina, Texas, and northern Mexico. Very variable in size, hairiness, and other details.

S. SAXATILIS (from 8 to 20 inches) has a rather weak, mostly smooth stem. The leaf-blades are ovate, with a few blunt teeth on either side, and long-stalked. The corolla is violet-and-white, about ½ inch long.

June to September: in rocky woods and on moist bluffs from Delaware to Indiana and southward to South Carolina and Tennessee.

B. The remaining species of group I have most leaf-blades forming a V at the tip of their stalk, not indented.

S. INCANA (from 1 to 4 feet) has ovate leaves which are white on the lower side with minute hairs; the stem and corolla are similarly hoary. The leaf-blades are scalloped, long-stalked. The corolla is blue, about an inch long. Some plants have heart-shaped leaves.

June to September: in dry woods and thickets from New Jersey and western New York to Iowa and southward to Florida, Alabama, Arkansas, and Kansas. *Plate 118*.

S. ELLIPTICA (from 6 inches to 2 feet) has a softly hairy stem bearing short-stalked leaves with ovate (*not* elliptic) scalloped blades, at rather wide intervals. The corolla is violet, less than an inch long.

May to August: in dry woods and thickets from southern New York to Missouri and southward to Florida and Texas. *Plate 118*.

S. SERRATA (from 8 to 28 inches) is smooth or nearly so. The leaf-blades are ovate or nearly elliptic, scalloped, on short stalks. The plant suggests *S. elliptica* except for its lack of hair. There is often only one inflorescence, at the summit of the stem. The corolla is blue, an inch or more long.

May and June: in woods from southern New York to Ohio and southward to Georgia and Alabama; and in Missouri. *Plate 119*.

S. INTEGRIFOLIA (from 1 to 2 feet) has densely hairy, slender stems bearing narrowly lanceolate leaves without stalks and without marginal teeth or notches (*integri-*). (Some of the lower leaves may have toothed blades on stalks, but these fall early.) The corolla is purplish-and-whitish (sometimes pink), about an inch long.

May to July: in the edges of woods and fields on the coastal plain from Massachusetts to Florida and Texas; inland from New York to Missouri and southward to Tennessee. *Plate 119*.

II. *Two species have flowers in several* one-sided *racemes growing from the axils of leaves from the middle part of the stem upward, with or without a short raceme at the summit of the stem. (Compare group III.)*

MAD-DOG SKULLCAP, S. LATERIFLORA (from 4 to 40 inches), spreads by slender creeping stems. The leaf-blades vary from ovate to lanceolate or even triangular, with toothed margins, on stalks about an inch long. The corolla is violet (or pink or white), about ⅓ inch long.

June to September: in wet woods and thickets

PLATE 119

Glecoma hederacea *Rickett*

Scutellaria epilobiifolia *Johnson*

Lamium purpureum *D. Richards*

Scutellaria integrifolia *Rhein*

Scutellaria serrata *Elbert*

Scutellaria lateriflora *Scribner*

Scutellaria parvula *Johnson*

and meadows from Quebec to British Columbia and southward to Florida and California. *Plate 119*. The English name is due to its supposed efficacy against rabies; at one time it was also reputed to furnish a remedy for hysteria!

S. CHURCHILLIANA resembles *S. lateriflora*, with a somewhat larger flower (the corolla nearly $\frac{1}{2}$ inch long) and flowers often borne singly in the axils of leaves. In the latter characteristic it resembles *S. epilobiifolia*, and it has been suspected to be a hybrid between this species and *S. lateriflora*.

III. *Species whose flowers are borne singly in the axils of leaves. Mostly small plants with leaf-margins indistinctly toothed.*

S. EPILOBIIFOLIA (4 to 40 inches) has slender creeping stems (or rhizomes). The erect branches bear lanceolate or ovate leaves, with no stalks or a very short one, their edges very indistinctly toothed. The corolla is violet with a whitish tube, from $\frac{3}{5}$ to 1 inch in length.

June to September: on gravelly or sandy shores and in meadows and swampy thickets from Newfoundland to Alaska and southward to Delaware, West Virginia, Missouri, New Mexico, and California. *Plate 119*. This is considered by some botanists to be the same as the Old-World *S. galericulata*.

S. PARVULA (3 to 12 inches) forms slim horizontal stems which are swollen at intervals into tubers. From these rise the floriferous branches bearing ovate, stalkless and toothless leaves scarcely more than $\frac{4}{5}$ inch long, with three, five, or more veins running lengthwise. The corolla is bluish, from $\frac{1}{5}$ to $\frac{2}{5}$ inch long.

May to July: in dry woods and prairies from Quebec to North Dakota and southward to Florida, Alabama, and Texas. *Plate 119*. Plants with narrow leaf-blades on short stalks are separated by some botanists as *S. leonardi*. The more southern plants with long glandular hairs are sometimes separated as *S. australis*.

S. NERVOSA (8 to 20 inches) has a weak, slender, smooth stem bearing short, broad leaves without stalks, the margins with a few notches. The corolla is pale blue, about $\frac{2}{5}$ inch long.

May to July: in moist woods from New Jersey to Ontario and Iowa and southward to North Carolina and Tennessee.

GLECOMA

One species of this Old-World genus has become established in North America.

GROUND-IVY, GILL-OVER-THE-GROUND, G. HEDERACEA, forms mats with its creeping, trailing stems. The leaf-blades are nearly round with scalloped or sometimes toothed edges, and stalked. Erect branches rise from the mat, bearing the bright violet flowers in the axils of their leaves. The corolla is about $\frac{3}{4}$ inch long, with a narrow upper lip and a wide, three-lobed lower lip.

April to July: on roadsides and in yards, woods, etc. from Newfoundland to Minnesota and southward to Georgia, Alabama, Missouri, and Kansas; and on the Pacific coast. *Plate 119*. This bitter herb was once used as hops now are in flavoring ale; it was called alehoof in England. It was also used in medicine. Some amusing variations of the name are Robin-run-in-the-hedge, runaway-Jack, Jenny-run-by-the-ground.

DEAD-NETTLES (LAMIUM)

The Old-World genus *Lamium* has contributed several species to the wild flowers of North America. Most of them are known as dead-nettles, having nettle-like leaves but no "sting." They have a characteristic corolla, erect, distended just below the lips, the upper lip erect and hooded, the lower extending forward and flanked by two side lobes. The leaves have broad and toothed or scalloped blades, all stalked except those just beneath flowers.

HENBIT, L. AMPLEXICAULE, has several stems from one root, spreading at various angles, some almost lying on the ground. The lower leaves have round, scalloped blades on long stalks; the upper ones have short stalks or none, and the edges may be coarsely toothed. Earlier flowers often do not open, fertilizing themselves inside the closed tube of the corolla. The corolla is purplish, dark-spotted, the upper lip crowned with a tuft of magenta hairs.

March to November: in lawns, waste land, and fields and on roadsides throughout our range and southward; commoner southward. *Plate 120*.

RED DEAD-NETTLE, L. PURPUREUM, grows erect to a height of from 4 to 12 inches. The leaf-blades are ovate, usually indented at the base, scalloped, the

PLATE 120

Lamium amplexicaule *Johnson*

Synandra hispidula *Johnson*

Lamium maculatum *Elbert*

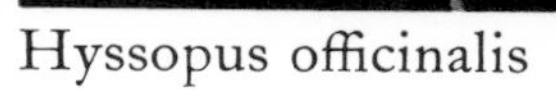

Hyssopus officinalis *Rhein*

Leonurus cardiaca *Johnson*

Perilla frutescens *Allen*

Trichostema dichotomum *Ryker*

lower ones on long stalks, the upper on short stalks. The flowers grow in the axils of the upper leaves, which are colored crimson underneath. The corolla is crimson, about ½ inch long.

April to October: in waste land and fields and on roadsides from Newfoundland to Michigan and southward to South Carolina, Tennessee, and Missouri. *Plate 119.*

L. HYBRIDUM resembles *L. purpureum* but has longer-stalked leaves and leaf-blades sharply toothed or even lobed, the teeth or lobes themselves cleft.

May to October: in waste places and cultivated ground from Newfoundland to New England and North Carolina. Supposed to be a hybrid between *L. purpureum* and *L. amplexicaule;* but this has not been proved.

WHITE DEAD-NETTLE, L. ALBUM, grows from 6 inches to 2 feet tall. The leaves have coarsely toothed, ovate blades tapering to long points, all stalked. The flowers are cream-colored, about an inch long.

April to October: in waste places and gardens and on roadsides from Quebec to Minnesota and southward to Virginia. *Plate 113.* An English name is Adam-and-Eve-in-the-Bower. Grigson writes, "turn the plant upside-down, and beneath the white upper lip of the corolla, Adam and Eve, the black and gold stamens, lie side by side, like two human figures." The plant had various medicinal uses, especially against the "King's evil."

L. MACULATUM is from 6 inches to 2 feet tall. The leaves have ovate blades usually with a white blotch (*maculatum* means "spotted"), often indented at the base, their edges coarsely scalloped, all stalked. The flowers are rose-purple, from ⅘ to 1 inch long.

April to October: on roadsides and in waste places from New England to Ontario and southward to North Carolina and Tennessee. *Plate 120.*

BALLOTA

One species from the Old World has become a weed in North America.

BLACK- or FETID-HOREHOUND, B. NIGRA, is a plant from 1 to 3 feet tall, with coarsely toothed, ovate, stalked leaf-blades; the whole plant more or less hairy. The flowers are clustered in the axils of the upper leaves. The calyx has five short teeth. The corolla is purplish, with a hooded upper lip and a broad, three-lobed lower lip.

June to October: in waste places from New England and New York southward to Maryland and Pennsylvania. It somewhat resembles *Marrubium*, horehound, but lacks the white wool and the longer calyx-teeth of that genus. It was once prescribed for ulcers and the bite of mad dogs.

MARRUBIUM

One species of *Marrubium* was formerly cultivated for medicinal use and is now found as a weed. Its bitter flavor is still found in certain candies esteemed by some as "cough-drops."

HOREHOUND, M. VULGARE, is from 1 to 2 feet tall, the leaves having ovate blades, edged with blunt teeth, on short stalks. The whole plant is covered with a white down. The flowers are in dense clusters in the axils of the leaves. The corolla is small, white, with an erect upper lip and a three-lobed lower lip.

May to September: in waste places practically throughout the United States. The name horehound is a corruption of Anglo-Saxon *har hune.*

MOTHERWORTS (LEONURUS)

The English name goes back to olden days when these plants were supposed to provide a remedy for the disease known as "mother." The botanical name means "lion's-tail," presumably from a fancied resemblance of the leaves of some species to the tufted tail of the king of beasts. The leaves are stalked and toothed, lobed, or cleft. The small flowers are in clusters in the axils of leaves.

COMMON MOTHERWORT, L. CARDIACA, grows up to 5 feet tall. The leaves have long stalks bearing blades generally palmately cleft or lobed into three sharply toothed parts. The corolla is pale purple, about ½ inch long, the upper lip hairy.

June to August: in waste land from Quebec to Montana and southward to North Carolina and Texas. *Plate 120.*

L. MARRUBIASTRUM has ovate, sharply toothed leaf-blades on short stalks. The corolla is almost white and scarcely longer than the calyx.

June to September: in waste land from Pennsylvania to Illinois and southward to Florida. The name refers to its likeness to *Marrubium*.

MELISSA

The Asian genus *Melissa* contributes one species that has been cultivated since earliest times, was brought to America for home use, and has escaped into roadsides and other places.

BALM, M. OFFICINALIS, is from 1 to 3 feet tall, with stalked leaves; the blades are ovate and scalloped. The crushed leaves have a lemon scent. The flowers are in small clusters in the axils of leaves. The calyx is two-lipped, with three short, broad teeth above and two longer and narrower teeth below. The almost white corolla curves upward; it is about ½ inch long.

June to September: on roadsides and in woods and waste land from Maine to Kansas and southward to Florida and Arkansas; and on the Pacific Coast. Balm tea "removes melancholy."

SYNANDRA

There is only one species of *Synandra*.

S. HISPIDULA, a native American, seems to have acquired no English name. It is a plant from 1 to 2 feet tall, with long-stalked leaves; the blades are ovate, coarsely toothed with blunt teeth, indented at the base. The large flowers are borne singly, without stalks. The corolla is yellowish or greenish, about 1½ inches long, with an inflated tube, an arched upper lip, and a three-lobed lower lip.

May to July: in woods and thickets from Virginia to Illinois and southward to Tennessee. *Plate 120*.

CONRADINA

The species of *Conradina* are southern, but one extends into our range. *C. verticillata* grows in Kentucky and Tennessee on sandy and gravelly shores and in open woods. It is really a shrub, having woody stems, but scarcely exceeds a foot in height, the branches being numerous and spreading out. The leaves are very narrow; other leaves appear in their axils so that they seem to be in circles rather than pairs. The flowers are in small clusters in the axils of the leaves and at the ends of the branches. The lavender flowers appear in May.

BLUE-CURLS (TRICHOSTEMA)

The blue-curls are distinctive and charming little plants, easily known by the very long, curved, blue-stalked stamens (to which the botanical and English names both refer). The blue corolla has four pointed lobes above and a broader, blunter lower lobe extending downward. The calyx also is two-lipped, the upper, three-toothed lip much longer than the lower, two-toothed lip. This calyx has the curious habit of becoming inverted as the corolla falls, so that in the fruiting stage the long lip becomes the lower.

BLUE-CURLS or BASTARD PENNYROYAL, T. DICHOTOMUM, is a delicate plant, much branched, from 4 to 30 inches tall. It is usually somewhat sticky. The leaves are narrow, without stalks and without teeth. The flowers are borne singly at the ends of slender branches.

August to October: in dry woodland from Maine to Michigan and southward to Florida and Texas. *Plate 120*. A variety with very narrow leaves grows from Louisiana and Florida northward to Connecticut, mostly near the coast, and inland to Pennsylvania and Ohio. It is by some considered a distinct species, *T. setaceum*.

THE SNAPDRAGON FAMILY (SCROPHULARIACEAE)

A flower of a snapdragon has five petals joined to form a corolla with two "lips," upper and lower. The opening between the lips is closed by an elevated part of the lower lip, called by botanists a "palate." This sort of corolla is found in many species of the snapdragon family; but in many others there is no "palate" and the corolla is open, and in a number of these there is little difference between upper and lower petals – the corolla may be almost radial in its symmetry. The number of stamens varies from five to two, but is most commonly four.

This is not an economically important family. It includes the foxglove (*Digitalis*), from which a valuable drug is obtained; and a number of species – snapdragons, beard-tongues, veronicas, and others – decorate our gardens. In the wild, however, we meet many attractive "scrophs," and a number of insignificant and puzzling small plants that may excite the curiosity of the amateur.

We place the genera in three groups by the number of fertile stamens (producing pollen) in each flower.

I. *Fertile stamens five:* Verbascum.

II. *Fertile stamens two:* Bacopa (*occasionally; see under III*); Lindernia (*sometimes; small plants with paired leaves, flowers less than ½ inch long, in axils*); Hemianthus (*creeping plant with minute paired leaves and flowers less than ⅛ inch long, without an upper lip*); Wulfenia (*long-stalked basal leaves with heart-shaped blades*); Veronicastrum (*leaves in circles; small flowers in dense, tapering spikes*); Veronica (*leaves borne singly; small blue, pink, or white flowers in axils or in racemes*); Gratiola (*flowers white or yellow, in axils*).

III. *Fertile stamens four. This, much the largest group, is divided by the manner in which the leaves are borne.*

A. Genera with leaves borne singly (compare B). The first four of these genera have "spurred" flowers – a hollow tube extends downward from the lower side of the corolla: *Linaria* (yellow-and-orange or blue-and-white flowers in a tight raceme); *Kickxia* (purple-and-yellow flowers in the axils; stems lying on the ground); *Cymbalaria* (blue-and-yellow flowers in the axils; stem trailing; leaves palmately lobed); *Chaenorrhinum* (leaves very narrow, less than an inch long; purple-and-yellow flowers in the axils); *Pedicularis* (yellow or red flowers with upper lip curved down, in dense heads on low stems; leaves pinnately lobed or cleft); *Castilleia* (flowers minute, green, more or less concealed by red or yellow bracts – or bracts with red or yellow tips); *Schwalbea* (narrow, yellowish-and-purplish flowers 1½ inches long, in axils).

B. Genera with four fertile stamens and leaves in pairs. These in turn may be separated into two groups according to the manner in which their flowers are arranged.

1. Flowers in the axils of leaves (singly or in clusters)*: these are arranged by the size of their flowers – in most species of the first two, more than ¾ inch long, in the next six, more than ⅓ inch but less than ¾ inch, in the last two less than ⅓ inch long: *Mimulus* (flowers blue or yellow, the corolla closed by a "palate"); *Agalinis* (flowers pink or purple, the corolla open; leaves narrow); *Tomanthera* (like *Agalinis* but with rough, lanceolate leaves, some of them lobed at the base); *Collinsia* (blue-and-white flowers; leaves narrow); *Melampyrum* (flowers narrow, white-and-yellow); *Euphrasia* (leaves less than an inch long, coarsely toothed; flowers white-and-blue with purple lines); *Lindernia* (flowers white or lavender); *Bacopa* (flowers white or blue; leaves roundish, palmately veined); *Mecardonia* (flowers white, on long stalks); *Leucospora* (flowers pale lavender; leaves pinnately cleft).

2. Flowers in an inflorescence: *Limosella* (leaves rod-shaped, bladeless, all basal; flowering stems leafless); *Buchnera* (blue purple flowers in a spike); *Rhinanthus* (yellow flowers all on one side of a spike); *Scrophularia* (brown or greenish flowers in a widely branched inflorescence); *Penstemon* (white or purple flowers in a narrow inflorescence); *Chelone* (white, pink, or purple flowers in a spike); *Dasistoma* (yellow flowers in a spike; corolla hairy inside); *Aureolaria* (yellow flowers in spikes; corolla funnel-shaped, almost radially symmetric).

*It is often difficult to decide whether a plant has an inflorescence – a flower-cluster which does not include ordinary foliage – or flowers in the axils of the foliage. In several species the flowers some distance from the tip of the stem are in the axils of leaves, but flowers nearer the tip are in the axils of smaller leaves that must be called bracts; and often the transition is gradual. In general such arrangements are called inflorescences – spikes, racemes, etc. – especially when there is a noticeable difference between the lower and upper leaves.

THE MULLEINS (VERBASCUM)

The mulleins are distinguished in this family by the formation of five functional stamens. The corolla consists of five almost equal petals joined only at the base; careful scrutiny will reveal that the lower petals are larger than the upper. The flowers are in tall spikes or racemes. All the following species have been introduced from the Old World.

COMMON MULLEIN, V. THAPSUS, is a familiar weed of dry fields and roadsides. A rosette of leaves is formed at the surface of the ground; from this rises a stem sometimes well over 6 feet tall, bearing more leaves and ending in a long spike of yellow flowers (the spike may be branched). The plant would be handsome if more flowers opened together; but there are usually only a few at a time in the tall, green, weedy spike. The whole plant is velvety with a white wool.

June to September: perhaps one of the most widely distributed of plants, originating in Europe (according to legend from Thapsus in the Mediterranean), and now found throughout the United States from the rocky fields of New England to the dry hillsides of Missouri and the burning canons of Arizona; evidently preferring situations which are not conducive to other large plants. *Plate 121*. In Europe many superstitions are connected with its tall wand; it had power to ward off evil. One English name is Aaron's-rod, since the spike suggests a "rod" that blossoms again. In Germany it is the king's candles. The hairs on the leaves repay examination with a microscope; they are branched, like minute trees. Their softness has led to such names as flannel-plant and beggar's-blanket. The plant had its medicinal uses in olden times.

MOTH MULLEIN, V. BLATTARIA, is a slender plant up to 5 feet tall, with leaves on the lower part of the stem, flowers on the upper part. The leaves are lanceolate and toothed. The flowers amusingly simulate a moth with the lower three hairy stamens as the tongue and antennae. The corolla is white or yellow, one form more abundant in some places, the other in other places. The hairs on flower-stalk, calyx, etc. are glandular. As the flowers open, from below upward, the flowering stem gets longer, until it is a tall, skinny, leafless wand with a flower or two at the top.

June to October: in fields and on roadsides practically throughout the United States. *Plate 121*.

Several other species are less abundant as wild flowers. *V. phlomoides* (*Plate 121*) resembles *V. thapsus;* it may be taller. The edges of the leaves do not run down on the stem as they do in *V. thapsus*. There may be gaps in the spike between the flowers. It has been found from Maine to Minnesota and southward to North Carolina, Kentucky, and Iowa. *V. lychnitis* has a more branched inflorescence with flowers in small clusters at intervals on the branches. The flowers are small, yellow or white. The leaves are smooth on the upper side. *V. phoeniceum* resembles *V. blattaria* but the flowers are purple. *V. virgatum* also resembles *V. blattaria* but is more densely covered with hairs.

FALSE PIMPERNELS (LINDERNIA)

Our species of *Lindernia* are small, branched plants of muddy places and wet shores, with paired, stalkless leaves scarcely more than an inch long. The flowers are recognizable by their very short, notched upper lip; the lower lip is much longer and three-lobed. They are borne singly in the axils of the leaves.

L. DUBIA has stems from 2 to 14 inches long, usually branched and spreading. The leaves are generally oblong or elliptic, with toothed or plain edges. The flowers are pale lavender, less than ½ inch long. They may fertilize themselves without opening.

June to October: in damp soil from Quebec to North Dakota and southward to Florida, Texas, and Mexico; and on the Pacific Coast.

L. ANAGALLIDEA is widely branched and not more than 8 inches tall. The leaves are ovate or elliptic, mostly about ½ inch long, usually with plain edges. The flowers are pale lavender, less than ½ inch long.

June to October: on damp shores and mud from Quebec to North Dakota and Colorado and southward to Florida and Texas.

HEMIANTHUS

The botanical name means "half-a-flower": the corolla of these plants has no upper lip. The lower lip is three-lobed; the middle lobe is much the largest and curls upward. This is a genus of the tropics, one species being North American.

H. MICRANTHEMOIDES is a creeping plant with leaves only about ⅕ inch long. The white flowers are single in the axils, the corolla minute.

August to October: on the tidal mud of estuaries from New York to Virginia.

WULFENIA

Of this western genus one species, *W. bullii*, is found rarely in prairies and on bluffs from Michigan to Minnesota and southward to Ohio and Iowa. It has long-stalked basal leaves with heart-shaped blades, and a spike of minute yellow or greenish flowers up to 16 inches tall.

CULVER'S-ROOT (VERONICASTRUM)

One species of *Veronicastrum* occurs in North America; the only other species, if it is indeed distinct, is in Siberia.

CULVER'S-ROOT or CULVER'S-PHYSIC, V. VIRGINICUM, grows from 2 to 6 feet tall. The leaves have lanceolate blades on short stalks; they are borne in circles. The many small white flowers are in several tapering spikes at the summit of the stem. The corolla has four lobes and two long stamens; the projecting stamens give the whole inflorescence a fuzzy appearance.

June to September: in meadows, prairies, and woods and on roadsides from Massachusetts and Vermont to Manitoba and southward to Georgia, western Florida, and Texas. *Plate 121*. The plant has been used in medicine, having rather violent cathartic and emetic properties. Just who Culver was, however, I have not discovered. In England culver is a dove or wood-pigeon; but this seems to have no relevance.

THE SPEEDWELLS (VERONICA)

Speedwells carpet fields in England and wish the traveler well with bright blue flowers. The corolla has four lobes, almost radially symmetric but the lowest narrower than the other three, all joined at the base. There are two stamens. The ovary develops into a small heart-shaped capsule, the shape of which is useful in identifying the species. Our species have blue, purple, or white flowers borne singly in the axils of leaves or in racemes. Some are native Americans, others introduced from the Old World.

I. *Species whose flowers are in racemes. (Compare II.)*

BIRD'S-EYE or GERMANDER SPEEDWELL, V. CHAMAEDRYS, has slender branches from 4 to 16 inches tall, rising from a creeping stem. The leaf-blades are ovate or heart-shaped, scalloped, on very short stalks. The flowers are in loose racemes. The corolla is blue with a white center, less than ½ inch across.

May to July: on roadsides and in fields from Newfoundland to Michigan and southward to Maryland, West Virginia, and Illinois; introduced from Europe. *Plate 122*. A belief of some English children is that if you pick this plant your mother's eyes will be picked out; hence one common name – pick-your-mother's-eyes-out! Also milkmaid's-eye, angel's-eyes, wish-me-well, and many others. The blue eyes look up at you from many a grassy field and path. Its magical qualities are suggested by such names as devil's-eyes.

COMMON SPEEDWELL or GYPSYWEED, V. OFFICINALIS, has stems that creep on the ground, forming mats. They bear small elliptic, toothed, rather thick leaves. Flowering branches are erect, ending in a raceme of pretty blue or lavender flowers about ¼ inch across.

May to July: in dry fields and woodland from Newfoundland to Wisconsin and southward to North Carolina, Tennessee, and farther west; also in Europe, and perhaps not native here but introduced from Europe. *Plate 121*.

V. LATIFOLIA has erect stems from 1 to 3 feet tall. The leaves are without stalks, ovate and toothed. The flowers are blue, about ½ inch across.

May to July: on roadsides and in waste land from New England to South Dakota and southward to Maryland and Indiana; introduced from Europe.

AMERICAN BROOKLIME, V. AMERICANA, is a rather succulent plant with creeping stems turning upward at their ends. The leaf-blades are lanceolate or ovate, toothed, on short stalks. The flowers are violet, only about ¼ inch across.

May to August: in shallow water and swamps and along streams from Newfoundland to Alaska and southward to North Carolina, Tennessee, Missouri, Nebraska, Mexico, and California. *Plate 122*. Brooklime, *V. beccabunga*, is an European plant occasionally found in this country. It is very like *V. americana*, the leaves more elliptic, the capsule less deeply notched. The name brooklime was once brok-lemoc or brooklemoc – lemoc being the name of a plant. *Beccabunga* means the same in old Norse.

PLATE 121

Gottscho

Verbascum thapsus

Verbascum phlomoides *Rhein*

Johnson

Veronicastrum virginicum

Rickett

Verbascum blattaria

Verbascum blattaria *Johnson*

Veronicastrum virginicum  *Rickett*

Veronica officinalis *Elbert*

BROOK-PIMPERNEL or WATER SPEEDWELL, V. ANAGALLIS-AQUATICA, resembles *V. americana* but has leaf-blades without stalks and extending around ("clasping") the stem, either toothed or plain. The flowers are lilac, about $\frac{1}{5}$ inch across.

May to October: in springs and brooks and on wet shores from New England to Washington and southward to North Carolina, Tennessee, Texas, and Arizona; also in the Old World.

MARSH SPEEDWELL, V. SCUTELLATA, is a weak plant with slim creeping stems bearing very narrow leaves, mostly without teeth. The flowers are lilac.

May to September: in swamps and other wet places from Labrador to Alaska and southward to Virginia, Illinois, Iowa, Colorado, and California; and in Europe.

V. LONGIFOLIA grows from 1 to 5 feet tall, with leaves paired or in threes, and a terminal raceme of deep violet flowers.

May to July: in woods and fields and on roadsides from Newfoundland to Ontario and Wisconsin and southward to North Carolina, Tennessee, and Illinois. *Plate 122.*

V. ALPINA is a little northern plant extending into the United States only on certain high mountains. It stands up to about a foot tall, with a dense raceme of blue flowers about $\frac{1}{5}$ inch across. The leaves are elliptic; the stem is hairy.

July and August: in wet moss and ravines at high altitudes in Maine, New Hampshire, and Colorado.

II. *Species whose flowers are borne singly in the axils of leaves. These leaves are themselves borne singly, as contrasted with the paired leaves of non-flowering stems, and are somewhat smaller than the latter; they may be termed bracts, and then the flowers may be said to be in a raceme. See, for instance,* V. serpyllifolia.

FIELD SPEEDWELL, V. AGRESTIS, has stems that lie on the ground with tips turning up. The flowers are on stalks nearly $\frac{1}{2}$ inch long. The corolla is blue or white, up to $\frac{1}{3}$ inch across.

May to September: in waste land and cultivated ground from Newfoundland to Michigan and southward to Florida and Texas; introduced from Europe. *Plate 122.*

BIRD'S-EYE, V. PERSICA, is similar to *V. agrestis* but larger, the leaves coarsely toothed. The corolla is bright blue with paler center, about $\frac{1}{2}$ inch across.

March to September: on lawns and roadsides and in waste land from Newfoundland to Alaska and southward to Florida, Indiana, and the Pacific states. *Plate 122.*

CORN SPEEDWELL, V. ARVENSIS, is a little plant not much more than a foot tall, with ovate, toothed leaves only about $\frac{1}{2}$ inch long. The flowers are blue, minute.

March to August: in waste land and open woodland from Newfoundland to Minnesota and south to the Gulf; and in the Pacific states; a native of Europe.

THYME-LEAVED SPEEDWELL, V. SERPYLLIFOLIA, has creeping stems which form mats, covered with small ovate or elliptic leaves only about $\frac{1}{2}$ inch long. The flowers are loosely disposed on erect branches in the axils of smaller leaves or bracts – thus forming a raceme or spike.

April to July: in fields and lawns and on roadsides throughout our range and beyond. *Plate 122.*

NECKWEED or PURSLANE SPEEDWELL, V. PEREGRINA, is a smooth, rather succulent plant from 2 to 18 inches tall, with inconspicuous white flowers in the axils of the upper leaves. The leaves are mostly oblong and blunt, and usually lack teeth.

March to August: in fields, lawns, and moist woodland and on roadsides from Quebec to Minnesota and southward to Florida and Texas. A variety, which extends farther westward, has glandular hairs.

IVY-LEAVED SPEEDWELL, V. HEDERAEFOLIA, is distinguished by its leaves which are as broad as or broader than long, with a few wide teeth or lobes. The minute blue or lavender flowers are in the axils of ordinary leaves in the upper parts of the branches, which curve upwards from a horizontal base.

March to June: in fields, woodland, and waste places from New York to Ohio and southward to North Carolina; not common.

HEDGE-HYSSOPS (GRATIOLA)

The hedge-hyssops are small plants with paired leaves and yellow or white flowers borne singly in their axils. In America they do not resemble hyssop and are not found in hedges. Our species grow in ditches, swamps, and wet mud. The botanical name is connected with "grace" or "thanks [to God]" and refers to a supposed medicinal potency. (Both English and botanical names were formerly applied to quite another plant, a species of skullcap in the mint family.) The leaves are without stalks and those of most species

PLATE 122

Veronica persica *Elbert*

Veronica longifolia *Elbert*

Veronica agrestis *D. Richards*

Veronica serpyllifolia *D. Richards*

Veronica chamaedrys *Elbert*

Veronica americana *Rhein*

Linaria vulgaris *Johnson*

lack teeth; there are generally three main veins. The calyx is composed of unequal teeth, separate almost to the base; and below the calyx there is in most of our species a pair of small bracts which resemble the sepals. The corolla apparently has four lobes at the end of the tube; the upper lobe, which is also the upper lip, is notched and is composed of two petals.

GOLDEN-PERT, G. AUREA, is the handsomest species, with yellow flowers ½ inch long. The stem may reach a height of a foot or more. The leaves are ovate or lanceolate, only about an inch long.

June to September: in wet sand and swamps from Newfoundland to North Dakota and southward along the coast to Florida, inland to New York and Illinois. *Plate 126*. There are forms with honey-colored and white flowers, and one that grows under water.

G. VISCIDULA has a clammy ("viscid") stem, up to 2 feet tall. The leaves are generally toothed. The corolla is cream-white with purplish lines and yellow center.

June to September: in marshes and wet woods and shores from Delaware and southern Ohio southward to Georgia and Tennessee.

G. RAMOSA is a southern species, extending north to Maryland and flowering all summer. The corolla has a yellow tube with white lobes.

G. NEGLECTA is the most widespread and abundant species, yet, because of its small size (rarely more than a foot tall, with leaves not more than 2 inches long), and because of the places where it grows, it is commonly overlooked (*neglecta*). The lobes of the corolla vary from yellowish to cream-white, the tube being honey-colored.

May to October: in wet mud practically throughout southern Canada and the United States. *Plate 126*. The flowers formed in autumn often pollinate themselves without opening.

G. VIRGINIANA is widely distributed but does not extend so far northward as *G. neglecta*. It is from 4 to 18 inches tall. The leaves are lanceolate or narrowly elliptic with teeth on the edges. The flowers are about ½ inch long, varying from white or pink to honey-colored.

March to October: in brooks, pools, and ditches, on wet mud, and in wet woods from New Jersey to Kansas and southward to Florida and Texas. *Plate 126*. Later flowers often do not open.

G. PILOSA, with a softly hairy stem and white flowers about ¼ inch long, is a southern species found northward to New Jersey and Kentucky.

TOADFLAXES (LINARIA)

The toadflaxes have the mouth of the corolla closed by a prominent ridge or "palate" rising from the lower lip. The corolla also is prolonged downward into a hollow extension, the "spur." The flowers are in racemes or spikes. The leaves are very narrow, without teeth, borne in circles or – on the upper part of the stem – singly.

BUTTER-AND-EGGS or COMMON TOADFLAX, L. VULGARIS, is now a common weed in America; it came from Europe. The flowers are yellow-and-orange.

May to October: on roadsides and in fields, gardens, and waste land practically throughout North America. *Plate 122*. In Europe it has been a weed in the flax (*Linum*); hence both botanical and English names. Many names refer to the orange and yellow: bacon-and-eggs, bread-and-cheese, chopped-eggs, bread-and-butter. It is a bad weed, once it is established in a garden, spreading underground. One occasionally finds radially symmetric flowers with five spurs or none.

Several other European species are occasionally found in this country.

OLD-FIELD TOADFLAX, L. CANADENSIS, has slender stems up to 3 feet tall, often forming large colonies in old fields. Often a circle of runners grows from the base. The leaves on these as well as on the flowering stems are short and narrow. The flowers are small but numerous, violet-and-white.

March to September: in dry soil practically throughout the United States and southern Canada. *Plate 123*.

KICKXIA

The plants that rejoice in this improbable name have trailing stems bearing small ovate leaves and, in their axils, small purple-and-yellow flowers. They were formerly included in *Linaria*, from which they differ only in certain details of the capsules. The flowers have much the same form. They grow from the

PLATE 123

Pedicularis canadensis *Rickett*

Pedicularis canadensis *Gottscho*

Chaenorrhinum minus *D. Richards*

Linaria canadensis *Rickett*

Linaria canadensis *Shuler*

Cymbalaria muralis *D. Richards*

Pedicularis lanceolata *Johnson*

axils of the leaves. Both species below come from Europe.

CANKER-ROOT, K. ELATINE, has more or less triangular leaf-blades with sharp basal lobes and sometimes a few teeth at the margins. The lower leaves are paired, the upper single.

June to October: in waste places, on roadsides, etc. from Massachusetts to Kansas and southward to Georgia and Louisiana. *Plate 180.*

FLUELLIN, K. SPURIA, has oval leaf-blades without teeth and without sharp lobes at the base.

June to September: in waste places and fields and on roadsides from Rhode Island to Missouri and southward to Florida and Alabama.

CYMBALARIA

One species of this Old-World genus is found as an infrequent escape from cultivation in America.

KENILWORTH-IVY, C. MURALIS, is a trailing plant with leaves borne singly; long stalks bear palmately lobed, round blades. The pretty flowers are on long stalks from the axils of the leaves. The corolla is violet with a yellow "palate," less than ½ inch long, and has a short "spur."

May to October: in waste places and on roadsides, mostly near gardens. *Plate 123.* It was introduced as a garden plant from Europe into England in the seventeenth century, and spread as a wild flower, growing readily on walls (*muralis*).

CHAENORRHINUM

One species of this Mediterranean genus is now a wild flower in America.

DWARF-SNAPDRAGON, C. MINUS, is a small, branched, glandular plant from 2 to 16 inches tall. The leaves are lanceolate or narrower, without teeth. The flowers are borne singly on long stalks from the axils of the leaves. The corolla is lilac with a yellow center, only about ½ inch long; it has a "spur" and a "palate."

June to September: on roadsides and railroad tracks and other waste places from Quebec and Ontario to Virginia and Missouri. *Plate 123.*

THE LOUSEWORTS (PEDICULARIS)

The elegant names, both English and Latin, derive from an old belief that cattle that grazed the European species of this genus would become infested with lice. (Probably they did, but not from grazing.) Our species are low plants with pinnately lobed or cleft leaves and flowers in dense heads or spikes. The corolla is two-lipped, the upper lip arching and forming a sort of beak, the lower three-lobed, spreading downward. The four stamens rise under the upper lip.

COMMON LOUSEWORT or WOOD-BETONY, P. CANADENSIS, grows from 6 to 16 inches tall. The leaf-blades are narrow, oblong, pinnately cleft into blunt lobes. The flowers are either yellow or red. The calyx is almost without teeth or lobes.

April to June: in meadows and open woodland from Maine to Manitoba and southward to Florida, Texas, and Mexico. *Plate 123.* Like many species of this family, the louseworts seem to be dependent on the roots of other plants, especially of grasses.

P. LANCEOLATA grows up to 3 feet tall. The leaves are pinnately lobed, the lobes scalloped or toothed. The flowers are pale yellow. A distinctive feature is the presence of distinct toothed lobes on either side of the calyx.

August to October: in meadows, swamps, and wet woods from Massachusetts to Manitoba and southward to Virginia, the mountains of North Carolina, Tennessee, Missouri, and Nebraska. *Plate 123.* *P. furbishiae*, a Canadian species with a five-lobed calyx, may be found in northern Maine.

THE PAINTED-CUPS (CASTILLEIA)

The genus *Castilleia* is one of those that make botanists wish they had embraced some easy branch of science such as theoretical physics. It is a western genus, with at least 250 species, on the exact characteristics of which no two botanists seem able to agree. Fortunately for the readers of the present volume,

PLATE 124

Mimulus alatus *Scribner*

Castilleia coccinea *Rickett*

Mimulus moschatus *Scribner*

Mimulus glabratus *Johnson*

Mimulus ringens *Johnson*

Castilleia sessiliflora *Johnson*

only four species are found in the northeastern states, and they are not hard to distinguish. The flowers are not often seen, being hidden behind the red or yellow bracts that give the genus its name. The corolla is a slender tube, divided into two lips at the end, the lower lip the shorter. The leaves and colored bracts may be cleft or lobed.

PAINTED-CUP or INDIAN PAINTBRUSH, C. COCCINEA, is the commonest eastern species. The leaves and red-tipped bracts are deeply cleft, mostly into three narrow lobes. The plants stand from 8 inches to 2 feet tall.

April to August: in meadows and fields from southern New Hampshire to southern Manitoba and southward to Florida, Louisiana, and Oklahoma. *Plate 124*. There is a form with yellow bracts. Like many other *Scrophulariaceae*, the painted-cups are parasitic on the roots of other plants, apparently most frequently on grasses.

DOWNY PAINTED-CUP, C. SESSILIFLORA, has leaves and bracts cleft into very narrow, yellowish green lobes. The flowers also are greenish or yellowish. The stems are usually clustered and up to a foot tall or slightly taller.

April to July: on prairies and dry hills from Wisconsin to Saskatchewan and southward to Illinois, Oklahoma, Texas, and Arizona. *Plate 124*.

NORTHERN PAINTED-CUP, C. SEPTENTRIONALIS, has mostly unlobed but narrow leaves. The bracts are somewhat fanlike, shallowly three-lobed, or unlobed, greenish, yellow, or red.

June to August: in rocky soil from Labrador to Alberta and southward to Maine, Michigan, Minnesota, South Dakota, and Utah. This species is variable and has been variously divided into species and varieties by different botanists.

C. purpurea, with bracts tipped with violet, is found in southwestern Missouri.

SCHWALBEA

There is only one species of *Schwalbea*.

CHAFFSEED, S. AMERICANA, is a plant from 1 to 3 feet tall, with lanceolate leaves borne singly and a spike of purple-and-yellow flowers at the top. The corolla is narrow, two-lipped, nearly 1½ inches long.

May to July: in woods from Massachusetts and New York to Florida and Texas; in Kentucky and Tennessee. The seeds have a loose, chaffy coat.

MONKEY-FLOWERS (MIMULUS)

The monkey-flowers are so named from the "grinning" or face-like corolla, which has an erect upper lip, a lower lip directed downward, and (in our species) a "palate" closing the tube. The flowers are borne singly on long stalks from the axils of the paired leaves.

I. *Species with blue flowers.*

M. RINGENS has a square stem from 16 inches to 4 feet or more tall, bearing leaves without stalks. The leaves are elliptic, oblong, or lanceolate, and toothed. The calyx is angular. The flowers are violet.

June to September: in swampy places, meadows, and water-sides from Quebec to Saskatchewan and southward to Georgia, Texas, and Colorado. *Plate 124*.

M. ALATUS resembles *M. ringens* but has stalked leaves. The stems have thin flanges along the angles.

June to September: in swamps and other wet places from Connecticut to Michigan and Nebraska and southward to Florida and Texas. *Plate 124*.

II. *Species with yellow flowers.*

MUSK-FLOWER, M. MOSCHATUS, is a somewhat clammy and woolly plant with weak stems that turn upwards at the tips. The leaves have ovate blades on short stalks.

June to September: along streams and in ditches from Newfoundland to Ontario and southward to North Carolina, West Virginia, and Michigan; and in the West. *Plate 124*.

M. GLABRATUS has weak, smooth stems that tend to lie on the ground, the tips turning up. The leaves are mostly without stalks, round, and toothed.

June to October: on wet shores and along streams from Ontario to Manitoba and southward to Michigan, Illinois, and Mexico. *Plate 124*.

M. GUTTATUS is a western species with round, toothed leaf-blades, the lower ones on stalks, the upper ones stalkless.

June and July: in brooks and meadows in the West and escaped from cultivation here and there in our range. *Plate 125*.

PLATE 125

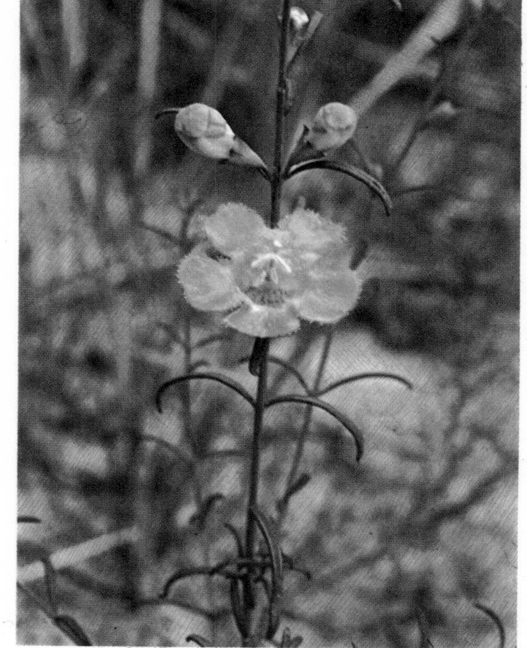

Agalinis paupercula *Elbert*

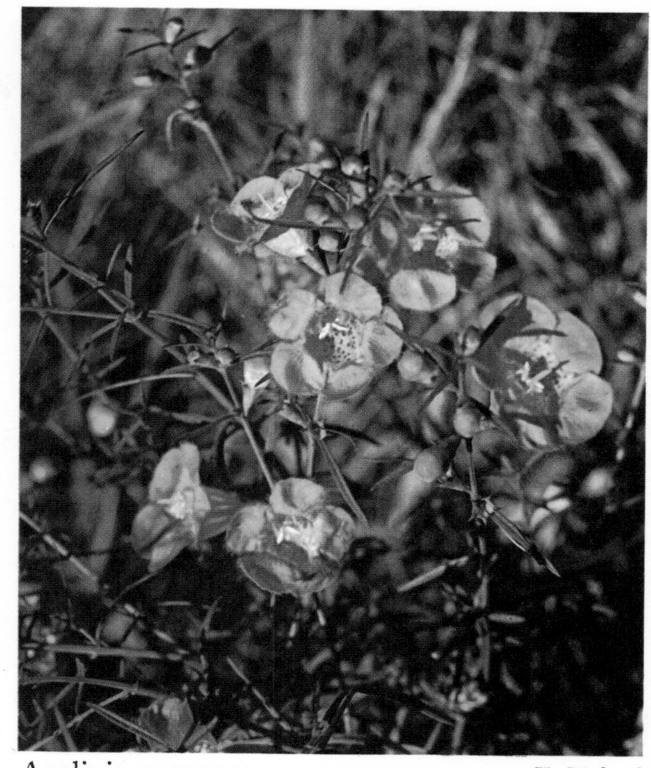

Agalinis purpurea *V. Richard*

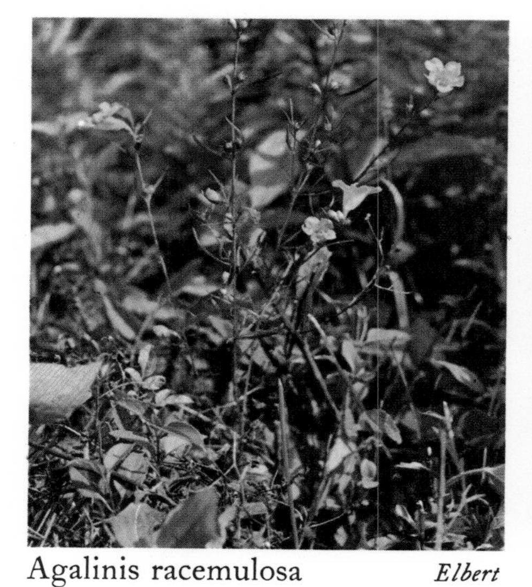

Agalinis racemulosa *Elbert*

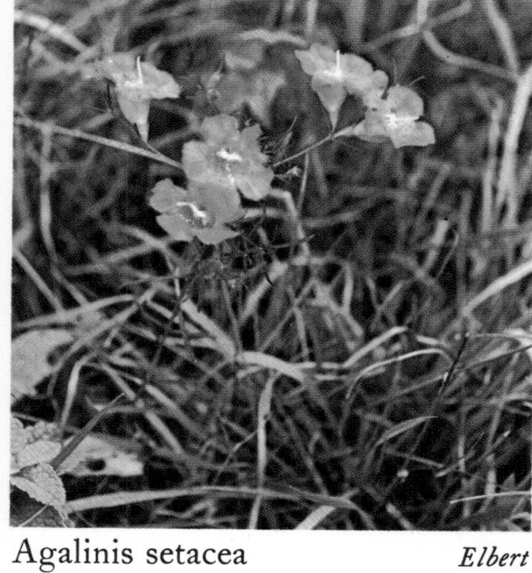

Agalinis setacea *Elbert*

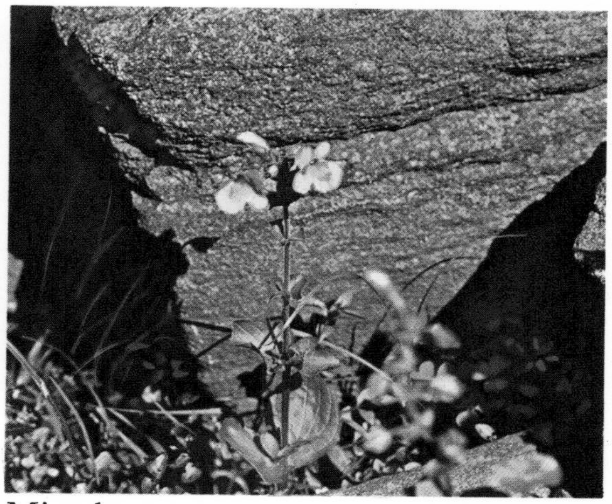

Mimulus guttatus *Smith*

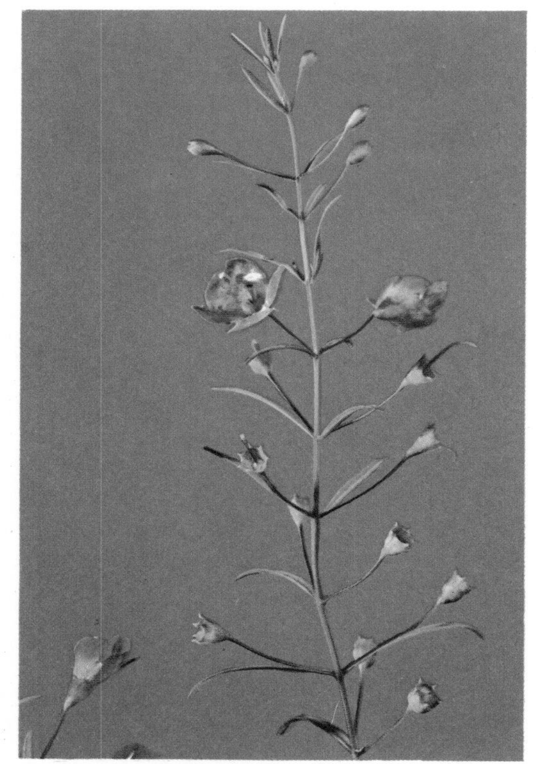

Agalinis tenuifolia *Johnson*

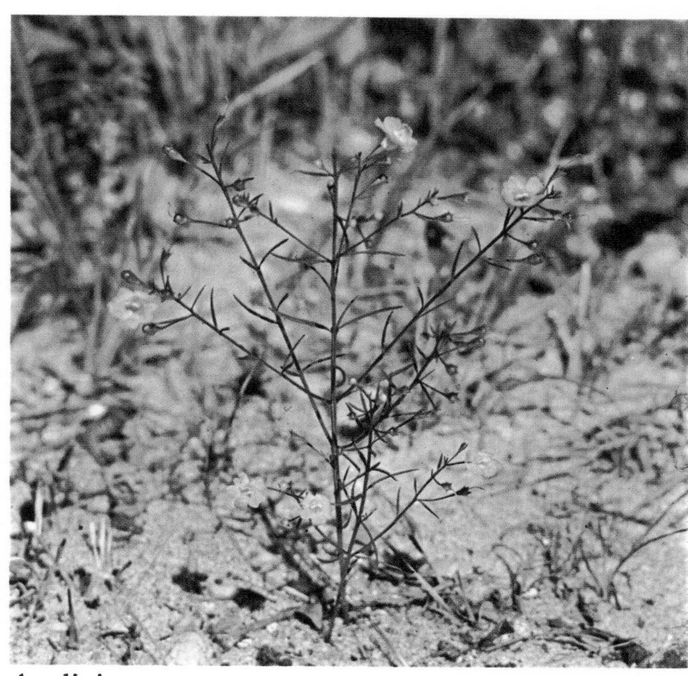

Agalinis acuta *Ryker*

GERARDIA (AGALINIS)

The gerardias are slender plants with wiry, branched stems and narrow leaves, and crimson or pink flowers on stalks that grow from the axils of the leaves, or from the axils of bracts thus forming a raceme. The corolla may appear almost radially symmetric, but closer scrutiny will show that its bell-like portion bulges more on the lower side, and of its five lobes the lower three differ slightly in shape or size from the upper two. These delicate plants all look much alike. The places where they grow, as stated below, will help in identifying them.

I. *Species with flower-stalks less than $\frac{1}{6}$ inch long.*

A. PURPUREA has leaves up to $\frac{1}{6}$ inch broad and corollas an inch or more long. The stem may be 4 feet tall. The flowers are in racemes.

July to September: in damp places from southern New England to Minnesota and Nebraska and southward to Florida and Texas. *Plate 125.*

A. RACEMULOSA has almost threadlike leaves and longer clusters of smaller flowers than *A. purpurea.*

September and October: in pine-barrens and bogs on the coastal plain from Long Island to South Carolina. *Plate 125.*

A. PAUPERCULA rarely reaches 3 feet in height. The leaves are very narrow, hairlike. The corolla is about $\frac{4}{5}$ inch long.

August and September: in damp soil and bogs from New Brunswick to southern Wisconsin and Iowa and southward to New Jersey and Indiana. *Plate 125.*

II. *Species with flower-stalks longer than $\frac{1}{4}$ inch.*

A. TENUIFOLIA has narrow leaves, and flowers about $\frac{1}{2}$ inch long. The upper lobes of the corolla are much longer than the lower. The teeth of the calyx are very short ($\frac{1}{25}$ inch or less).

August to October: in dry woods and fields from Maine to Manitoba and southward to Georgia, Louisiana, and Texas. *Plate 125.*

A. SETACEA has threadlike leaves and flowers from $\frac{1}{2}$ to 1 inch long. The teeth on the calyx are minute, scarcely visible.

August to October: in dry sandy woods on the coastal plain and piedmont from Long Island to Georgia and Alabama. *Plate 125.*

A. ACUTA resembles *A. tenuifolia;* the lobes of the corolla are notched.

August and September: in sandy soil from Cape Cod to Long Island and inland to central Massachusetts and Connecticut. *Plate 125.*

A. GATTINGERI and A. SKINNERIANA are similar to *A. acuta. A. skinneriana* has rough angles on the stem; the corolla has two yellow lines inside. *A. gattingeri* has smooth stems and somewhat longer teeth on the calyx (up to $\frac{1}{12}$ inch). They grow on dry bluffs and in open barren places from Ontario to Wisconsin and (*A. gattingeri*) Minnesota and Nebraska and southward to Alabama and Texas (*A. skinneriana* to Ohio and Oklahoma).

A. ASPERA is a stiff plant up to 3 feet tall with rough narrow leaves. The teeth on the calyx are up to $\frac{1}{8}$ inch long. The corolla is about an inch long.

August and September: in sandy and rocky places from Wisconsin and Minnesota to Manitoba and southward to Illinois and Oklahoma.

A. LINIFOLIA has a creeping stem which sends up an erect stem from 3 to 5 feet tall. The leaves are all paired, about $\frac{1}{8}$ inch wide. The calyx-teeth are minute. The corolla is more than an inch long.

August to October: in damp places on the coastal plain from Delaware southward; not common.

A. MARITIMA is somewhat succulent and green or purplish, with leaves up to $\frac{1}{10}$ inch wide. The calyx-teeth are about $\frac{1}{16}$ inch long and the corolla from about $\frac{1}{2}$ inch to nearly 1 inch.

July to September: in salt marshes along the coast from Nova Scotia to Florida and Texas; often making a colorful show.

TOMANTHERA

There is one species in our range; there is another farther west.

T. AURICULATA is a rough plant with an angular stem from 12 to 30 inches tall. The leaves are lanceolate, broadbased, with no stalk; there are often two lobes at the base, these being the "little ears" or "auricles" to which the name refers. The flowers are borne in the upper axils, without stalks. The corolla is purple, bell-shaped, with five large, nearly equal lobes – much like that of *Agalinis.*

August and September: in prairies and open woodland from New Jersey to Minnesota and southward to Virginia, Alabama, Tennessee, and Texas.

COLLINSIA

The genus *Collinsia* is characterized by its peculiar corolla. Four lobes are at first sight visible, two upper and two lower; but the fifth lobe is the middle one of the lower lip, folded vertically so as to disappear between and underneath the other two. In this pouch are the style and four stamens. This is an American genus, mostly western.

BLUE-EYED-MARY, C. VERNA, is our most widespread species. The upper lip of the charming flower is white, the lower bright blue. The flower is about ½ inch long.

April to June: in woods and thickets, especially on banks, from New York to Michigan and Iowa and southward to West Virginia, Kentucky, and Arkansas. *Plate 126.*

C. VIOLACEA might be called violet-eyed-Mary; its lower lip is violet rather than blue. The lobes of the corolla are notched.

April to June: in moist woods from Illinois and Kansas to Texas.

C. PARVIFLORA has flowers only ¼ inch long, both lips blue; the upper lip has yellow marks at the base.

May to July: in gravelly soil from Vermont and Ontario westward to Alaska and southward to South Dakota and California.

MELAMPYRUM

The mostly Old-World genus, *Melampyrum* has one species in North America.

COW-WHEAT, M. LINEARE, is an inconspicuous little plant from 2 to 20 inches tall, usually but not always branched, and with leaves of very various shapes from very narrow to lanceolate or even ovate, the upper ones often more or less toothed near the base. The flowers are borne singly in the upper axils. The slim corolla has something the appearance of a minute snake's head, from ¼ to ½ inch long, white with a yellow "palate."

May to September: in dry or moist woods and in bogs and rocky places from Labrador to British Columbia and southward to Virginia, the mountains of Georgia and Tennessee, Indiana, Wisconsin, Montana, and Idaho. *Plate 126.* The more northern plants tend to be less branched and to have very narrow leaves without teeth. The broadest, ovate leaves are found through the Atlantic states. The English name really belongs to an European species, also called poverty-weed. It was formerly common in wheatfields, was difficult to separate from the wheat, and imparted a bitter taste to the flour. It is now rare in England.

THE EYEBRIGHTS (EUPHRASIA)

The eyebrights are plants of our northern boundaries, the species apparently not yet well understood by botanists. They are small plants, not more than a foot tall, with paired leaves not an inch long, stalked, and toothed at the edges. The flowers grow almost without stalks in the axils of the leaves. They have a four-lobed calyx and a two-lipped corolla, the three lobes of the lower lip often notched or two-lobed.

The name belongs primarily to an European species which was reputed useful in curing diseases of the eye and even in restoring lost sight.

E. AMERICANA has a corolla with a pale blue upper lip and a lower lip white with violet lines. The flowers are formed in the axils of the upper leaves.

June to September: common in fields and on roadsides from Newfoundland to Maine. *Plate 127.*

E. CANADENSIS differs from *E. americana* in forming flowers almost throughout the stem. They are smaller, about ¼ inch as against ⅓ inch.

July to September: in fields and on roadsides from Quebec to Maine and western Massachusetts.

Several Canadian species reach our northern borders. *E. hudsoniana* (*Plate 126*) is found in Minnesota; it has downy leaves. *E. randii*, *E. oakesii*, and *E. williamsii* occur on the highest mountains of Maine and New Hampshire; they have tiny purple or white corollas, ⅛ inch long or less. *E. arctica*, with a white corolla from ⅕ to ⅓ inch long, extends to Maine, northern Michigan, and Minnesota. For the characteristics of these species the technical manuals must be consulted.

BACOPA

The plants of *Bacopa* are small inhabitants of wet mud, shores, and shallow water. Most of the species are natives of tropical America; a few stray into temperate North America. Only one is at all widespread in our range.

WATER-HYSSOP, B. ROTUNDIFOLIA, grows in shallow water with most of the stems floating. The leaves (*-folia*) are round (*rotundi-*), with no stalks or marginal teeth, in pairs, about an inch across, with several veins diverging from the point of attachment. The flowers are white, the corolla almost radial in symmetry and less than ½ inch long. There are typically two pairs of stamens, but one pair may fail to develop.

June to September: in muddy water from Indiana to Montana and southward to Mississippi and Texas. Five other species have been found in southern Virginia and one in southern Missouri.

MECARDONIA

This tropical American genus closely resembles *Bacopa* and the one species of our range is by some botanists treated as *Bacopa acuminata*.

M. ACUMINATA has small white flowers, less than ½ inch long, on long stalks from the axils of the leaves. The leaves are widest towards the tip and toothed near the base. The veins branch pinnately from the midrib. It flowers from July to September in damp places from Delaware and Maryland to Tennessee and southern Missouri and southward to Florida and Texas.

LEUCOSPORA

There is only one species.

L. MULTIFIDA is a much-branched, spreading little plant only 6 or 8 inches tall. The leaves are stalked, the blade pinnately cleft into narrow lobes. The flowers are borne on long stalks from the axils. The corolla is greenish-white, only about ⅙ inch long.

June to October: along streams from southern Ontario to Kansas and southward to Georgia and Texas.

MUDWORTS (LIMOSELLA)

One species extends along the Atlantic Coast, and a western species is found in Minnesota.

L. SUBULATA consists mainly of tufts of threadlike leaves about 2 inches long, growing from a mat of creeping stems. From among them rise stems not so tall as the leaves, each bearing one tiny white flower. The five lobes of the corolla are radially symmetric.

June to October: in tidal mud or sand from Newfoundland to Virginia. *Limus* in Latin means "mud." *Subula* is an "awl"; *subulata*, "awl-shaped," refers to the leaves.

L. AQUATICA has small elliptic blades on the ends of the leaf-stalks. The flowers are pink. It grows in Canada and is found in Minnesota.

BUCHNERA

This tropical genus has four species in North America, one of them in the Northeast.

BLUE-HEARTS, B. AMERICANA, is a roughly hairy plant with an unbranched stem from 1 to 3 feet tall and toothed, paired, mostly lanceolate leaves without stalks. The flowers form a loose spike, growing in the axils of bracts on the upper few inches of the stem. They are from ½ to 1 inch long, the corolla purple. The lobes of the corolla are almost equal.

June to September: in moist sand in prairies and woodland from western New York to Michigan and Kansas and southward to Florida and Texas. *Plate 126.*

PLATE 126

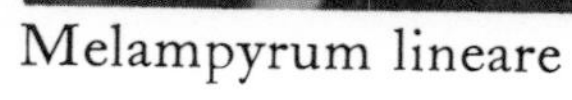

Melampyrum lineare *Rickett*

Gratiola aurea *Gottscho*

Buchnera americana *D. Richards*

Gratiola neglecta *D. Richards*

Gratiola virginiana *Phelps*

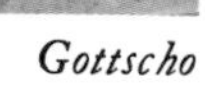

Collinsia verna *Gottscho*

Euphrasia hudsoniana *Johnson*

YELLOW-RATTLE (RHINANTHUS)

The yellow-rattles form a genus of arctic and northern-temperate plants, with one species (and perhaps several others) in New England and New York.

R. CRISTA-GALLI (the name means "snout-flower cock's-comb") grows from 4 to 40 inches tall, bearing paired, oblong or lanceolate, toothed leaves without stalks. The yellow flowers form a one-side spike at the summit of the stem. The four-toothed calyx is flattened vertically, as if squeezed from the sides. The corolla is about ½ inch long, the upper lip forming a long hood, the lower lip much shorter. When the fruit develops the calyx enlarges also, forming a sort of bladder.

May to September: in fields and thickets from Labrador to Alaska and southward to New England and New York. *Plate 127*. Like many *Scrophulariaceae*, these plants obtain nourishment parasitically from the roots of other plants. Several other species, differing only in minute details, have been listed from New England and northern New York; they are treated by some botanists as part of *R. crista-galli*.

FIGWORTS (SCROPHULARIA)

The figworts are tall plants with ovate, toothed leaves and many small greenish or brownish flowers in a large branched inflorescence. The corolla is characteristic: it stands almost erect, with a two-lobed upper lip extending straight up, two small lobes at the sides, and a lower lip bent straight down. There are four fertile stamens and a rudimentary stamen.

The English name has nothing to do with figs, but with the disease once known as "fig" (piles); the European species were used in medieval medicine as a remedy. It was also prescribed for scrofula, the "King's evil."

CARPENTER'S-SQUARE, S. MARILANDICA, may reach 10 feet in height. The leaves are stalked, the stalks being often half or a third as long as the blades. The corolla is brown; the rudimentary stamen is brown or purplish.

June to October: in woods from Quebec to Minnesota and southward to northern Georgia, Louisiana, and Oklahoma.

S. LANCEOLATA has leaf-blades that run down onto the stalk as two narrow flanges ("wings"). The corolla is greenish-brown; the rudimentary stamen is greenish-yellow, often wider than long.

May to July: in open woodland and thickets and on roadsides from Quebec to British Columbia and southward to South Carolina, Indiana, Oklahoma, New Mexico, and northern California. *Plate 127*.

THE BEARD-TONGUES (PENSTEMON)

The beard-tongues are named for the presence of a fifth stamen that forms no pollen; in many species it wears a "beard" – a tuft of hairs. The botanical name is a contraction of two Greek words meaning "five stamens." In the West this is an enormous and baffling genus. In the East we have fewer species – about sixteen – but some of them are not easy to distinguish. The corolla forms a distinct tube with five spreading lobes at the end; these are distinctly two-lipped in some species, almost radially symmetric in others. The flowers are in small clusters (cymes) that spring from the axils of mostly small leaves or bracts in the upper part of the stem, forming a complex inflorescence, generally called a panicle. The leaves are paired and without stalks. Several western species are cultivated in "rock gardens." Some of the eastern species also are worthy of a place in the border.

We may first separate our species into two groups by the size of their flowers.

I. *Species with flowers from 1½ to 2 inches long.* (*Two species in group II,* P. canescens *and* P. calycosus, *may reach a flower-size of 1½ inches, but not more and usually less.*)

P. GRANDIFLORUS is the handsomest of all our eastern species, standing up to 4 feet tall and bearing large flowers colored lavender. The broad roundish leaves, which "clasp" the stem by two lobes at the base, are distinctive. So is the bluish waxy "bloom" of the whole plant.

May and June: in prairies from Wisconsin to Wyoming and southward to Illinois and Texas; and escaping from cultivation farther eastward. *Plate 127*.

P. COBAEA grows up to 30 inches tall, with a downy stem. The leaves are more or less lanceolate and toothed and may "clasp" the stem. The flowers vary from nearly white to pale or deep purple.

PLATE 127

Penstemon grandiflorus *Johnson*

Penstemon tubaeflorus *Johnson*

D. Richards

Scrophularia marilandica

Euphrasia americana *Scribner*

Scrophularia lanceolata *Horne*

Rhinanthus crista-galli *Johnson*

May and June: in prairies and on rocky bluffs from Missouri westward and southward, and sometimes escaping from cultivation farther eastward.

II. *Species with flowers generally not more than an inch long (often exceeding 1 inch in* P. canescens *and* P. calycosus).

A. Of these we may distinguish two species that have a markedly two-lipped corolla, the opening or "mouth" almost closed by ridges on the lower lip.

P. HIRSUTUS is from 8 inches to 3 feet tall. The stem is covered with a fine, whitish down (the hairs are often tipped with glands). The flowers are pale, dull violet outside. The upper lip is two-lobed, erect, the lower three-lobed and directed forward.

May to July: in dry soil, often in rocky places, from Quebec to Wisconsin and southward to Virginia and Tennessee. *Plate 128.*

P. TENUIFLORUS closely resembles *P. hirsutus* but has white flowers. The leaves are more apt to be softly hairy.

April to June: in dry woodland from Kentucky to Alabama.

B. Other species have a wide-open corolla, the five lobes often nearly equal.

1. Some of these species may be recognized by the sharp difference in width between the lower part of the corolla and the upper: a slender tube is suddenly dilated to twice its width. The first three of these have hairs on the pollen-bearing tips of the stamens.

P. DIGITALIS has shining smooth stem and leaves, up to 5 feet tall. The corolla is white, often with purple lines inside, an inch or more long. The stem is glandular in the flowering part.

May to July: in meadows, prairies, and open woodland from Maine to South Dakota and southward to Virginia, Alabama, and Texas. *Plate 128.* A handsome species worthy of cultivation.

P. DEAMII, in Indiana and Illinois, and P. ALLUVIORUM, from Ohio to Missouri and southward, differ from *P. digitalis* chiefly in their smaller flowers, scarcely exceeding $\frac{4}{5}$ inch. The stem and leaves of *P. alluviorum* are more or less downy. *P. deamii* does not exceed 3 feet in height.

P. CANESCENS stands from 1 to 3 feet tall, the stem being covered with a minute grey down (or in a variety with longer hairs). The flowers are violet or purple outside with darker lines inside, from $\frac{4}{5}$ to $1\frac{1}{2}$ inches long, the lower lobe pointing straight out, the upper erect.

May to July: in woods and thickets chiefly in the hills from Pennsylvania to Indiana and southward to Georgia and Alabama. *Plate 129.*

P. CALYCOSUS often has a downy stem. The flowers resemble those of *P. canescens* in size and color: pale purple outside, white inside. The botanical name refers to the large calyx, often $\frac{1}{3}$ inch long.

May to July: in woods and meadows and on rocky hillsides from Ohio to Michigan and Illinois and southward to Georgia and Alabama; spreading eastward to New England, Pennsylvania, and Maryland. *Plate 129.*

P. LAEVIGATUS is similar to *P. canescens*, but has smaller flowers, scarcely an inch long.

May and June: in meadows and wooded bottomlands from New Jersey and Pennsylvania to Florida and Mississippi.

2. The remaining species of group B have a gradually flaring corolla without a sharp difference in width between lower and upper halves. The first two and the last two have white flowers.

P. PALLIDUS has a downy stem up to 3 feet tall and hairy leaves which are rarely toothed. The flowers are from $\frac{3}{5}$ inch to almost an inch long, with purple lines inside.

April to July: in woods and fields and on roadsides from Maine to Michigan and Iowa and southward to Georgia, Tennessee, Arkansas, and Kansas. *Plate 128.*

P. ARKANSANUS is very like *P. pallidus* and by some botanists is included in that species. The leaves are smooth and the flowers very small, not exceeding $\frac{4}{5}$ inch in length.

May and June: in rocky woods from Missouri to Arkansas and Texas. *Plate 128.*

P. BREVISEPALUS has a minutely downy stem from 15 to 30 inches tall. The leaves tend to be sharply toothed but blunt at the tip. The flowers are from $\frac{3}{5}$ to 1 inch long, pale violet with purple lines inside.

April to July: in open woodland and on rocky bluffs from Virginia to Ohio and southward, chiefly in the mountains, to Georgia and Tennessee. *Plate 128.*

P. GRACILIS is a slender plant not 2 feet tall, with very narrow leaves. The leaves are generally finely toothed and smooth. The flowers are pale violet, from $\frac{3}{5}$ to $\frac{4}{5}$ inch long.

June and July: in dry woods and prairies from Wisconsin to Alberta and southward to Iowa and New

PLATE 128

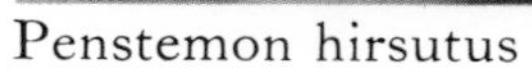
Penstemon hirsutus 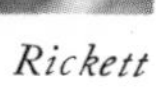 *Rickett*

Penstemon digitalis  *Johnson*

Penstemon pallidus *Johnson*

Penstemon brevisepalus *Elbert*

Penstemon arkansanus *Johnson*

Mexico. *Plate 129*. A variety in Wisconsin has downy leaves. The species is found also farther eastward.

P. ALBIDUS grows from 6 to 16 inches tall, several stems usually together forming a tuft. The leaves are narrow and downy. The inflorescence is densely glandular and the corolla is minutely glandular inside. The flowers are short-stalked. The corolla is white or lightly violet-tinged, less than an inch long.

May and June: in prairies from Minnesota to Alberta and southward to Iowa and New Mexico.

P. TUBAEFLORUS grows from 1 to 3 feet tall. The stem and rather blunt leaves are smooth; the leaves are rarely toothed. The corolla is much like that of *P. albidus*.

May to July: in open woods and fields and on roadsides from Indiana to Wisconsin and Nebraska and southward to Mississippi and Texas; also in the Atlantic states. *Plate 127*.

P. australis is a southern species with a red-purple corolla; it is found in southeastern Virginia.

TURTLE-HEADS (CHELONE)

The flower of the turtle-heads simulates the head of those beasts to a remarkable degree; the two lips – a half-open mouth – terminate a large egg-shaped corolla.

TURTLE-HEAD, SNAKE-HEAD, or BALMONY, C. GLABRA, is the common species, a plant from 1 to 7 feet tall with lanceolate or ovate, toothed leaf-blades on short stalks or without stalks. The flowers are generally an off-white, but vary to being tipped with purple or yellow.

July to October: in moist ground, along streams, and in thickets from Newfoundland to Minnesota and southward to Georgia, Alabama, and Missouri. *Plate 129*. Variable in flower-color and in shape and hairiness of leaves.

C. LYONI, with crimson corolla and round-based leaf-blades on distinct stalks, a southern species, has been cultivated and has escaped in thickets and moist ground in New England. *Plate 129*.

C. OBLIQUA, with pink corolla and lanceolate leaf-blades tapering down to distinct stalks, grows from Maryland and Indiana to Minnesota and southward to Florida, Mississippi, and Arkansas. *Plate 129*.

DASISTOMA

The name of the genus means "woolly mouth," and the wool inside the corolla characterizes the genus – which has only one species. (By some botanists, however, this is placed in the larger genus *Seymeria*. *S. cassioides*, whose corolla is smooth inside, is found in southeastern Virginia.)

MULLEIN-FOXGLOVE, D. MACROPHYLLA, is a downy plant from 3 to 5 feet tall, with only slight resemblance to either mullein or foxglove. The lower leaves are pinnately divided and the segments are pinnately cleft. The upper leaves are undivided. The corolla is yellow; the five lobes are almost equal.

June to September: in woods from Ohio to Wisconsin and Nebraska and southward to Georgia, Alabama, and Texas. *Plate 130*.

THE FALSE FOXGLOVES (AUREOLARIA)

The false foxgloves are named for their narrowly bell-shaped corolla which somewhat resembles the purple or white corolla of the foxglove, *Digitalis*. The yellow bell flares into five almost equal lobes, the upper two joined for a slightly greater part of their length than the lower three. The larger leaves are paired; the upper ones, borne singly, are generally smaller and may be called bracts, since the flowers appear in their axils. Some species have pinnately lobed, cleft, or divided leaves.

To distinguish the species one must give attention to apparently unimportant details.

I. *A species with relatively long-stalked flowers (the stalks from $\frac{2}{5}$ to 1 inch long) and a hairy or downy stem with some of the hairs gland-tipped. (See also* A. grandiflora *under* II.)

A. PEDICULARIA stands from 1 to 4 feet tall. Practically all the leaves are pinnately cleft and even the teeth of the calyx are usually pinnately lobed or bluntly toothed.

August and September: in dry woods from Maine to Minnesota and southward to Florida, Kentucky, and Missouri. *Plate 129*. The Missouri plants

PLATE 129

Chelone lyoni *Lee*

Chelone glabra *Johnson*

Penstemon canescens *Elbert*

Chelone obliqua *Johnson*

Penstemon calycosus *Johnson*

Penstemon gracilis *Lee*

Aureolaria pedicularia *Rhein*

are by some botanists placed in a separate species, *A. pectinata;* their leaves are more sharply toothed, the glands more abundant, the flower-stalks shorter.

II. *Species with short-stalked flowers, but the stalks not less than $\frac{1}{6}$ inch; no glands.*

A. FLAVA may reach a height of 8 feet. The stem is smooth. The lower leaves are deeply cleft pinnately, the upper leaves merely lobed or toothed.

July to September: in woods and at the margins of woods from Maine to Minnesota and southward to Florida and Louisiana. *Plate 130.* Many of the western and southern plants have a corolla 2 inches or more long and downy leaves; these have sometimes been treated as a separate species, *A. macrantha.*

A. GRANDIFLORA has a downy stem up to 5 feet tall. The lower leaves are pinnately lobed or deeply cleft, the upper sometimes only toothed. In spite of its name (*grandi-* means "large"), the flowers are no larger than those of other species; but the flower-stalks may be $\frac{3}{5}$ inch long.

July to September: in dry, often rocky woods from Indiana, Wisconsin, and Minnesota to Arkansas and Texas. *Plate 130.*

III. *Species with practically no flower-stalks, the stalks when visible not more than $\frac{1}{6}$ inch long.*

DOWNY FALSE FOXGLOVE, A. VIRGINICA, is from 1 to 5 feet tall, with stem and leaves downy. The lower leaves are pinnately lobed or cleft or sometimes only with a wavy edge; the upper leaves are less lobed or not at all. There is generally a short leaf-stalk. The flower-stalks do not exceed $\frac{1}{8}$ inch.

June to September: in dry woods from New Hampshire to Michigan and southward to Florida and Louisiana. *Plate 130.*

A. LAEVIGATA is about 3 feet tall. Stem and leaves are quite smooth and the leaves are generally unlobed and without stalks.

July to September: in woods, chiefly in the mountains from Pennsylvania to Ohio and southward to Georgia and Tennessee. *Plate 130.*

THE UNICORN-PLANT FAMILY (MARTYNIACEAE)

The *Martyniaceae* are a small family native to the warm parts of the Americas, with several species in the Southwest and one in the East.

PROBOSCIDEA

We have a single species.

UNICORN-PLANT, P. LOUISIANICA, is a sticky, bushy plant standing about 3 feet tall. The leaves have broad, heart-shaped blades on long stalks. The flowers are large and handsome, in racemes at the summit of the stem and branches. The corolla is about 2 inches long, the five lobes forming two lips; the color varies from almost white to purplish-pink, often mottled with purple and yellow. The curious and distinctive feature is the fruit, which is at first fleshy, the flesh later disappearing and leaving an inner hard shell prolonged into a sharp, upcurved spine – the "unicorn." This splits lengthwise, the spine splitting also.

June to September: in waste places and fields and on stream-banks from West Virginia to Illinois and Minnesota and southward to Georgia and Mexico. *Plate 130.* The fruits have been made into pickles, and the plant has occasionally escaped from cultivation farther northeastward.

THE LOPSEED FAMILY (PHRYMACEAE)

The lopseed family has the distinction of including only one genus.

PHRYMA

And the genus *Phryma* is generally considered to have only one species (but recently two others have been reported in Asia).

LOPSEED, P. LEPTOSTACHYA, is a slender, branched plant from 1 to 3 feet tall, with ovate, coarsely toothed leaf-blades on short stalks. The flowers are in

PLATE 130

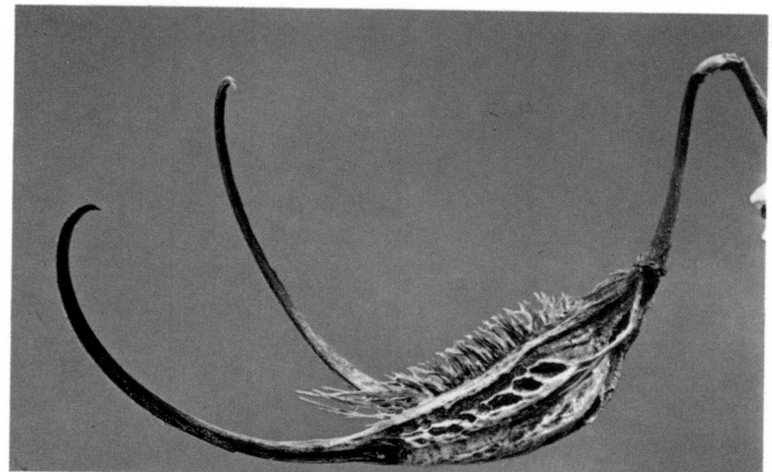

Proboscidea louisianica *Allen*

Aureolaria flava *Gottscho*

Dasistoma macrophylla *Rickett*

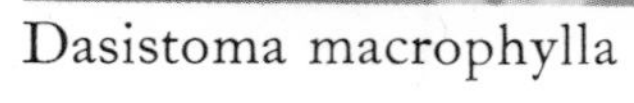

Aureolaria laevigata *Elbert*

Aureolaria grandiflora *Dobbs*

Proboscidea louisianica *Allen*

Aureolaria virginica *Rhein*

long spikes. The corolla is much like that of the snapdragon family, two-lipped, the lower lip longer and three-lobed. (The species is separated from other families by the technical characteristics of the pistil.) The feature by which the plant is easily recognized is the curious habit of the calyx with its enclosed achene, after the corolla has fallen, of lying downward along the stem (the "seed" "lops").

June to September: in woods from Quebec to Manitoba and southward to Florida and eastern Texas. *Plate 131.*

THE BROOM-RAPE FAMILY (OROBANCHACEAE)

The *Orobanchaceae* are all parasites. All are often called cancer-roots. They lack the green pigments, chlorophyll, necessary to the making of food; instead they attach themselves underground to the roots of other plants, from which they draw nourishment. Their leaves are whitish scales. They have more or less two-lipped flowers. There are four stamens in two pairs.

BROOM-RAPES (OROBANCHE)

The broom-rapes have unbranched stems bearing one or a few flowers at the tip. An Asian species with many yellow-and-blue flowers sometimes occurs.

CANCER-ROOT, O. UNIFLORA, has, as the Latin name indicates, a single flower at the summit of each stem; the stems usually grow several together, up to 10 inches tall. Its tube flares into five nearly equal lobes. The pretty corolla is pale lavender or lilac, with two yellow bands inside on the lower side.

April to June: in woods from New Brunswick to British Columbia and southward to Florida, Texas, and California. *Plate 131.*

O. FASCICULATA forms stems from 2 to 6 inches tall, bearing small scales, and, at the summit, from four to ten flowers on stalks an inch or more long. The corolla is purple, with five nearly equal lobes.

April to August: in dry open soil from Michigan to Yukon and southward to Indiana, Nebraska, New Mexico, and California. *Plate 131.*

O. LUDOVICIANA forms a dense spike of many flowers on a sticky stem from 4 to 12 inches tall. The corolla is purplish with two yellow bands inside on the lower side. It spreads at the tip into five pointed lobes.

July to September: in prairies from Minnesota to British Columbia and southward to Ohio, Illinois, and Texas. *Ludoviciana* means "of Louisiana" – the name having been given when everything west of the Mississippi was Louisiana. This species is parasitic mostly on *Compositae.*

O. MINOR has flowers in a rather loose spike, on a stem from 4 to 18 inches tall. The purplish corolla is more markedly two-lipped than those of our native species. This comes from Europe, having become established from New Jersey to North Carolina. It lives chiefly on the roots of clovers. English farmers named it hell-root and devil's-root when they found it in their clover. Two other species, *O. ramosa* and *O. purpurea*, have been introduced from the Old World, but are uncommon in our range.

EPIFAGUS

There is only one species.

BEECH-DROPS, E. VIRGINIANA, is found under beech trees, on whose roots it feeds. It is a brownish, much-branched plant, rising to a height of from 6 to 18 inches. The branches bear small scales (the leaves; borne singly), in the axils of which are the flowers. The corolla is a slim tube, whitish with purplish-brown stripes. The upper ones are two-lipped at the tip. The lower ones do not open, but they alone form fruits (capsules). As the fruit develops, it wears the unopened corolla for a time on its tip.

August to October: in beech woods from Quebec to Wisconsin and southward to Florida and Louisiana. *Plate 131.*

PLATE 131

Epifagus virginiana *Rhein*

Orobanche uniflora *Rhein*

Epifagus virginiana *Elbert*

Phryma leptostachya *Horne*

Conopholis americana *Elbert*

Orobanche fasciculata *Johnson*

Pinguicula vulgaris *Johnson*

CONOPHOLIS

One species of *Conopholis* occurs in our range, the other three ranging far to the southwest.

SQUAW-ROOT, C. AMERICANA, consists of several thick, pale brown or yellowish stems from 4 to 10 inches tall, covered with scales. The scales (leaves) are at first soft and thick, finally becoming dry and hard, when the whole branch resembles a small pine-cone. In the axils of the upper scales are the flowers. The corolla is about $\frac{1}{2}$ inch long, the tube curved, the upper lip forming a narrow hood, the lower lip three-lobed.

April to July: in woods, commonly under oaks, from Nova Scotia to Wisconsin and southward to Florida and Lousiana. *Plate 131*.

THE BUTTERWORT FAMILY (LENTIBULARIACEAE)

The butterworts and their relatives, the bladderworts, form a family of small plants which capture insects in various ways. The corolla is two-lipped, in some species with an elevated part or "palate" closing the opening, and a hollow projection or "spur" extending downward from the lower lip. There are only two stamens. In many features these plants resemble the snapdragon family; they differ in details of the pistil.

There are two genera in our range.

PINGUICULA

We have only one species, the others being more southern.

BUTTERWORT, P. VULGARIS, is a little plant with a rosette of leaves at the surface of the ground. These are elliptic, without stalks or teeth, yellowish-green and greasy (*pinguis* means "fat"). Small insects that alight on this surface are unable to take off again, and are digested by the plant. The flowers are borne singly on leafless stems from 2 to 6 inches tall; they are violet, about $\frac{1}{2}$ inch long.

June to August: on wet rocks and in meadows and bogs from Greenland to Alaska and southward to Vermont, Michigan, and Washington. *Plate 131*.

THE BLADDERWORTS (UTRICULARIA)

The bladderworts are small plants, the flowering stems seldom a foot tall. They grow on wet soil or in water, their leaves being at the base of the flowering branch and on creeping stems, in many species under the water and divided into hairlike segments. On these segments are minute bladders of intricate construction in which small, even microscopic insects are trapped; the victims of these traps are digested and contribute to the nourishment of the plant.

There are many species widely distributed, most numerous in the tropics. They can be distinguished only by technical characteristics, but the following brief classification may help the reader to arrive at a name.

I. *Species with purple or deep pink flowers.*

U. PURPUREA has long stems which float beneath the surface of water, bearing leaves in circles. The leaves are long-stalked, with many hairlike segments at the end of the stalk, and bladders at the ends of some of the segments. The flowers are on stems from 1 to 6 inches tall.

June to September: in quiet waters from Quebec to Minnesota and southward to Florida and Louisiana; and in the West Indies and Central America. *Plate 132*.

U. RESUPINATA grows on wet soil (or occasionally in shallow water), the slender horizontal stems creeping below the surface bearing mostly undivided narrow leaves. The flowering branches are from 1 to 6 inches tall, each bearing one flower.

July to September: at the margins of ponds, lakes, and streams from Quebec to Wisconsin and southward to Delaware and Illinois; and from South Carolina to Florida. *Plate 132*.

II. *Species with yellow or yellowish flowers.*

A. Species with a circle of leaves, at the base of the flowering stem, with inflated stalks or "floats" some 2 or 3 inches long.

PLATE 132

Lobelia dortmanna *Scribner*

Utricularia cornuta *Allen*

Utricularia subulata *Johnson*

Utricularia inflata *Rhein*

Utricularia purpurea *DeVoe*

Utricularia resupinata *Voss*

Utricularia vulgaris *Johnson*

U. INFLATA has, besides the floats, submerged stems on which are many leaves, borne singly, divided into threadlike segments which bear small bladders. The flowers are borne several together at and near the tip of a stem up to 10 inches tall.

May to November: in ponds and ditches on the coastal plain from Maine to Florida and Texas; inland from New England to Pennsylvania and Tennessee; South America. *Plate 132*. Over most of this range the plants have smaller flowers and shorter stems. They are sometimes treated as a separate species, *U. radiata*.

B. Species without "floats" at the base of the flowering stem.

1. Some of these have scarcely any visible leaves. They are usually found on wet soil.

U. CORNUTA has slender underground stems which bear threadlike, undivided leaves, on which are bladders; these leaves are difficult to find. The flowering stems are from 1 to 12 inches tall and usually bear from one to three flowers, sometimes more. The corolla is from ½ to 1 inch tall.

June to September: on wet shores and in bogs from Newfoundland to Minnesota and southward to Delaware, Ohio, and Illinois; from North Carolina to Florida; and in Texas. *Plate 132*.

U. JUNCEA resembles *U. cornuta* but has smaller flowers, from ⅖ to ½ inch tall. It grows on the coastal plain from New York to Florida and Mississippi.

U. SUBULATA also resembles *U. cornuta*. The stems may bear up to twelve flowers, on longer stalks. The bracts at the base of the flower-stalks are attached by their middle. On the coastal plain from Nova Scotia to Florida and Texas and inland to Arkansas.

2. The remaining species have leaves divided into hairlike segments, borne on horizontal stems which are generally in shallow water.

U. INTERMEDIA has flowers with the lower lip about twice as long as the upper. The stem generally creeps on the bottom of the pool or other water. The flowering stems are from 2 to 12 inches tall.

May to September: in shallow pools and mud from Greenland to Alaska and southward to Delaware, Indiana, Iowa, and California.

U. MINOR is like *U. intermedia*. A hand lens will show that the spur, which in *U. intermedia* is nearly ½ inch long, is in this species very short or almost lacking. The ranges of the two species are much the same.

U. GIBBA has upper and lower lips about equal. The spur is short and broad, shorter than the lower lip. The flowering stem rises from 1 to 4 inches, with one or several flowers.

June to September: in shallow pools and bogs from Quebec to Minnesota, southward to Florida, Texas, and Mexico, and in California; also in the West Indies and Central America.

U. BIFLORA is like *U. gibba*, but larger. The spur is pointed and nearly as long as the lower lip. The flowering stem grows from 1 to 6 inches tall.

July to October: in shallow pools from New England and Oklahoma southward to Florida, Texas, and New Mexico.

U. FIBROSA has a flowering stem from 4 to 16 inches tall, with up to seven flowers. The lips of the corolla are equal, rather broad; the spur is narrow and as long as or longer than the lower lip.

May to November: in pools and wet peat on the coastal plain from Massachusetts to Florida and Texas.

U. VULGARIS has floating stems covered with much-divided leaves bearing numerous bladders. The flowering stems are from 4 to 30 inches tall and bear from six to twenty or even more flowers. The palate of the corolla is marked with brown lines.

May to September: in quiet waters from Labrador to Alaska and southward to Virginia, Ohio, Texas, and Mexico; also in the Old World. *Plate 132*.

U. GEMINISCAPA is a small edition of *U. vulgaris*, with flowers only about ¼ inch tall on branches up to 6 inches tall.

June to September: in quiet waters from Newfoundland to Michigan and Wisconsin and southward to Virginia and Pennsylvania.

THE LOBELIA FAMILY (LOBELIACEAE)

Only one genus, *Lobelia*, is comprised in this family. Botanists now place *Lobelia* with *Campanula* in the bluebell family, on the basis of technical characteristics. This is undoubtedly correct, but in visible and obvious characteristics the lobelias are very distinctive and for the purposes of this book it is convenient to keep them in a family of their own.

LOBELIA

The flowers of the lobelias are bilaterally symmetric, with distinct upper and lower lips – as is evident from even a glance at *Plate 133*. The peculiarity of the genus is that the upper lip is split, each half ending in an erect or horizontal pointed lobe in front; and through the split emerges a curved rod or column consisting of the joined stamens around the style. This object is further distinguished by a tuft of hairs, a "beard," at the tip. The juice of the plants is milky and more or less poisonous.

The genus comprises a number of species, among them two or three of our best-known and most striking wild flowers.

I. *Species with flowers (the corolla-tube) about an inch long or (usually) longer.*

GREAT BLUE LOBELIA, L. SIPHILITICA, stands from 8 inches to 5 feet tall. The leaves are lanceolate, tapering to both ends, toothed. Much of the stem is occupied by the large, brilliant blue or blue-violet flowers. The lower lip is composed of three large pointed lobes directed downward, with white marks where they begin. The upper lip is two erect teeth. The underneath of the tube is white or pale blue with darker ribs. The buds stand erect, displaying this ribbed under surface.

August and September: in wet land, roadside ditches and banks, and moist woods from Maine to Manitoba and southward to North Carolina, Alabama, and Texas. *Plate 133*. The plants west of the Mississippi are generally smaller than the more eastern plants. The unpleasant name of this strikingly beautiful plant refers to its use as a supposed cure for syphilis; "antisiphilitica" would seem more appropriate. It was said to be a secret remedy of the American Indians, the secret having been purchased from them by Sir William Johnson in the eighteenth century. But no mention of it in this connection appears in medical books of the nineteenth century, and Sir William's money would seem to have been wasted. The root (which was the part used) does contain alkaloids and, if taken internally, will cause violent vomiting and other unpleasant symptoms. (See *L. inflata* below.)

CARDINAL-FLOWER, L. CARDINALIS, is even more striking than the great blue lobelia, the flowers being a brilliant dark red. The brush ("beard") at the end of the stamens is glistening white. The tube of the corolla is from $1\frac{1}{5}$ to $1\frac{4}{5}$ inches long. The lower three lobes are lanceolate and sharp; the upper two nearly as long, wider towards their tips and spreading horizontally.

July to September: in wet soil along streams, in swamps, and in wet woods from Quebec to Minnesota and southward to Florida and Texas. *Plate B*. Forms are known with pink or white flowers. In the West, including a bit of Missouri, grows *L. splendens*, like *L. cardinalis* but with narrower leaves and flowers scarcely more than an inch long.

L. PUBERULA may be mentioned here; in flower-size it is on the borderline between the two groups – the corolla-tube from $\frac{3}{5}$ to 1 inch long. The stem is very finely downy or hairy. The leaves are mostly elliptic or lanceolate. The small blue flowers often grow all on one side of the stem.

August to October: in woods and damp soil from New Jersey to Missouri and southward to Florida and Texas. *Plate 133*. A variable southern species. Several other southern species just enter our range, with flowers up to an inch long. *L. elongata*, with narrow, sharp-toothed leaves and deep blue flowers, from Delaware southwards; *L. georgiana*, with elliptic, sharp-toothed leaves and pale blue flowers, from Virginia and Kentucky southward; and *L. glandulosa*, with almost threadlike leaves and blue-and-white flowers, from southeastern Virginia southward.

II. *Species with flowers rarely reaching $\frac{4}{5}$ inch in length of corolla-tube.*

INDIAN TOBACCO, L. INFLATA, has a slender, branched, hairy stem from 6 inches to 3 feet tall. The leaves are ovate or elliptic, rather bluntly toothed. The flowers are small, not $\frac{1}{2}$ inch long, not close together. The botanical name derives from the lower part of the flower (receptacle), which expands into a round sac as the corolla falls. The corolla is pale violet or pinkish.

June to October: on roadsides and in fields, woodland, and waste land from Quebec to Minnesota and southward to Georgia, Alabama, Arkansas, and Kansas. *Plate 133*. The plant contains appreciable quantities of the alkaloids characteristic of the genus. It has certain medical uses, but the plant, especially the root, should not be eaten or chewed; it can have violent emetic and other effects and can even, taken in any quantity, be fatal. The symptoms are similar to those due to nicotine, whence the name Indian tobacco.

PALE-SPIKE LOBELIA, L. SPICATA, grows in two forms. In one, most of the leaves are at the surface of the ground, with only bracts farther up the stem; in the other they are scattered up the stem. They vary from ovate to lanceolate, often with the broadest part between middle and tip. The flowers are small, about $\frac{1}{2}$ inch long or less, pale blue or white, rather scattered in the upper part of the stem.

June to August: in meadows and fields from New Brunswick to Alberta and southward to Georgia, Alabama, Louisiana, and Kansas. *Plate 133*. A rather common little weed, the flowers individually pretty.

L. KALMII has a slender, often branched stem from 6 to 24 inches tall, with very narrow, almost hairlike leaves. The flowers stand at rather wide intervals on the stem, from $\frac{2}{5}$ to $\frac{3}{5}$ inch long. The corolla is blue with a white center.

July to September: in moist soil and bogs and on wet rocks from Newfoundland to Mackenzie and southward to New Jersey, Ohio, Illinois, Iowa, South Dakota, and Colorado. *Plate 133*.

WATER LOBELIA or WATER-GLADIOLA, L. DORTMANNA, grows under water, only the leafless flowering stem above the surface. The leaves form a rosette at the base: they are narrow, succulent, and hollow. The flowering stem is also hollow; it rises from 2 to 40 inches, depending on the depth of the water. It bears rather few, slender flowers up to $\frac{4}{5}$ inch long.

July to October: in water at the margins of ponds, usually on a sandy bottom, from Newfoundland to Minnesota and southward to New Jersey, Pennsylvania, and Wisconsin; also from British Columbia to Oregon. *Plate 132*.

L. NUTTALLII has a very slender, threadlike stem with leaves not much wider, the plant from 8 to 30 inches tall. The flowers are blue with a white center, only about $\frac{2}{5}$ inch long.

July to October: in sandy or limy soil on the coastal plain from southern New York to Florida and Alabama and inland from Pennsylvania to Kentucky and Alabama.

L. canbyi and *L. boykinii* are two other species of the southern coastal plain extending northward to New Jersey and Delaware respectively. They have very narrow, often threadlike leaves and blue flowers about $\frac{1}{2}$ inch long. *L. appendiculata* resembles *L. spicata* but has leaves with a round instead of tapering base. It has been reported from Illinois and Missouri.

THE MILKWORT FAMILY (POLYGALACEAE)

The milkwort family contributes only one genus to our wild flowers: *Polygala*.

THE MILKWORTS (POLYGALA)

The milkworts must not be confused with the milkweeds (*Asclepias*). The milkweeds derive their name from the white sap or latex which oozes from any cut surface. The milkworts have no such milk. Their name comes from an old belief that certain of their species, if eaten by cattle, increase the flow of milk. (The botanical name is from two Greek words meaning "much" and "milk"; compare *Ornithogalum*.)

The flower has an intricate structure, best seen in *P. paucifolia*. There are five sepals, two of which are larger than the others and colored like petals; these are the "wings." There are three petals, partly joined with each other and with the stamens. The lower petal generally bears a fringe. The stamens mostly number eight. The fruit is a small pod (capsule).

Many species also form smaller flowers (some even underground) which do not open but which form fruits.

In many of our species these parts are hard to see, the wings generally concealing the other parts, the flowers being densely clustered, and the head or spike suggesting a clover. In one species, however, the comparatively large flowers are borne singly on long stalks from the axils of leaves.

FRINGED MILKWORT or FLOWERING-WINTERGREEN, P. PAUCIFOLIA, has creeping stems (both above ground and underground) from which erect branches rise from 2 to 4 inches tall. Near the summit of these branches are several leaves with short stalks and ovate blades. The beautiful rose-magenta flowers are on long stems from the axils of these leaves. The two sepals called wings and the fringe on the lowest petal are conspicuous. *Plate 134*.

May to July: in woods and on banks from Quebec to Manitoba and southward to Connecticut, the mountains of Georgia, Tennessee, northern Illinois, and Minnesota. The second English name above is unfortunate: the plant has nothing in common with wintergreen.

All our other species of *Polygala* have numerous small flowers in spikes or heads.

I. *Species with rose-pink or white or greenish flowers.*

A. Some of these have their flowers in very narrow spikes; the lower few flowers are likely to be separated. These species have mostly white flowers.

PLATE B

Lobelia cardinalis

Gottscho

SENECA SNAKEROOT, P. SENEGA, with white flowers, is recognizable by its comparatively broad, lanceolate or even ovate leaves. It grows from 6 to 18 inches tall. *Plate 134.*

May to July: chiefly in dry soil, often in rocky places, from Quebec to Alberta and southward to Georgia, Tennessee, Arkansas, and South Dakota.

P. POLYGAMA grows up to 18 inches tall. The leaves are narrow, tending to widen towards their tips. The flowers are rose-pink or white, the wings being the conspicuous part.

June and July: in sandy soil from Maine to Minnesota and southward to Florida along the coast (and from Florida to Texas), and inland southward to West Virginia, Indiana, and Iowa. *Plate 134.*

WHORLED MILKWORT, P. VERTICILLATA, is distinguished in this group by having its narrow leaves mostly in circles, from three to seven at each level. The spikes of white or greenish (or sometimes purplish) flowers taper towards their tips.

June to October: in open places from Maine to Manitoba and southward to Florida and Texas. A variable species, some varieties having only the lower leaves in circles. One variety is sometimes treated as a distinct species, *P. ambigua;* it differs chiefly in the relative size of the wings.

B. Species with flowers in dense heads and spikes, no flowers distinctly separate from the others. These species have mostly pink flowers.

In the first two species the leaves are in circles; in the others they are scattered along the stem, mostly borne singly.

CROSS MILKWORT, P. CRUCIATA, is so called from its several circles of four narrow leaves each; but there may only be three in a circle. Branches may spring from the axils of the upper leaves, in pairs. The stem is from 4 to 20 inches tall. The rose-purple or whitish flowers are in dense heads a little longer than wide. The wings, which are sharp-pointed, of each flower spread to either side, disclosing the small petals within.

July to October: in damp soil, including bogs, peats, and marshy borders on the coastal plain, from Maine to Florida and Texas and inland from Ohio to Minnesota and southward to Alabama. *Plate 134.*

SHORTLEAF MILKWORT, P. BREVIFOLIA, resembles *P. cruciata* but is smaller, rarely more than a foot tall. The leaves are commonly in fours and fives, and the branches from their axils are long, often overtopping the main stem. The flowers are pale purplish rose.

July to September: in sandy swamps on the coastal plain from New Jersey to Florida and Mississippi. *Plate 135.*

P. SANGUINEA has flowers varying from rose-purplish to pink, white, and greenish, in dense short spikes or long heads. The stem is from a few inches to a foot or more tall, bearing very narrow leaves. When the flowers fall, small bracts may be seen on the stem where they grew.

June to October: in meadows and fields from Nova Scotia to Minnesota and southward to South Carolina, Louisiana, and Oklahoma. *Plate 134.*

P. CURTISSII is similar to *P. sanguinea*, the flower-heads shorter and the rose-purple petals tipped with yellow.

June to October: in sandy soil, chiefly pinelands, on the coastal plain, from Delaware to Georgia and Louisiana, and inland from western Virginia to Ohio and Kentucky. *Plate 133.*

P. NUTTALLII is another species similar to *P. sanguinea* but smaller, from 2 to 10 inches tall. The flowers are greenish-white or dull-purplish.

June to October: in dry open places and pinelands from Massachusetts to Georgia and inland in Kentucky, Arkansas, and Mississippi.

P. INCARNATA has rather pale rose-pink flowers in spikes two or three times as long as they are thick. The stem grows from 6 inches to 2 feet tall, bearing scattered narrow leaves. The petals are more conspicuous than in the preceding species, forming a tube which projects beyond the sepals (wings).

June to November: in dry open places from southern Ontario to Michigan and Kansas and southward to Long Island, Florida, Texas, and Mexico. *Plate 134.*

P. MARIANA resembles *P. incarnata*. The flower-heads may taper to their tips, and the flowers are not so tightly packed. The wings and petals are purplish, rose, or greenish, the petals not projecting beyond the wings.

June to October: in sandy and peaty places chiefly on the coastal plain from New Jersey to Florida and Texas, and inland in Kentucky and Tennessee.

II. *Species with yellow flowers.*

YELLOW MILKWORT or YELLOW BACHELOR'S-BUTTONS, P. LUTEA, grows from a rosette of blunt leaves, its stem or stems from 3 to 18 inches tall. The leaves on the stems are lanceolate, often with the broadest part outermost. The flowers are orange-yellow, in a dense, almost spherical head, each head usually on a long leafless stem.

May to October: in wet sandy soil and bogs on the coastal plain from Long Island to Florida and Louisiana. *Plate 135.*

PLATE 133

Polygala incarnata *Johnson*

Lobelia kalmii *Johnson*

Lobelia spicata *Johnson*

Polygala curtissii *Elbert*

Lobelia puberula *Elbert*

Lobelia inflata *Williamson*

Lobelia siphilitica *Rickett*

P. RAMOSA means "branching polygala," and this species has more branches and more flower-heads than any other species of our range except the following one. The stem is from 4 to 18 inches tall, branching repeatedly near the summit to form an inflorescence about 6 inches across. There is a rosette of leaves at the base of the stem and very small leaves on the stem. The flower-spikes are less than ½ inch wide, not very dense. The bracts stay on the stem of the spike after the flowers fall.

July to September: in damp fields and pinelands on the coastal plain from New Jersey to Florida and Texas. *Plate 135*.

P. CYMOSA is similar to *P. ramosa* but much larger – up to 4 feet tall, with a branched inflorescence up to 8 inches across.

July to September: in wet woodlands and open places on the coastal plain from Delaware to Florida and Louisiana. *Plate 135*.

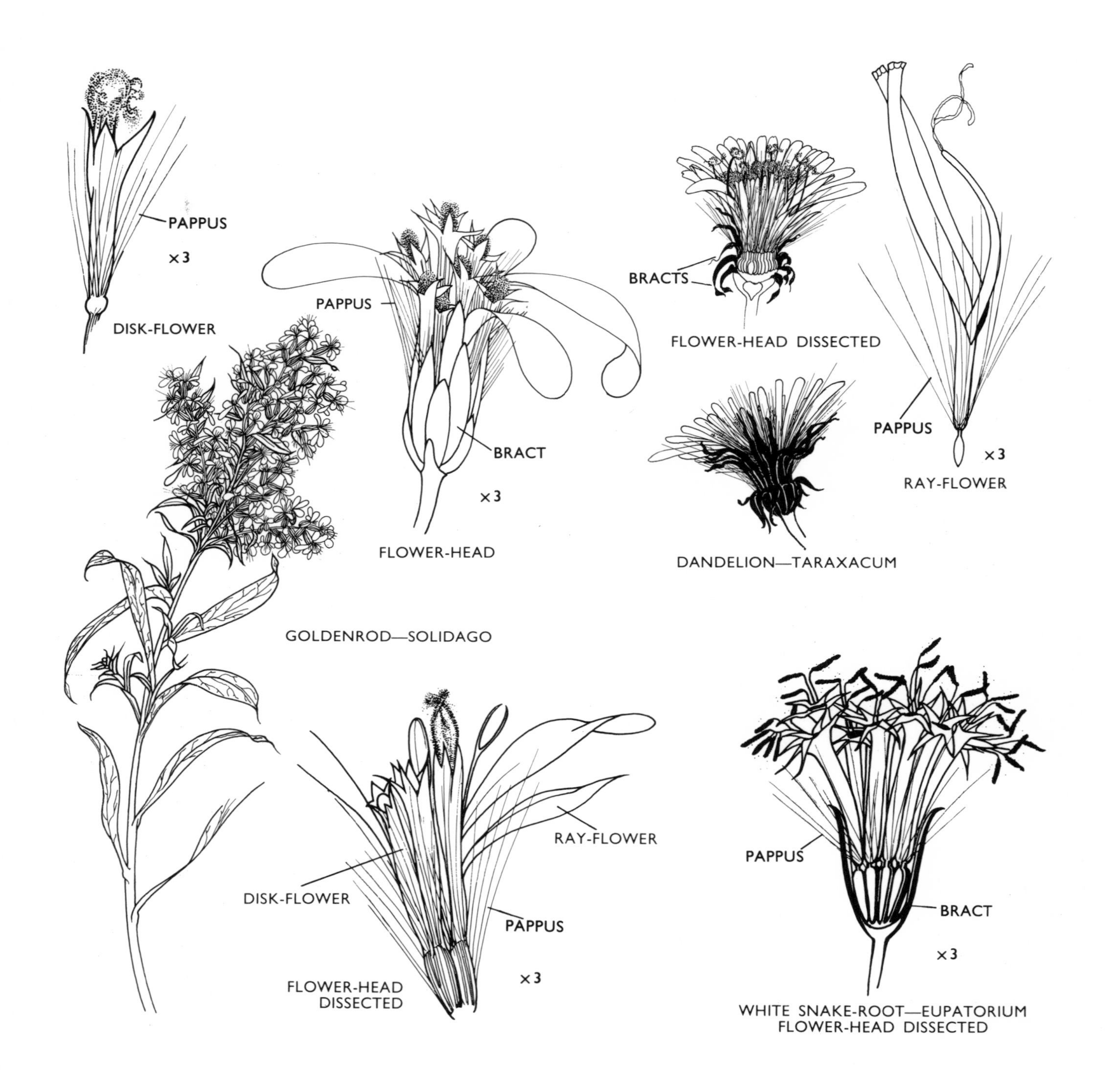

PLATE 134

Polygala paucifolia *Gottscho*

Polygala sanguinea *Johnson*

Polygala polygama *V. Richard*

Polygala cruciata *V. Richard*

Polygala senega *Johnson*

GROUP XIV

FLOWERS small, in dense heads which simulate single flowers, surrounded by bracts which may be mistaken for a calyx. Sepals various, petals four or five; petals joined, radially or bilaterally symmetric. Stamens four or five.

This group consists almost entirely of the very large daisy family, whose distinguishing characteristics are described below. Associated with it (because sometimes mistaken for one or more of its genera) is the very small teasel family.

THE TEASEL FAMILY (DIPSACACEAE)

The teasels and their relatives have small flowers packed into heads which bear a superficial resemblance to those of the daisy family – even to the involucre of bracts around each head. The flowers in our species have four petals and four stamens; there is nothing corresponding to the "rays" of a daisy or sunflower. The stamens are separate, not joined as they are in the daisy family.

DIPSACUS

The genus *Dipsacus* is known by the prickly stems and the long, spinelike bracts around the compact, oval flower-heads. The flowers, also, are mixed with spinelike bracts which make the whole head a sort of pincushion.

TEASEL, D. SYLVESTRIS, is a tall plant (20–80 inches) with paired lanceolate leaves on the prickly stem. The small flowers, which are immersed in a mass of spiny bracts, have a tubular four-lobed, radially symmetric corolla.

July to October: naturalized from Europe and established in old fields and waste land and on roadsides from Quebec to Michigan and southward to North Carolina, Tennessee, and Missouri. *Plate 135.*
Fuller's teasel, *D. fullonum*, closely related to the above species and perhaps only a cultivated variety of it, differs in having hooked spines. The dried heads are placed on spindles and revolved against cloth to *tease* it – that is, to raise the nap; not in *fulling* (cleaning and beating) cloth, as some botanists have written.

KNAUTIA

We have only a single species of this Old-World genus.

BLUEBUTTONS, K. ARVENSIS, like teasel has flowers with four-lobed, radially symmetric petals. They are not, however, mixed with spinelike bracts (merely with hairs). The bracts around the head do not extend much beyond it. The paired leaves are pinnately divided into from five to fifteen narrow segments.

June to September: in dry fields and waste places from Newfoundland to North Dakota and southward to Pennsylvania; naturalized from Europe. *Plate 135.*

PLATE 135

Polygala brevifolia *Allen*

Dipsacus sylvestris *Roche*

Polygala lutea *Johnson*

Dipsacus sylvestris *Murray*

Knautia arvensis *Elbert*

Polygala ramosa *Allen*

Scabiosa atropurpurea *Horne*

Polygala cymosa *Johnson*

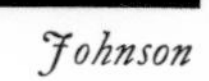

SCABIOSA

The species of *Scabiosa* differ from those of *Dipsacus* and *Knautia* in having the marginal flowers of a head larger than those in the center and somewhat bilateral in their symmetry. The lobes of the corolla are quite long and brightly colored and the plants have a decorative appearance which has earned them a place in gardens. There are numerous species in the Old World, a few of which have escaped from cultivation in North America. The species named below is the only one at all widespread or abundant.

S. AUSTRALIS has a branching stem from 1 to 3 feet tall or slightly taller. The leaves are paired and narrow, lanceolate or wider towards the tip. The flowers are light blue.

June to October: in fields and waste land from New England to Pennsylvania. *Plate 135* illustrates sweet scabious, *S. atropurpurea*, with pinnately divided leaves and dark purple or rose or white flowers. It escapes from cultivation but is rarely established as a wild flower.

THE DAISY FAMILY (COMPOSITAE)

The daisy family is easily the largest family of wild flowers in the United States or in any temperate country; that is, it includes the most species. Indeed it is usually considered to be the largest family of any plants anywhere; but when we include the flowers of the tropics with those of temperate regions, the orchids also are in the running for the title. In the United States the *Compositae* (or "composites") are almost all herbaceous; in the tropics many of them are trees and shrubs. Many herbaceous species are cultivated in our gardens for ornament: chrysanthemums, dahlias, asters, zinnias, daisies of various kinds, sunflowers, golden-glow, and others. Several are well known in the herb-garden: southernwood, wormwood, feverfew, costmary, chamomile, tansy. A few come to our tables: artichoke, lettuce, chicory, endive, salsify. Several are among our most objectionable weeds: ragweeds, mugwort, cocklebur, wild lettuce, dandelions, and the thistles.

The flowers of the family are distinctive. They are very small, but several or many are associated in a head which can be mistaken for a single large flower. (See page 422). Careful examination discloses that a daisy or sunflower – the flower-head – is composed of two kinds of flowers: those around the margin, the white or yellow "rays"; and those in the center, the "disk." Each disk-flower has five petals joined to form a tubular, radially symmetric corolla, generally with five teeth. Each ray-flower has a bilaterally symmetric corolla tubular at the base, but with the joined petals expanded on one side into the "ray." Each flower of either kind stands on its own small stalk (within which is the inferior ovary). All these stalks are close-packed, side by side, on the enlarged end of the main stem. Other composites have only flowers of the "disk" or tubular kind, no "rays"; such are the thistles, ironweeds, Joe-Pye-weeds, and many others. Still others have only "strap-shaped" flowers resembling the "rays" of a sunflower, and none of the "disk" or tubular type; among these are dandelions and chicory.

The entire head is surrounded by one or several circles of bracts. If the head is mistaken for one flower and the rays for its petals, these bracts may be taken for a calyx.

The fruit of each tiny flower is commonly miscalled a seed. A sunflower "seed" for example, is a small hard fruit, developed from the inferior ovary of one disk-flower, and containing one seed. It does not split open at maturity, and is of the type named achene. The ray-flowers of sunflowers are sterile, form no fruit; but each disk-flower may form a fruit, so the disk comes to be a mass of small achenes. Some composites, on the other hand, form fruits from their ray-flowers and none from the disk-flowers; while in still other genera all flowers are fertile.

So much for the family as a whole; it is easily recognized. But because of the large size of the family and the small size of the flowers, identification of genera and species is more difficult than in most other families, and when one has in hand an unknown composite, it may be best to turn the pages until something suggestive is seen in a photograph or drawing. However, with the aid of a hand magnifier and careful attention to one or two details, most genera in our area may be discovered through the guide below.

One of the details referred to is what is called the pappus. This is a circle of fine hairs or a few minute scales or teeth or bristles or spines situated at the base of each corolla, on the summit of each small individual flower-stalk (see the drawings). It corresponds to the calyx of other flowers. In some genera it is lacking. It may best be seen by cutting the flower-head in half vertically so that the small individual flowers are wholly visible. If the plant is passing into the fruiting stage ("going to seed"), the pappus, if any, may often be quite easily seen on the summit of each seedlike fruit (achene).

There are also, in some genera, narrow green or papery or spiny or bristly bracts among the disk-flowers. These are called, collectively, chaff. The presence or absence and character of chaff sometimes aids in identification.

Botanists divide the family into "tribes" which represent their ideas of natural relationships among the genera. These tribes are distinguished chiefly by minute characteristics of stamens and style, by the small bracts or chaff mixed with the flowers, and other details not easily visible. For easier identification, the genera of composites growing wild in our area are here arranged in three purely artificial groups, each group divided into several subordinate groups. The botanical tribes are not mentioned further, but may be found in the standard botanical manuals and floras.

Guide to Genera of Compositae

I. *Plants with both ray-flowers and disk-flowers in each flower-head (compare II, p. 428, and III, p 428).* CAUTION: *The ray-flowers are very small, scarcely projecting beyond the bracts of the involucre, in* Parthenium, Dyssodia, Galinsoga, *and* Madia. *The disk-flowers may be overlooked in* Solidago.

In this group we may distinguish two smaller groups by the color of the rays.

A. Plants with yellow, orange, or reddish rays (compare B).

1. Among these, the following genera have an easily seen pappus of numerous fine bristles: *Solidago* (numerous small flower-heads in large clusters; short rays); *Chrysopsis* (rays long; pappus-bristles in two rings; plants hairy or woolly); *Heterotheca* (rays short; pappus-bristles in two rings, the outer shorter; plants hairy and glandular); *Senecio* (bracts all equal, except for a few short outer ones); *Inula* (heads large, rays numerous and narrow); *Arnica* (leaves in pairs or all at the base); *Tussilago* (no leaves at flowering time).

2. The following genera have a pappus of a few (mostly from one to four) short, stiff teeth or spines (best seen on the achenes): *Coreopsis* and *Bidens* (bracts of two kinds in each head; leaves paired, in circles, or at the base); *Actinomeris* and *Verbesina* (leaf-edges "running down" on the stem, the stem therefore "winged"; leaves mostly borne singly).

3. These genera with yellow rays have a pappus of a few bristles, or scales that may taper up into bristles: *Helianthus* (rays long, toothed at the end; some leaves paired); *Grindelia* (plant sticky); *Dyssodia* (leaves pinnately divided into narrow segments; rays very small; ill-scented); *Helenium* (rays three-lobed; disk dome-shaped).

4. In the following genera there is no evident pappus: *Chrysogonum* (about five rays); *Rudbeckia* (disk conical or hemispherical; leaves borne singly); *Ratibida* (disk columnar; leaves pinnately cleft or divided); *Heliopsis* (disk conical; leaves paired); *Hymenoxys* (all leaves basal, narrow, without teeth or lobes); *Silphium* (disk-flowers sterile, with unbranched styles; heads large); *Polymnia* (leaves coarsely and irregularly lobed or cleft); *Madia* (plant sticky; leaves narrow, borne singly).

B. Plants with rays of colors other than yellow or orange – white, pink, purple, lavender, etc. (See also *Coreopsis rosea* and *Verbesina virginica.*)

1. Of these the following have a pappus of numerous fine bristles: *Aster* (bracts of several lengths, overlapping; disk yellow turning reddish; flowering in summer and fall); *Sericocarpus* (rays few, white; disk cream-colored); *Erigeron* (bracts all of the same length; disk yellow; flowering mostly in spring and summer).

2. Two genera with white or purple rays have a pappus of a few bristles or teeth (the bristles are really scales cut into fine parts): *Boltonia* (leaves borne singly; rays white or purple, about $\frac{1}{4}$ inch long); *Galinsoga* (leaves paired; rays white, three-lobed, less than $\frac{1}{4}$ inch long).

3. The remaining genera of group B have no easily visible pappus: *Echinacea* (rays drooping, long; spiny chaff obscures the disk-flowers); *Parthenium* (leaves borne singly; usually five rays, no longer than the bracts; disk white); *Eclipta* (leaves narrow, paired; rays white, very short); *Chrysanthemum* (rays from ten to twenty-five, up to an inch long; leaves toothed); *Achillea* (rays short; disk white; leaves divided into many fine segments); *Anthemis* (rays up to ½ inch long; disk yellow; chaff present; leaves divided into fine segments); *Matricaria* (like *Anthemis* but no chaff). See also *Polymnia*, whose rays may be whitish; and *Verbesina*.

II. *Plants with no ray-flowers, only disk-flowers.* NOTE: *The marginal flowers of* Centaurea *are somewhat intermediate, the corolla having large and unequal lobes.*

A. Genera with bracts that are whitish and papery at least at the edges; leaves attached singly; plants mostly woolly; *Antennaria* (leaves mostly basal); *Gnaphalium* (leaves on the stem; outer flowers of each head without stamens); *Anaphalis* (leaves on the stem; pistillate and staminate flowers on different plants).

B. Genera with bracts tipped with sharp spines: *Cirsium* (pappus of numerous feathery bristles); *Carduus* (pappus of plain bristles); *Onopordon* (pappus plain; stem prickly – winged); *Arctium* (bracts with hooked spines); *Centaurea* (leaves without prickles, pinnately cleft).

C. Genera with bracts not whitish and papery nor spine-tipped.

1. Of these, the following have a pappus of numerous bristles: *Mikania* (a vine; leaf-blades indented at the base); *Eupatorium* (leaves mostly in pairs or circles; bracts in most species nearly all of one length; *Vernonia* (leaves borne singly; bracts of several lengths, overlapping); *Liatris* (flower-heads short-stalked, in a spike); *Kuhnia* (pappus-bristles feathery); *Pluchea* (leaves borne singly, toothed; rank smelling); *Cacalia* (heads numerous, small; bracts all of one length; leaves stalked); *Erechtites* (heads small; bracts all of one length; leaves without stalks, sharply toothed); *Filago* (leaves narrow and covered with white wool); see also *Senecio* (group I), some of whose species may lack rays.

2. The following genera of group II have a pappus of a few bristles or small scales, or no evident pappus: *Marshallia* (leaves at or near the base of the long stalks of the flower-heads); *Sclerolepis* (leaves narrow, in circles; heads single); *Elephantopus* (heads very small, in dense clusters each of which simulates a single head; flowers all bilateral in symmetry, the lobes on one side); *Centaurea* (bracts tipped with a toothed appendage); *Adenocaulon* (heads small; bracts all of the same length; leaves more or less triangular, toothed, woolly on the under surface); *Tanacetum* (leaves much divided, aromatic; flowers yellow); see also *Matricaria*, *Polymnia*, and some species of *Bidens* (group I).

III. *Genera with no disk-flowers of the tubular type; all the flowers are "strap-shaped," more or less like the rays of group I. These may be separated by the color of their flowers.*

A. Flowers white, pink, purplish, blue – not yellow or orange: *Cichorium* (flowers blue or sometimes white; heads borne singly on very short stalks); *Prenanthes* (flower-heads small, hanging in clusters); *Lactuca* (heads small, numerous; stem leafy; tall weeds with milky juice).

B. Flowers yellow or orange. The pappus of some is feathery; i.e. each bristle bears smaller bristles along its sides.

1. Pappus feathery: *Tragopogon* (very smooth; leaves narrow, on the stem; pappus on very long beak); *Hypochoeris* (leaves all at base, pinnately lobed or cleft; chaff present); *Leontodon* (leaves toothed or pinnately lobed or cleft, all at the base; no chaff);

2. Pappus plain or none: *Pyrrhopappus* (leaves all at or near the base, mostly toothed or pinnately lobed or cleft; stems branched; pappus reddish, on long beak); *Lapsana* (leaves stalked, blades ovate, toothed); *Lactuca* (heads small, numerous; stem leafy; tall weeds with milky juice); *Sonchus* (like *Lactuca*, but leaves mostly prickly and clasping the stem); *Krigia* (heads terminating long leafless stalks; leaves at or near the base; bracts all of one length; pappus not on beak); *Serinia* (like *Krigia* but no pappus); *Crepis* (bracts in two circles, the outer ones much shorter; heads small on long stalks; leaves narrow, the lower ones toothed); *Hieracium* (mostly hairy plants; leaves not lobed, in most species all at the base); *Taraxacum* (leaves toothed or pinnately lobed or cleft, all at the base; pappus on long beak; flower-heads large, each on a leafless stem).

THE GOLDENRODS (SOLIDAGO)

The common and beautiful plants known as goldenrods have numerous small flower-heads clustered in inflorescences of various types. Rays and disk-flowers are both yellow, except in one common species and some forms of other species, in which the rays are white. Rays are not numerous – seldom more than ten; and in many species quite small. The plants are perennial, with rhizomes, runners, or thick crowns. In many species there is a marked difference between the leaves at the base of the stem (or on an offset) which often form a rosette which may last through the winter, and the leaves on the erect stem. All the goldenrods flower in late summer and autumn.

This is chiefly an American genus, reaching its greatest complexity in the eastern United States. There is one European species, called farewell-summer and woundwort. The reason for the first name is obvious; the latter refers to the use of the plant in healing wounds. Gerard, the herbalist, tells an amusing story of fantastic prices paid for the imported herb – which dropped to nothing when the plants were found growing at the gates of London.

Goldenrods are commonly believed to be responsible for the distressing allergies known as "hay fever." Actually not much goldenrod pollen is carried by the wind (the flowers are pollinated by butterflies and other insects) and in the flowering season of these plants the air is likely to contain plenty of ragweed and other pollen known to be the principal cause of the disease. However, though it has never been proved that goldenrods are dangerous to hay-fever sufferers, it is possible that they play some small part in the affliction.

The great genus *Solidago* cannot be recommended to the amateur naturalist for the easy identification of its species. The goldenrods as a group are easy to recognize. Some species also are fairly distinctive. But most are separated by rather technical characteristics, and, since they are very variable, the easily seen features of leaves, stems, and flowers of various species overlap. Moreover, some species certainly hybridize in nature, yielding intermediate plants which add to the confusion. Even professional botanists do not agree on just how many species there are nor how they may be separated. One man's species is another man's variety. One man's species will cover the United States; another man will use the same name for plants of a restricted range.

The photographs and drawings, with the brief descriptive notes, will enable the amateur to recognize and name at least some of the more distinctive goldenrods of the northeastern United States. But he will find many plants that are impossible to classify without some technical training and the use of the botanical manuals and floras – and even then it will be difficult.

It is impossible even to illustrate some of these variable species so as to show all their forms. The photographs were made from plants fairly typical of their species, and the drawings represent common or "average" characteristics of leaves and flower-heads.

To make identification a bit easier, we here place all the goldenrods in three groups according to the way in which their flower-heads are arranged. The distinctions between these groups are not, however, sharp; some species have plants in more than one group; if trouble is encountered in identifying a plant, all three groups should be searched.

In group I the flower-heads are borne *along the upper sides of several or many branches which curve outward and often downward.* These branches may be long or short. In many species they are clustered towards the top of the main stem, and this stem may also bend and bear flower-heads along the upper side near its tip. The form of the whole inflorescence is generally broadly or narrowly conical.

This group includes the tall weeds so familiar in the autumn along roadsides and in old fields. Many exceed 6 feet in height.

In Group II the flower-heads are clustered in a *cylindric* inflorescence at the top of the stem, usually with small clusters also *in the axils of the leaves below.*

Most of these are smaller than those of group I, seldom exceeding 4 feet. Many are woodland plants.

Group III contains plants whose flower-heads are clustered *at and near the tips of a number of more or less erect branches*, so as to form a more or less *flat or domed inflorescence.* The leaves of most species in this group are narrow and sharp-pointed.

I. *Flower-heads along the upper side of branches that curve outward.*

Plants in this group may be further separated into two smaller groups by the veins in their leaves.

A. At least the lower leaves with three principal veins running lengthwise; the leaves lack stalks.

S. MISSOURIENSIS (4–32 inches) has numerous very narrow leaves mostly without teeth. The whole plant is smooth. This is mainly a western species, with a variety in our range.

In prairies and open gravelly places from Ontario to British Columbia and southward to Tennessee, Oklahoma, and Arizona; also in New Jersey.

S. CANADENSIS (1–5 feet) and the two following species are the commonest tall goldenrods of eastern fields and roadsides. They are difficult to distinguish, many intermediate forms are seen, and indeed they are considered by some botanists as merely varieties of one species. *S. canadensis* has typically narrow, sharp-pointed leaves with prominent sharp teeth along the edges; the surface may be rough. The involucre is tiny, mostly only about $\frac{1}{12}$ inch tall.

In fields and open woodland and on roadsides from Newfoundland to Saskatchewan and southward to North Carolina, Tennessee, South Dakota, and Colorado. *Plate 136.*

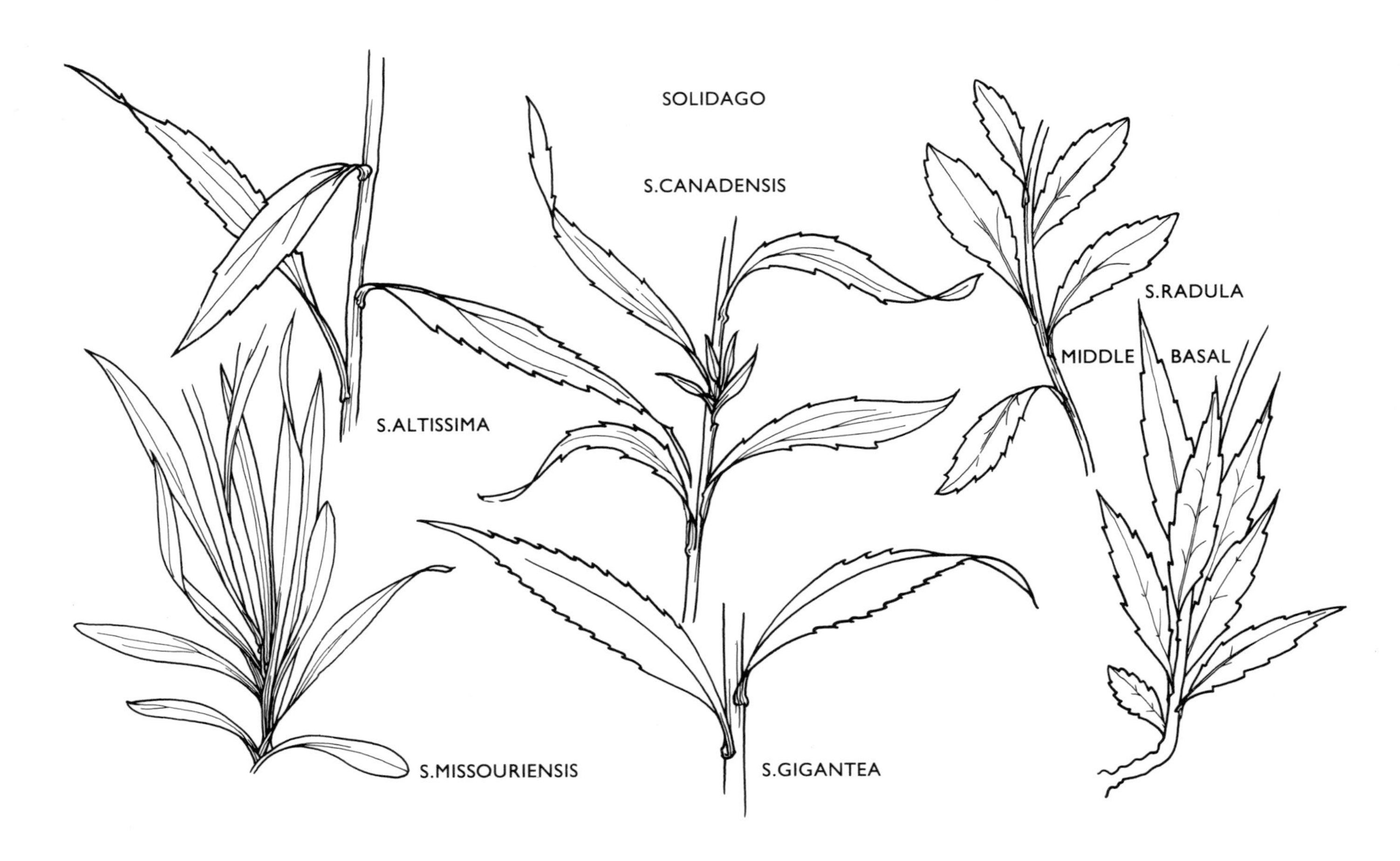

S. GIGANTEA (20–100 inches) differs from *S. canadensis* chiefly in having a stem generally (but not always) covered with a whitish bloom; leaves with less prominent teeth which are situated chiefly in the terminal parts of the blade; and a larger involucre (about $\frac{1}{6}$ inch).

In moist open places and thickets from Quebec to British Columbia and southward to Florida and Texas. *Plate 136.*

S. ALTISSIMA (28–80 inches) has typically a stem covered with a fine grayish down. The leaves may lack teeth, or they may be toothed towards their tips. The involucre is much as in *S. gigantea.*

In old fields, on roadsides, etc. from Quebec to North Dakota and southward to Florida and Arizona. *Plate 136.*

S. RADULA (8–40 inches) is grayish with a fine but rather rough down. The leaves are rough and stiff, with low teeth towards the tip, or without teeth. They are oblong, or broader above the middle. Runners are present at the base. A southwestern species.

In open woods and rocky places from Illinois to Missouri and southward to Louisiana and Texas.

B. The leaves of the remaining species of group I have one principal vein, with or without visible veins branching from it, but none running parallel with it.

Some of these plants have stalked leaves; some have leaves without stalks. Compare 1, 2, and 3 below.

1. Plants whose lower leaf-blades are sharply distinct from their stalks.

S. ARGUTA (20–80 inches) has leaf-blades on stalks that have thin margins or "wings"; each margin bears a row of short hairs. The leaves are sharp-pointed and finely toothed, smooth or somewhat rough.

In meadows and open woodland from Maine to Ontario and southward to North Carolina, Alabama, and Illinois.

S. BOOTTII (20–60 inches) has ovate or lanceolate basal and lower leaf-blades, finely and sharply toothed. The whole plant is smooth. The inflorescence is generally loose. A southern species much like the preceding.

In woodland from Virginia to Missouri and southward to Florida and Texas.

S. SPHACELATA (16–60 inches) is easily recognized if basal leaves are present. The blades are heart-shaped (indented where the stalk is attached), and

PLATE 136

Solidago nemoralis *Johnson*

Solidago gigantea *Johnson*

Solidago nemoralis *Johnson*

Solidago canadensis *Johnson*

Solidago altissima *Johnson*

Solidago ulmifolia *Johnson*

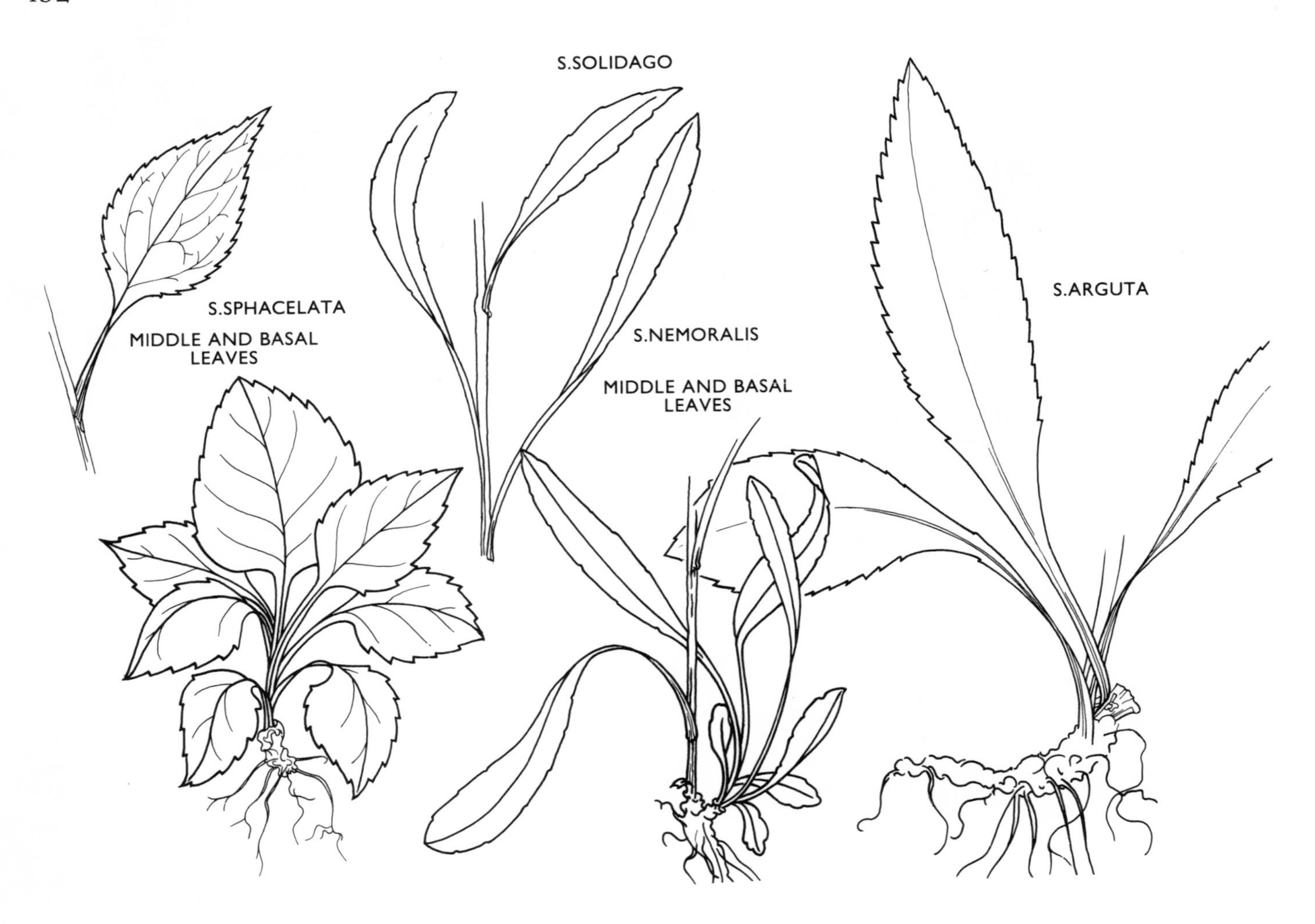

coarsely toothed. The lower leaves on the stem are either similar to the upper or have ovate, coarsely toothed blades. The inflorescence is narrow, the flower-heads crowded.

In woods and rocky places from Virginia to Illinois and southward to Georgia and Alabama.

2. Plants whose lower leaf-blades taper gradually down into their stalks.

S. NEMORALIS (6–50 inches) has basal and lower leaves that are generally broader and toothed or scalloped towards their tips. The upper leaves may be narrow and pointed. The stem is finely downy. The inflorescence is generally narrow and one-sided, but may be composed of several erect branches curving outwards. This is often a tall and slender plant.

In dry woods and open places from Quebec to North Dakota and southward to Florida and Arizona. *Plate 136*. One of the first goldenrods to bloom.

S. ULMIFOLIA (1–5 feet) has coarsely toothed or jagged leaves, elliptic and sharp-pointed. Through a lens one can see hairs on the under surface along the veins. The stem is smooth. The leaves diminish in size from base to tip of the stem, those in the inflorescence being mere bracts.

In rocky thickets and open woodland from Vermont to Minnesota and southward to Georgia and Texas. *Plate 136*.

S. JUNCEA (4 inches–4 feet) has finely toothed lower leaves, or they may be scalloped. The basal leaves are very long-stalked; their blades are broader towards the tip. The upper leaves are very narrow, smaller than the lower. The leaf-margins are rough, otherwise the plant is smooth.

In dry open places from Nova Scotia to Saskatchewan and southward to Georgia and Missouri. *Plate 137*. An early-flowering species.

S. RUGOSA (16–80 inches) is a very variable species, some forms of it resembling the preceding. The leaves are sharply toothed and sharp-pointed. Stem and leaves may be rough and hairy or smooth; the stem is usually hairy in the inflorescence. The stem and the inflorescence are both typically very leafy, the leaves of the inflorescence being smaller than those on

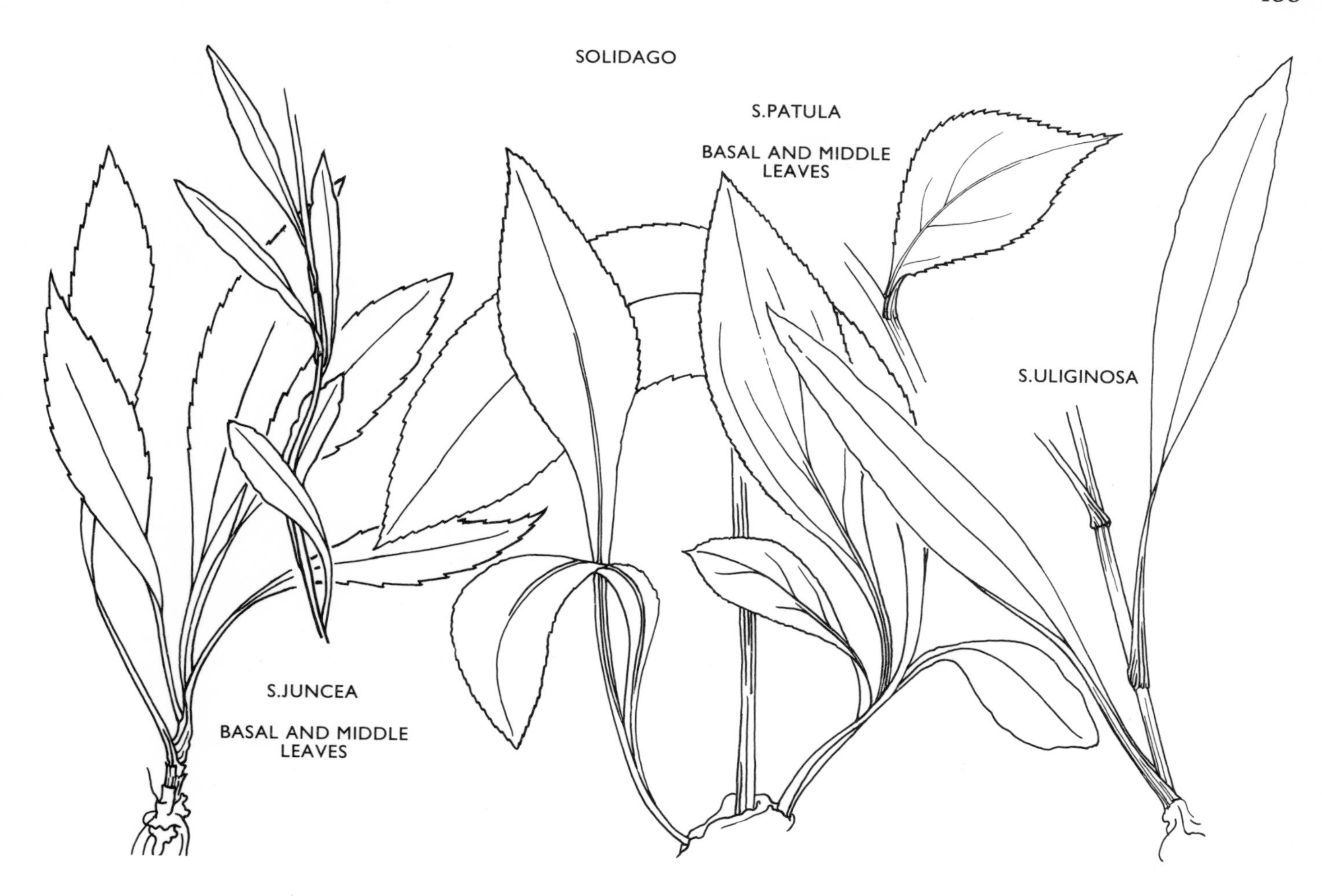

the stem. Typically a number of flowering branches spread widely from near the tip of the main stem.

Mostly in open places, wet and dry, from Newfoundland to Ontario and Michigan and southward to Florida and Texas. *Plate 137*. This is the species that Dr Fernald has called "hopelessly variable"; over such a vast range much variation is to be expected. Yet it is possible to become acquainted with at least some characteristic forms of the species, as illustrated and briefly described here.

S. ULIGINOSA (16–60 inches) is another quite variable species. The leaves are smooth, rather thick, and very long, tapering to a margined ("winged") stalk which embraces the stem at the point of attachment. The blades may or may not be toothed.

In bogs, swamps, and wet meadows from Newfoundland and Quebec to Minnesota and southward to North Carolina, Ohio, and Indiana. *Plate 137*.

S. PATULA (20–80 inches) has leaves which are very rough on the upper surface. Their teeth are often prominent and sharp. The stem may be square, or it may bear several lengthwise ridges. The lower leaves are large, elliptic. The inflorescence is loose.

In wet woods and swamps, also on ledges, from Vermont to Minnesota and southward to Georgia and Louisiana. *Plate 137*.

SEASIDE GOLDENROD, S. SEMPERVIRENS (8–100 inches), has thick, succulent leaves, as is so often the case with seaside plants. The basal leaves have long stalks. Each leaf has one strong central vein, with few visible veins branching from it.

In wet places, mostly salt, near the coast from Newfoundland and Quebec to Virginia. *Plate 137*. The southern *S. mexicana*, which grows in similar situations, is sometimes considered a variety of *S. sempervirens*.

3. Plants none of whose leaves are stalked.

S. ODORA (20–40 inches) is a very narrow-leaved species. All the leaves lack teeth; the upper are very small. The plant when crushed emits the odor of anise.

In dry open woods from Massachusetts and New Hampshire to Oklahoma and southward to Florida and Texas. *Plate 180*.

S. ELLIOTTII (20–80 inches) has elliptic leaves mostly without teeth, often crowded on the stem, all smooth. Runners are present at the base of the stem.

In swamps and wet woods near the coast from Nova Scotia to Massachusetts and Florida.

Besides all these, the southern species *S. pinetorum*, *S. gattingeri*, *S. drummondii*, *S. fistulosa*, *S. strigosa*, *S. ludoviciana* may be found along our southern borders. *S. shortii*, with rather leathery, long-stalked leaves, grows in Kentucky. The western species variously known as *S. elongata* or *S. lepida* or as a variety of *S. canadensis* reaches Minnesota and Michigan.

II. *Flower-heads in the axils of leaves and/or in a long, cylindric, terminal cluster.*

A. In this group, one species has white rays, a yellow disk.

SILVER-ROD, S. BICOLOR (8–32 inches), is almost unique among species of *Solidago* in having white rays. The stems are whitened with a very fine down.

In dry soil in the open or in thin woods from Quebec to Wisconsin and southward to Georgia and Arkansas. *Plate 137*.

B. In one species of group II the tips of the bracts – the small scales of each flower-head – are bent sharply outwards and downwards (they are "squarrose").

S. SQUARROSA (8–70 inches) has lower leaves whose blades taper down into stalks; the upper leaves mostly lack stalks. All are more or less sharply toothed.

In rocky woods from Quebec to Ontario and southward to North Carolina and Kentucky. *Plate 138*.

C. In two species of group II the leaves have blades on distinct stalks; that is, one can see clearly where stalk ends and blade begins (the blade may have a *small* V-shaped portion where it joins the stalk).

S. MACROPHYLLA (4 inches–5 feet) has long-stalked basal leaves with large, sharp-pointed teeth. The flower-heads are the largest of any formed by our goldenrods: nearly $\frac{1}{2}$ inch tall, with ten or fifteen rays. The flower-cluster is usually leafy.

In damp woods from Labrador and Newfoundland to Hudson Bay and southward to New York and Ontario.

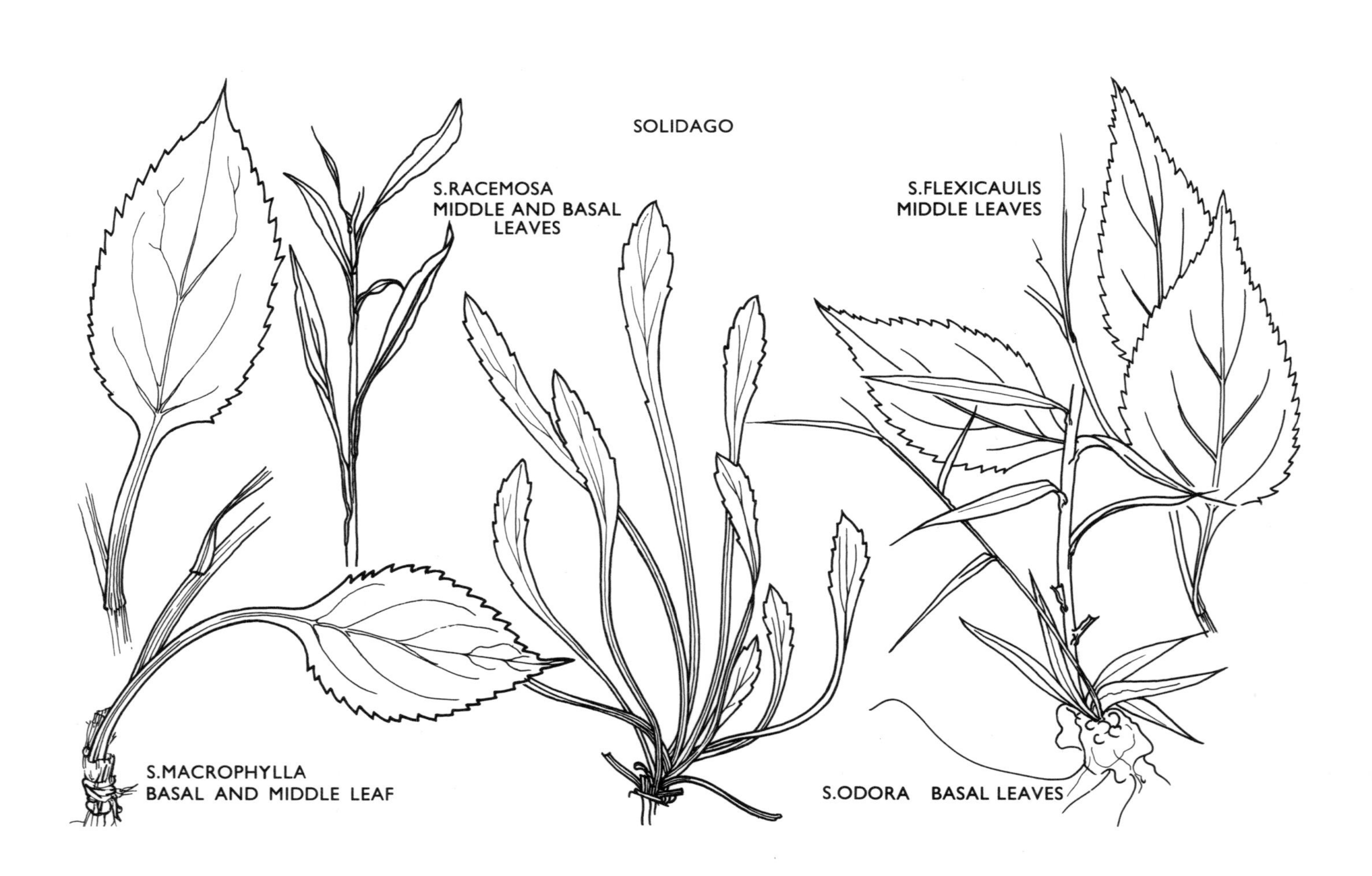

PLATE 137

Solidago sempervirens *Ryker*

Solidago juncea *Johnson*

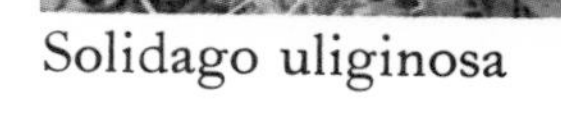

Solidago uliginosa *Gottscho*

Solidago bicolor *Rickett*

Solidago patula *Rickett*

Solidago rugosa *Rickett*

S. FLEXICAULIS (8–40 inches) has long-stalked leaves on the stem as well as at the base; the blades are coarsely and sharply toothed. The stem (*caulis*) is generally more or less zigzag. The flower-heads are mostly in the axils.

In woodland from Quebec to North Dakota and southward to Georgia, Arkansas, and South Dakota. *Plate 138.*

D. In group II three species have leaves that lack stalks or almost lack them.

BLUE-STEM GOLDENROD or WREATH GOLDENROD, S. CAESIA (8–40 inches), has purplish stems covered with a waxy bloom. The leaves taper to sharp points and are often sharp-toothed. The stems are apt to arch and grow horizontally, whence the name wreath goldenrod. The name is particularly appropriate when, as is often the case, the flowers are all distributed along the stem in the axils.

In woodland from Nova Scotia and Quebec to Wisconsin and southward to Florida and Texas. *Plate 138.*

S. CURTISII (16 inches–5 feet) resembles *S. caesia* but the stem is not waxy. It is more southern.

In upland woods from Virginia to Kentucky and southward to Georgia and Alabama.

S. MOLLIS (6–20 inches) has a stem whitened with a fine down. The lower leaves generally have three main veins running lengthwise. Runners are usually present at the base of the stem.

In dry open places from Minnesota to Saskatchewan and southward to Iowa, Oklahoma, and New Mexico.

E. The remaining species of group II have leaves that taper gradually downwards into stalks or stalklike portions.

The first five of these have, in general, rather narrow leaf-blades, often broadest between the tip and the middle.

S. ERECTA (12–52 inches) has blunt-toothed, scalloped, or plain-edged leaves. The flower-cluster is loose. The bracts of each flower-head have broad midribs and round ends.

In woods and thickets from Massachusetts to Georgia and Alabama.

S. RANDII (4 inches–3 feet) has numerous leaves on the stem. The basal leaves are short-stalked, variously toothed or scalloped. The flower-heads form a dense cylinder.

In rocky places from Quebec and Nova Scotia to Minnesota and southward to New York.

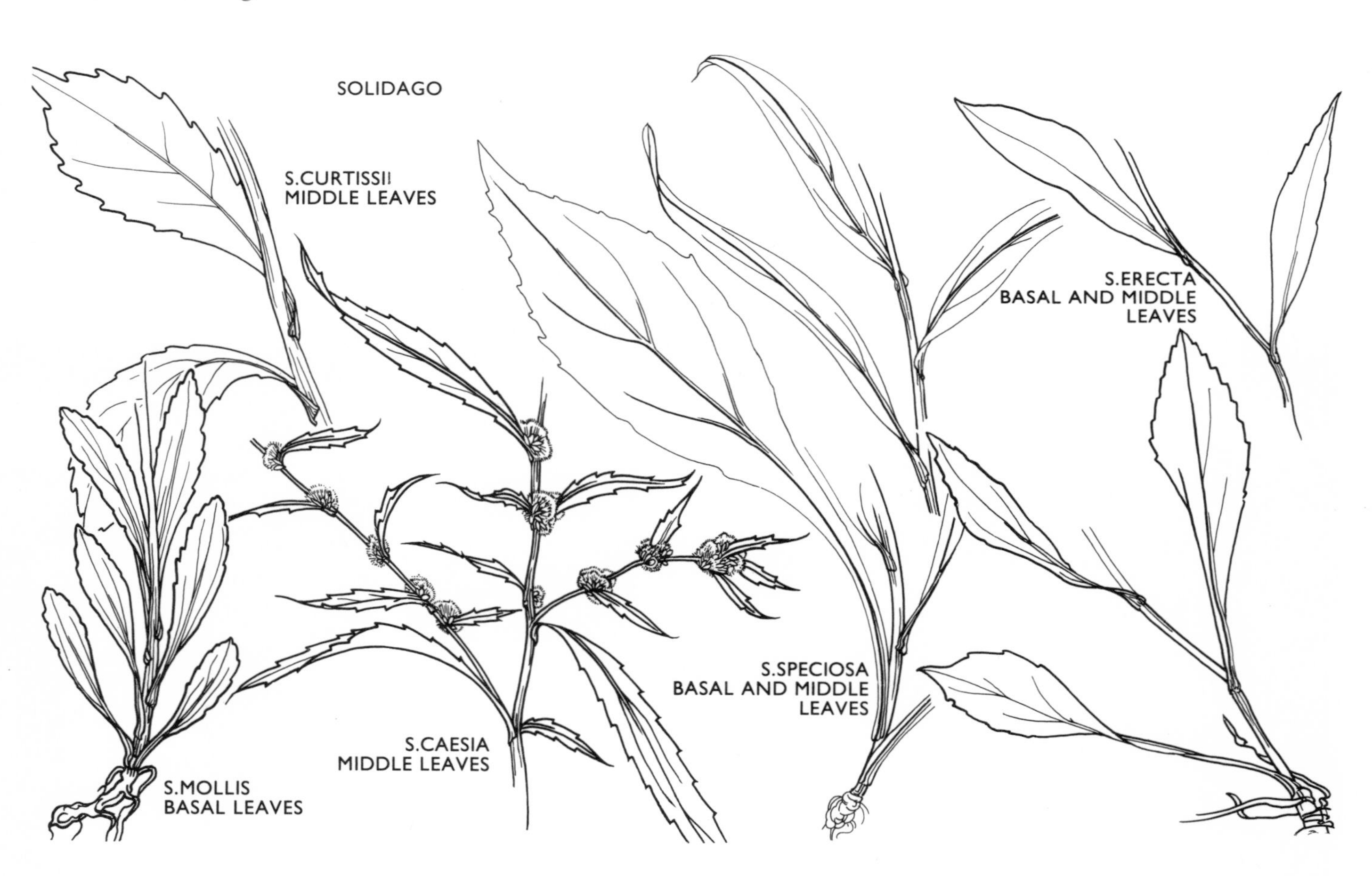

PLATE 138

Solidago squarrosa *Rhein*

Solidago flexicaulis *Johnson*

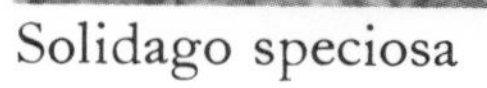

Solidago speciosa *Johnson*

Solidago puberula *Rickett*

Solidago puberula *Ryker*

Solidago jejunifolia *Johnson*

Solidago caesia *V. Richard*

S. RACEMOSA (4 inches–2 feet) is recognized by its very narrow, sharp-pointed leaves; they may be toothed or plain. The inflorescence is rather loose. A form is known with white rays.

On rocky ledges and banks from Quebec to Ontario and Wisconsin and southward to New York and Kentucky.

S. PUBERULA (8–40 inches) is a small but beautiful species, with large rays of a deep yellow color. There are from twelve to sixty leaves crowded on the stem.

In dry open places, in sand, on rock ledges, etc. from Quebec to Florida and Mississippi. *Plate 138.*

S. PURSHII (8–40 inches) has long-stalked basal leaves, the lanceolate blades about equal to the stalk. The leaves are rather fleshy and quite smooth. There are from four to six rays per head and up to fifteen disk-flowers.

In peat and damp places from Labrador to Manitoba and southward to Pennsylvania and West Virginia, Indiana, and Minnesota.

S. JEJUNIFOLIA (16–32 inches) has many narrow leaves without teeth (*jejunus* means "hungry, poor") and with almost spinelike tips.

In sandy places from Ontario to Minnesota and southward to Indiana and Illinois. This is sometimes considered to be a variety of *S. speciosa*, poverty-stricken by the places in which it grows. *Plate 138.*

The last three of these species tend to have broader leaf-blades, more or less elliptic.

S. SPECIOSA (2–7 feet) is a beautiful species (*speciosa* means "beautiful"). The stem bears from twenty to forty leaves. The flower-heads form a large terminal inflorescence; each head has usually five large rays.

In thickets, open woods, prairies, and on rocky ledges from Massachusetts to Wyoming and southward to Georgia and Texas. *Plate 138.*

S. HISPIDA (2–40 inches) is in general hairy ("hispid"), sometimes with conspicuous long hairs; but the species is extremely variable and some plants are merely downy or almost smooth. The leaves have

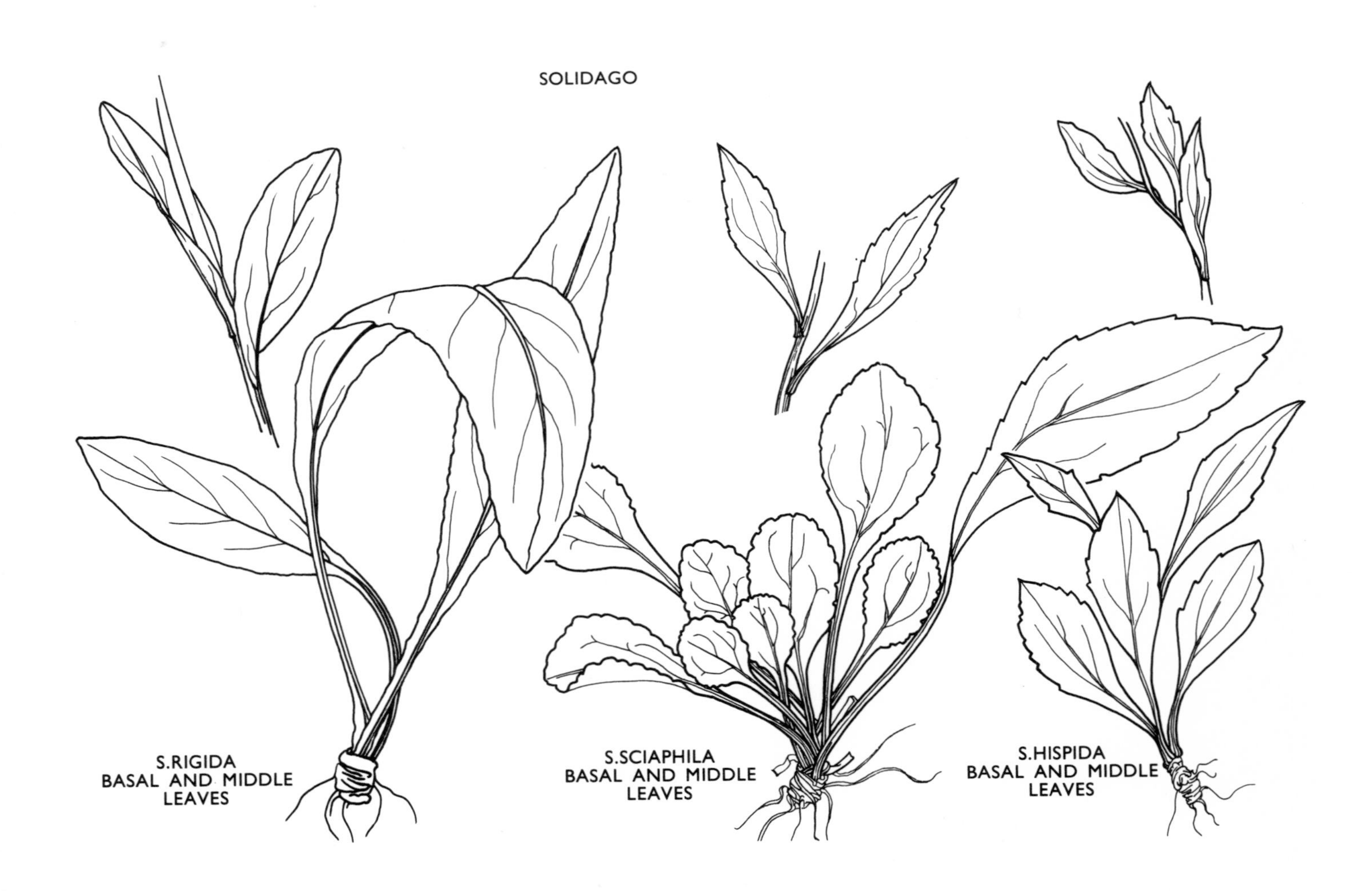

PLATE 139

Solidago graminifolia *Rickett*

Solidago rigida *Johnson*

Solidago riddellii *Johnson*

Solidago cutleri *Rhein*

Solidago graminifolia *Johnson*

Solidago hispida *Johnson*

blunt teeth. The upper leaves are much smaller. The inflorescence is often loose.

In rocky and peaty places from Quebec and Labrador to Manitoba and southward to Georgia and South Dakota. *Plate 139.*

S. SCIAPHILA (16–32 inches) has rather large, sharply toothed leaves. The flower-cluster is often leafy.

On cliffs and in sand from Ontario to Minnesota and southward to Illinois and Iowa.

On the southern borders of our range one may also find *S. buckleyi* and *S. flaccidifolia*. *S. calcicola* is a far-northern species found in northern New England. On the high mountains of Maine and New York one may find *S. cutleri*, only a few inches tall, with a few leaves (from two to seven) on the stem and an inflorescence of a few flower-heads, or even of only one. The basal leaves form a tuft. *Plate 139.*

III. *Flower-heads at the ends of more or less erect branches, forming a rather flat or dome-shaped inflorescence.*

The species may be separated into two groups according to the shape of their leaves.

A. Plants whose basal leaves are composed of definite blades on long stalks. The upper leaves are much smaller and mostly lack stalks.

S. RIGIDA (10–60 inches) has elliptic leaf-blades, the basal on definite stalks, the upper stalkless and much smaller. The blades of the basal leaves may be quite large and bluntly toothed; the upper leaves lack teeth. Leaves and stem are rough, generally with a harsh covering of short gray hairs (easily visible with a hand lens). The outer branches of the inflorescence often bend outward to a horizontal position, but the flower-heads are not restricted to their upper sides as in group I. The round-ended bracts are easily seen.

In prairies and dry open places from Connecticut to Alberta and southward to Georgia and New Mexico. *Plate 139.*

S. OHIOENSIS (16 inches–3 feet) is smooth. The basal leaves have narrow blades which taper very gradually into long stalks; the blades are commonly about equal in length to the stalks. The leaves on the stem are much smaller, stalkless, and tend to be lanceolate. The flower-heads are numerous.

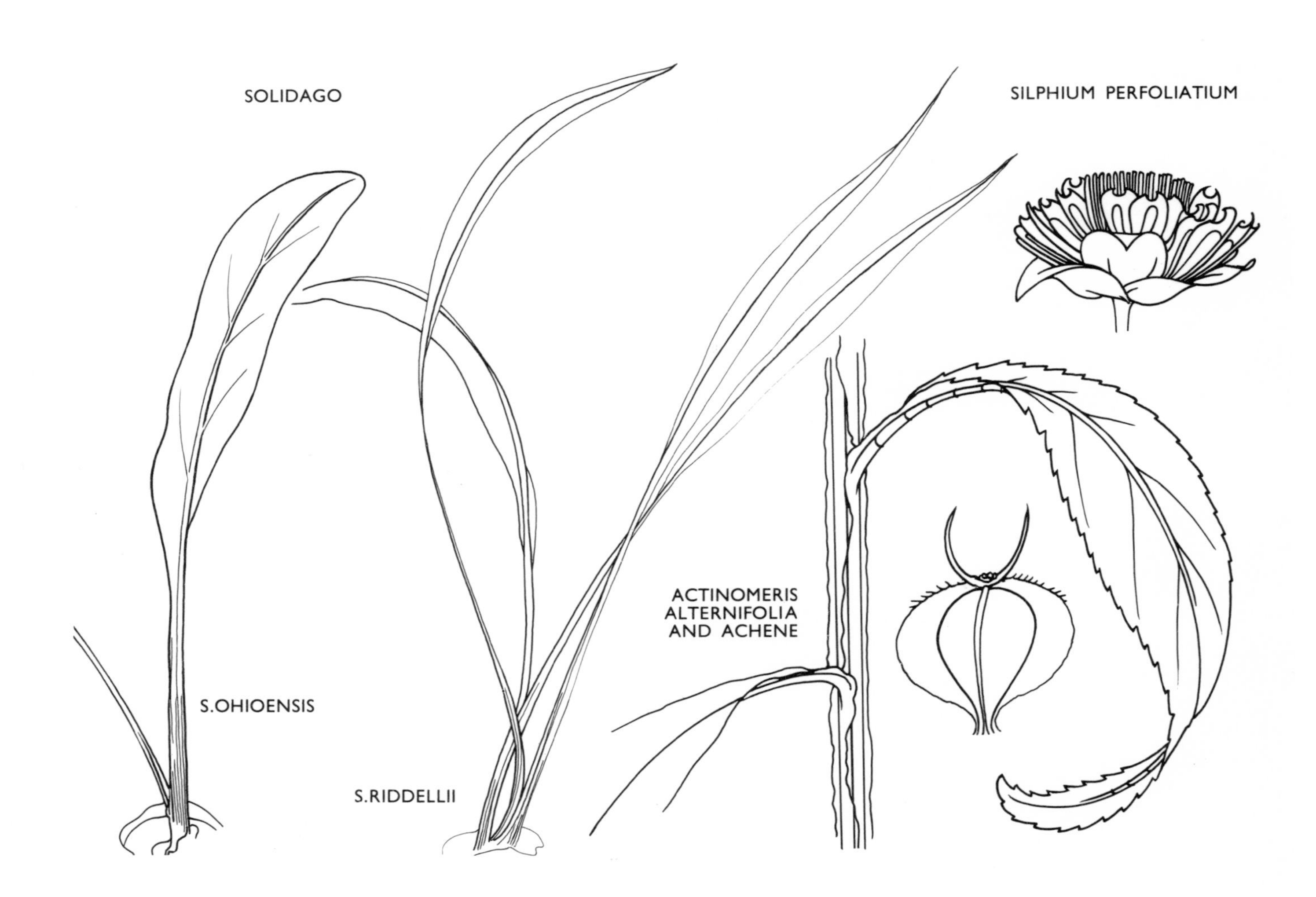

In swamps and bogs and on wet shores from New York and Ontario to Wisconsin and Illinois.

S. RIDDELLII (16–40 inches) is smooth. Its basal leaves have short and narrow blades which taper gradually to the much longer stalk. The leaves on the stem are quite long and usually folded lengthwise and bent outwards to the form of a sickle.

In swamps and wet meadows from southern Ontario to Minnesota and southward to Virginia, Ohio, and Missouri. *Plate 139.*

B. Plants whose basal leaves, like the upper leaves, are grasslike, without stalks.

S. GRAMINIFOLIA (1–5 feet) is the most widespread of the grass-leaved species. The foliage may be smooth or it may bear short hairs visible only with a lens. The lower leaves are usually broader than the upper, and tend to have from three to five ribs running lengthwise.

In open ground and thickets and on roadsides from Newfoundland and Quebec to British Columbia and southward to Florida and New Mexico. *Plate 139.*

S. GYMNOSPERMOIDES (16–40 inches) is very like *S. graminifolia*, but is far less common in our range. The leaves average narrower (up to $\frac{1}{5}$ inch wide) and generally have only one rib. The flower-heads contain fewer flowers: from three to eight disk-flowers as against from five to ten in *S. graminifolia*.

In open ground from Wisconsin to Colorado and southward to western Illinois, Arkansas, and Texas; also in Virginia. *Plate 140.*

S. TENUIFOLIA (12–30 inches) is smaller than the preceding species, with weak, very narrow leaves (not over $\frac{1}{6}$ inch wide). The flower-heads are small and numerous. The ray-flowers are minute. There are only from five to seven disk-flowers in a head.

In sandy soil near the coast from Nova Scotia to Florida and Louisiana; also in southern Michigan and northern Indiana. *Plate 140.*

S. MICROCEPHALA (16 inches–3 feet) resembles *S. tenuifolia*. The leaves are folded lengthwise and almost bristle-like. There are only three or four disk-flowers in a head.

In sandy soil from New Jersey to Florida and thence to Mississippi. *Plate 140.*

Besides these there are two species with very restricted ranges: *S. houghtonii* is a very slender plant with only fifteen flower-heads, or fewer. It grows in wet places from Ontario and New York to Michigan. *S. remota* resembles *S. microcephala*, but has five or six disk-flowers to a head. It is found in sandy soil and on prairies from northern Ohio to southern Wisconsin and northern Illinois. *S. leptocephala*, similar to *S. graminifolia*, grows in open places and thin woods from Kentucky to Missouri and southward to Louisiana and Texas.

GOLDEN-ASTERS (CHRYSOPSIS)

The species of *Chrysopsis* ("gold-look" or perhaps "golden face") somewhat resemble asters but have yellow rays instead of the white, pink, lavender, or purple rays of the latter. They are close to the goldenrods, but have longer rays and broader disks. The difference is that there are two circles of pappus, the outer of minute bristles or scales, the inner of long bristles or hairs. Recent study shows that *Chrysopsis* should be merged with *Heterotheca*.

I. *In some species the leaves, which are narrow, have a single central vein (midrib) with no visible branches, or several such veins running parallel through the length of the leaf.*

SILK-GRASS, C. GRAMINIFOLIA (1–3 feet), has straight leaves, the length from three to five times the breadth. The stem is silky-white.

July to September: in dry woods from Delaware to Ohio and southward to Florida and Mexico. *Plate 140.* Most plants of our range are distinguished by slender runners and by silky-webbed bracts; they are sometimes treated as a distinct species, *C. nervosa*.

C. FALCATA (4–16 inches) is distinguished by curved leaves (*falcata* means "sickle-shaped"). The stem, at least when it is young, is covered with white wool.

July to October: in sand on the coastal plain from Massachusetts to New Jersey. *Plate 140.*

II. *In other species the veins form a net, the midrib giving off numerous branches.*

C. MARIANA (8–28 inches) has silky stems but smooth leaves (when they are fully grown). The leaves on the stem are lanceolate; those at the base are widest between middle and tip.

August to October: in sandy and rocky woods and open places from New York to Ohio and southward to Florida and Texas. *Plate 140.*

C. VILLOSA is a name given to what is perhaps a group of interbreeding species (*C. angustifolie*, *C. bakeri*, *C. berlandieri*, *C. stenophylla*, *C. ballardi*, *C. camporum*). They are all characterized by more or less hairy stems from 8 to 40 inches tall, not woolly, and narrowly elliptic or lanceolate leaves.

July to October: in dry places from Indiana to Minnesota and British Columbia and southward to Missouri, Texas, and Arizona. *C. pilosa* is a western species seen in Missouri. It has a hairy and often sticky stem up to 4 feet tall.

HETEROTHECA

We have only one species of this genus, which belongs to Mexico and the South.

CAMPHORWEED, H. SUBAXILLARIS, has a group of stems from 1 to 3 feet tall bearing single ovate leaves, the upper indented and "clasping" the stem, the lower with stalks. The heads are numerous, with yellow rays and disk. The pappus is of bristles, with an outer circle of minute bristles.

July to November: in open sandy places from Delaware to Kansas and southward to Florida and Mexico. *Plate 141.*

THE RAGWORTS AND GROUNDSELS (SENECIO)

The genus is named from the Latin *senex*, "old man," in reference to the copious white hairs of the pappus (but this would as well do for a dozen other genera). Disk and rays are both yellow. The bracts of the involucre are all about equal in length (there may be a few minute bracts in an outer circle). Several species yield a yellow dye.

This is an enormous genus in the tropics. Many of the species there are woody; in our range they are all herbaceous. Some species lack rays. The ragworts, of course, are not ragweeds. The latter unpleasant weeds are avoided in this book.

I. *Species with the principal leaves in a tuft at the base of the stem, the leaves on the stem being much smaller and generally different in shape.*

A. Species with the principal leaves, in a tuft at the base of the stem, with long stalks and blades with round or indented base (compare B).

GOLDEN RAGWORT or SQUAW-WEED, S. AUREUS (1–4 feet), sends up its flowering stems from creeping branches. The basal leaves have blunt, heart-shaped blades on slender stalks; they are usually edged with blunt teeth. The leaves on the stem are much smaller, mostly without stalks, often pinnately cleft. The whole plant is smooth when mature. The flower-heads are golden-yellow, about an inch across.

April to August: in meadows, swamps, and moist woodland from Quebec to North Dakota and southward to Florida and Arkansas. *Plate 141.* A variable species; several varieties have been distinguished, differing in size and shape of leaves and other details. The young stems and leaves are often covered with loose tufts of wool.

S. ROBBINSII (1–3 feet) somewhat resembles *S. aureus*, but the blades of the long-stalked basal leaves are narrow and pointed – lanceolate or almost heart-shaped. The flowers are a paler yellow.

May to August: in moist meadows and woods from Quebec to New York and southward to North Carolina and Tennessee.

S. PLATTENSIS (4–16 inches) has one or more stems, permanently covered with a felt of hairs. The basal leaves have ovate, bluntly toothed blades on long stalks. The lower leaves on the stem may be pinnately cleft.

May to July: on dry bluffs and prairies from Vermont to Saskatchewan and southward to western Virginia, Ohio, Louisiana, Texas, and Arizona.

B. Species with the principal leaves, in a tuft at the base of the stem, with blades tapering down into the stalks.

S. SMALLII (8–30 inches) forms many stems in a cluster, each crowned by many flowers only about ½ inch across. The stems are woolly, at least in the lower part. The blades of the basal leaves are narrow, more or less elliptic, toothed, and taper gradually into the stalks. The leaves on the stem are mostly pinnately cleft.

May and June: in meadows and open woodland from New Jersey to Indiana and southward to Florida and Alabama. *Plate 141.* A very abundant species in the South.

S. OBOVATUS (6 inches–2 feet) has slender runners from which the flowering stems rise. The basal leaves have narrow blades widest above the middle and tapering downward to the stalks ("obovate"; i.e. ovate upside-down). The leaves may have some tufts of wool when young. The stem-leaves are pinnately cleft.

April to June: in rocky woods and on shaded banks from New Hampshire to southern Ontario and

PLATE 140

Solidago tenuifolia *Becker*

Solidago gymnospermoides *Johnson*

Chrysopsis mariana *V. Richard*

Chrysopsis graminifolia *Gottscho*

Chrysopsis falcata *V. Richard*

Solidago microcephala *Ryker*

Michigan and southward to Florida, Alabama, and Kansas. *Plate 141.*

S. PAUPERCULUS (2 inches–2 feet; the Latin name means "poor" or "small") often has tufts of white wool on its stems. The basal leaves are much like those of *S. obovatus* but smaller, rarely over 4 inches long. The leaves on the stem are narrow, pointed, and pinnately cleft or lobed. The flowers are deep yellow and handsome, nearly an inch across.

May to August: in meadows and bogs and among wet rocks from Labrador to Alaska and southward to Virginia, Ohio, Missouri, and British Columbia. Under this name several varieties are included.

S. INTEGERRIMUS is a northern species which extends to Minnesota and Iowa. Its basal and lower leaves are narrow, pointed, stalked; the upper leaves similar but smaller.

S. INDECORUS usually has no rays. Its basal leaves have narrow, toothed blades tapering to long stalks. The stem-leaves are pinnately cleft, the lobes sharply toothed or again cleft. This is a northern species which reaches Michigan.

S. TOMENTOSUS (8–30 inches) has several stems in a cluster, covered at least when young with white wool. The basal leaves have narrow, pointed, toothed blades tapering into long stalks. The leaves on the stem are few, much smaller, and without stalks. The flower-heads are bright yellow, nearly an inch across.

April to June: in dry open woodland mostly on the coastal plain from New Jersey to Florida and Texas, and in Arkansas. *Plate 142.*

S. ANTENNARIIFOLIUS (4–16 inches) has numerous basal leaves with small elliptic blades toothed towards the end and tapering to the stalk. Stem-leaves are few and narrow. Stem and leaves are white with wool.

April and June: in barren open places from Maryland to Pennsylvania, Virginia, and West Virginia. *Plate 141.*

II. *Species with the principal leaves borne along the flowering stems.*

GROUNDSEL, S. VULGARIS (4 inches–2 feet), has rather soft and thick leaves which are wavy-edged, coarsely toothed, or pinnately lobed or cleft. The stem and leaves are smooth or nearly so. The small flower-heads have no rays.

March to October: an Old-World plant now established as a weed throughout most of North America. *Plate 142.* The English name is from two Old-English words which mean "ground-swallower."

RAGWORT or STINKING-WILLIE, S. JACOBAEA (1–4 feet), has an unbranched stem which bears many pinnately divided or cleft leaves; the segments or lobes are themselves pinnately lobed or coarsely toothed. The young plant is covered with a thin wool but this disappears by the time it is mature. The flower-head bears yellow rays ¼ inch long or longer.

July to October: an Old-World weed now found from Newfoundland to New Jersey, on the Pacific coast, and probably elsewhere. *Plate 180.* In England, where it is very common, this is called ragweed as well as ragwort (that country being spared *our* ragweed, *Ambrosia*). The appearance of the leaves explains the name. The botanical name commemorates St. James, the plant being in full bloom (in Europe) on July 25, the day consecrated to that saint. Apparently this ragwort was much used by fairies and witches for transportation, in the way that broomsticks are said to have served.

BUTTERWEED, S. GLABELLUS (4 inches–3 feet), has a smooth, hollow stem bearing pinnately cleft or lobed leaves, the end lobe often much larger than those at the sides. The flower-heads are often numerous, with showy rays.

April to July: in wet woods and ditches from Ohio to South Dakota and southward to Florida and Texas. *Plate 142.*

MARSH-FLEABANE, S. CONGESTUS (6–24 inches) is a plant of the far north, some forms of which are found in Minnesota and Iowa. It has a hollow, hairy stem and narrow leaves with coarse teeth. The rays are pale yellow and short.

S. SYLVATICUS and S. VISCOSUS are similar to *S. vulgaris.* They are both hairy, the first slightly and the second markedly sticky; both have a fetid odor. The rays are short, curved, inconspicuous. These Old-World weeds are found in waste places near the northern Atlantic Coast.

ARNICA

The species of *Arnica* are characteristic of far northern regions and high mountains; there are many in the West. Only two species are likely to be found in the northeastern states, though one or two others touch our boundaries. They are known among other composites with yellow rays and pappus of hairs by their paired leaves and the lack of chaff.

PLATE 141

Senecio obovatus *Core*

Chrysopsis villosa *Johnson*

Senecio smallii *Johnson*

Senecio antennariifolius *Murray*

Heterotheca subaxillaris *Uttal*

Senecio aureus *Rickett*

To many persons "arnica" means a medicinal preparation made from *A. montana.*

A. MOLLIS grows from 2 inches to 2 feet tall, the hairy stem bearing from three to five pairs of ovate leaves; the lowest of these may be stalked, and the basal leaves have blades tapering into long stalks. The rays are ½ inch long or longer. The pappus is brownish.

July to September: along streams and on wet rocks from Quebec and New Brunswick to the mountains of New England and New York; and in the West. *Plate 142.*

A. ACAULIS has a hairy stem from 1 to 3 feet tall. The leaves are mostly at the base, those on the stem being much smaller. The basal leaves are hairy, ovate, with several veins running lengthwise and scarcely any stalks. The rays are about an inch long.

April to July: in sandy woodland from Pennsylvania and Delaware to Florida. *Plate 142.*

INULA

One species of this Old-World genus has become established in this country.

ELECAMPANE, I. HELENIUM, grows from a root deep underground, the stem from 3 to 5 feet tall. The leaves are large, toothed, mostly ovate, woolly on the under surface, the lower ones with stalks. The plant is easily recognized by its numerous very narrow yellow rays; the disk also is yellow.

May to September: on roadsides and in waste places from Quebec and Ontario southward. *Plate 143.* The species probably came originally from Asia (where there are other members of the genus), and was brought into cultivation in Europe in ancient times. It possessed many supposed medicinal and magical properties, and was, like other such plants, brought to America by the colonists. The name *Inula* is classical. In medieval times it became *Enula campana*, the "inula of the fields," which in turn was corrupted to elecampane. The name underwent many changes; Thackeray, for instance, spelled it alycompaine. The *helenium* of the botanical name is said to refer to Helen of Troy; but just what plant was called after her, and why, I do not know.

TUSSILAGO

There is only one species of *Tussilago*, a native of the Old World now widespread in the northeastern United States.

COLTSFOOT, T. FARFARA, has a horizontal underground stem (rhizome), which in spring sends up a stem that bears flowers but no green leaves; this grows from 2 to 20 inches tall. Numerous scales clothe its sides, and a single flower-head is formed at its tip. The numerous narrow rays and the disk are yellow. After the flowers disappear several leaves appear, with a large, round, indented and toothed blade – the "colt's foot" – on a long stalk.

March to June: in damp soil from Newfoundland to Minnesota and southward to New Jersey and West Virginia. *Plate 142. Farfarus* was a classical name for the plant; as was the name of the genus. The latter comes from the word for "cough." Coltsfoot tea, coltsfoot "rock" – a candy, and coltsfoot syrup are still used to relieve coughing and colds, and the leaves, with those of other herbs, form herbal tobacco, which is helpful in cases of asthma and (says Mrs. Grieve) "does not entail any of the injurious effects of ordinary tobacco."

THE TICKSEEDS AND BEGGAR-TICKS (COREOPSIS AND BIDENS)

The two genera *Coreopsis* and *Bidens* are quite easily distinguished from most other genera of the daisy family but rather hard to separate from each other. Both have two distinct circles of bracts around each head, the outer circle quite different from the inner; the outer bracts are generally green, like small leaves, the inner broader and thinner and often yellowish.* The seedlike fruits (achenes) of both genera are tipped with rigid spines (the pappus); these are barbed with minute hooks along their sides, generally pointing upwards in *Coreopsis* and downward in *Bidens* (but there are many exceptions). The achenes of *Coreopsis* are generally flat, with two spines, and a thin marginal flange or "wing." (*Coreopsis* is from Greek words meaning "buglike," referring to these achenes, which

* This characteristic is shared by some Mexican species of *Cosmos* widely cultivated and sometimes found growing wild; and by the western genus *Thelesperma*, species of which occur along our western boundaries.

PLATE 142

Arnica mollis *Clark*

Senecio tomentosus *Johnson*

Senecio vulgaris *Fischer*

Tussilago farfara *Rhein*

Senecio glabellus *Johnson*

Arnica acaulis *Johnson*

simulate small ticks or beetles.) The achenes of *Bidens* (which means "two teeth") may have from one to eight spines, but generally from two to four; the achene itself may be flat or square. Because of their backward-pointing barbs, the spines adhere to cloth or wool that they penetrate, and the achenes become the "beggar-ticks" which decorate the clothing of many a hiker in late summer. *Coreopsis* generally has eight showy rays, in some species toothed at the end; several find a place in gardens. Many species of *Bidens* have no rays and are unattractive weeds; but some are as handsome as the tickseeds and merit cultivation.

TICKSEEDS (COREOPSIS)

I. *Species with leaves generally undivided and unlobed except sometimes for a pair of small lobes or segments at the base of the blade.* (*Compare II.*)

C. LANCEOLATA is a smooth plant (there is a hairy variety) with most of the long narrow leaves growing from the base of the stems or from its lower few inches. The stems are from 8 to 30 inches tall. The flower-heads are borne on long leafless stalks.

May to July: in dry soil from New England to Michigan and Wisconsin and southward to Florida and New Mexico. *Plate 143.*

C. PUBESCENS has stems from 2 inches to 4 feet tall bearing ovate or lanceolate leaves, the lower stalked; the leaves are generally hairy. (Some may be divided, with a pair of small segments at the base of a much larger end segment.)

June to September: in dry open woods and grassy places from Virginia to Missouri and Oklahoma and southward to Florida and Louisiana; and occasionally farther north as an escape from cultivation. *Plate 144.*

C. AURICULATA is so named from the frequent presence of a pair of small lobes or segments at the base of the leaf-blades (these are the "little ears" or *auriculae*). The stems are from 6 to 24 inches tall, with the leaves mostly at and near the base.

April to June: in open woods from Virginia to Kentucky and southward to Florida and Louisiana. *Plate 143.*

C. ROSEA is distinguished by its pink or white rays. The leaves are very narrow (sometimes cleft). The plant is from 6 to 30 inches tall.

July to September: in wet soil from Nova Scotia to Georgia.

II. *Species with leaf-blades deeply cleft or divided pinnately or palmately.* (Caution: *a pair of leaves with no stalks and blades palmately divided can be mistaken for a circle of undivided leaves.*)

A. Of these, several species have rays not toothed at the end, or only inconspicuously so, and yellow throughout.

TALL TICKSEED, C. TRIPTERIS, grows up to 3 feet tall or taller. The leaves have blades divided palmately into three or five narrow segments on a stalk often about an inch long. The flower-heads are anise-scented.

July to September: in the borders of woods, fields, etc. from southern Ontario to Wisconsin and southward to Florida, Louisiana, and Kansas; and escaped from cultivation elsewhere. *Plate 143.*

C. PALMATA, from 2 to 3 feet tall, may be described as having leaves without stalks, the lower half of the blade (which might, however, be called the stalk) narrow, the upper half deeply cleft into three narrow lobes.

June and July: in prairies and open woods and on bluffs from Michigan to Manitoba and southward to Indiana and Texas. *Plate 143.*

C. VERTICILLATA has leaves palmately divided into very narrow, almost hairlike segments; the central segment may be again divided. Since there is commonly no stalk below the point of division, the segments may seem to be a circle of narrow undivided leaves (a "verticil"). The stem varies from 6 inches to 4 feet tall.

June to August: in dry woods from Maryland to Florida, Alabama, and Arkansas; like other species, escaped from cultivation farther north.

C. MAJOR is really no greater (*major*) than all other species, being from 2 to 3 feet tall; the stem is downy. The leaves are without stalks, the blades divided palmately into three lanceolate narrow segments; those of each pair seem to be six undivided leaves in a circle.

June and July: in open woodland from Pennsylvania and Ohio southward to Florida and Mississippi.

B. One species with divided leaves has rays evidently toothed at the end and colored red at the base.

C. TINCTORIA is a western native much cultivated and often running wild in the east. It is from 1 to 4 feet tall. The leaves are mostly pinnately divided into very

PLATE 143

Coreopsis palmata *Rickett*

Coreopsis palmata *Johnson*

Coreopsis tinctoria *Johnson*

Coreopsis tripteris *Gerard*

Coreopsis auriculata *Johnson*

Coreopsis lanceolata *Rickett*

Inula helenium *Johnson*

narrow segments, and these segments themselves sometimes pinnately cleft.

June to September: in moist ground from Minnesota to Washington and southward to Louisiana and California; elsewhere escaped from gardens. *Plate 143.*

Besides these species, several southerners may be found in Virginia and Missouri, and escaped from cultivation farther north: *C. heterogyna*, with narrow undivided and unlobed leaves; *C. onisci-carpa*, with long-stalked leaves widest near the tip; *C. grandiflora*, with leaves deeply cleft pinnately into very narrow lobes; *C. delphinifolia*, with leaves divided as in *C. major*.

BEGGAR-TICKS (BIDENS)

The beggar-ticks may be grouped by the presence or absence of rays and by the division or lobing of leaves.

I. *Species with rays (with* radiate *heads). (In several of these the rays are* occasionally *lacking.) About half of these have undivided leaves.*

A. Species with rays and undivided leaves.

B. CERNUA is from 4 inches to 6 feet tall. The leaves vary from narrowly lanceolate to ovate, with toothed or plain margins. The flower-heads are the largest of our species of *Bidens* except for those of the following species which equal or exceed them; the disk is from ½ to 1 inch across and the six or eight yellow rays up to ⅝ inch long (they may be lacking). The heads are generally directed sideways by a bend in the stalk, or they may hang slightly downward (they are "cernuous").

August to October: in wet places throughout southern Canada and southward to North Carolina and California; also in the Old World. *Plate 144.*

B. LAEVIS closely resembles *B. cernua*, growing from 1 to 3 feet tall. It holds its heads erect. The disk is from ½ to 1 inch wide, the yellow rays from ⅝ inch to more than an inch long.

August to November: in wet places from New Hampshire mostly near the coast, to Florida; inland to West Virginia and Indiana; also on the Pacific Coast and in South America. *Plate 144.*

B. HYPERBOREA is a Canadian species which extends southward to Massachusetts. The flower-heads often lack rays; when they are present they are pale yellow and less than ½ inch long. The leaves are narrow, with or without stalks.

B. Species with rays and divided (or deeply cleft) leaves.

TICKSEED-SUNFLOWER, B. POLYLEPIS, is a plant from 1 to 5 feet tall, with stalked leaves, the blades divided pinnately into narrow toothed segments, and these often themselves pinnately lobed. The flower-heads have a disk up to ½ inch wide and rays up to an inch long.

August to October: in fields and prairies, along depressions in which water may run, from Indiana to Colorado and southward to Tennessee and Texas; occasionally found farther to the east. When I first saw Missouri the fields were covered with the gold of this and the following species. They are used by bees to make a dark and pungent honey, much esteemed by connoisseurs. Both species are known to the country folk as Spanish needles, though this is a book-name for *B. bipinnata*.

B. ARISTOSA closely resembles *B. polylepis*, the difference being chiefly in the bracts. The outer bracts of *B. aristosa* are fewer, not more than ten, and shorter, only up to ½ inch long. It grows in low fields from Delaware to Minnesota and southward to Virginia, Alabama, and Texas, and occasionally farther to the northeast.

B. CORONATA grows up to 5 feet tall. The leaves are pinnately divided into from three to seven narrow, sharply toothed segments. The distinction between this and the two preceding species is chiefly in the achenes; those of *B. coronata* are narrow, with straight sides, and lack the narrow margins of the other species.

August to October: in prairies, bogs, etc. from Massachusetts to Maine and southward to Virginia, Kentucky, and Nebraska. *Plate 144.*

B. MITIS resembles *B. coronata;* the leaves are pinnately divided and the segments themselves often pinnately divided into hairlike narrow segments. The end segment is often the broadest.

July to October: in swamps near the coast from Maryland to Florida and Texas.

C. One species grows in water.

WATER-MARIGOLD, B. BECKII, has some leaves under water, others, with the single flower-head, above the surface. The under-water leaves are masses of hairlike segments. The aerial leaves are undivided but sharply cleft, lobed, or toothed. The flower-head is handsome, with yellow rays ½ inch long.

July to October: in ponds and sluggish streams from Quebec to Manitoba and southward to New Jersey, Pennsylvania, Indiana, and Missouri. *Plate*

PLATE 144

Coreopsis pubescens 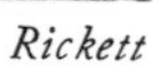 *Rickett*

Bidens connata *Johnson*

Bidens coronata *Johnson*

Bidens cernua *Johnson*

Bidens laevis *Gottscho*

145. Because the achene is not flat and because it may have more than four spines, this species is by some botanists placed in a separate genus, *Megalodonta* ("big-tooth").

II. *Species regularly without rays (but see also* B. cernua, B. hyperborea *in group I, which sometimes lack rays). These are the weedy, mostly unattractive species, to which the name beggar-ticks seems most appropriate. They grow usually in damp places and waste land through most of the United States, flowering in late summer and autumn.*

SPANISH NEEDLES, B. BIPINNATA, is the easiest of this group to recognize – and the easiest to acquire when one is out walking. The leaves are "bipinnate" – the segments again divided or cleft pinnately. The achenes are the "Spanish needles," four-angled, ½ inch long or longer, very narrow, with from two to four spines at the end. They form an efficient means of dispersing the species.

B. CONNATA may have leaves deeply cleft into three parts but scarcely divided, or not at all lobed; all being coarsely toothed. There are from two to six outer bracts like small leaves. The achenes have from two to five spines which have barbs pointing up or down. (See the note under *B. comosa.*) *Plate 144.*

B. COMOSA has leaves mostly unlobed and undivided, coarsely toothed. The outer bracts (from six to ten) are like narrow leaves rising round the flower-head and overtopping it. The achenes are flat with usually three spines with down-pointing barbs. *Plate 145.*

This and the preceding species are very variable and often difficult to identify. Perhaps they interbreed. One botanist assigns them both to the European *B. tripartita* (which is occasionally found in this country).

B. FRONDOSA has pinnately divided leaves, the end segment on a long stalk. There are usually about eight outer bracts, often unequal in size, all more or less leaflike. Rays are sometimes present. *Plate 145.*

B. VULGATA has pinnately divided leaves, the segments sometimes cleft. The outer bracts are numerous (more than ten), unequal, as long as the disk which they surround, and edged with short stiff hairs.

B. DISCOIDEA is a slim plant with a red stem bearing leaves divided into three. The outer bracts are few – from two to five – narrow and smooth.

B. BIDENTOIDES has an undivided leaf-blade tapering down into a long stalk, toothed and sometimes jagged. There are from three to five narrow outer bracts about 2 inches long. The achenes are long and narrow and two-horned.

ACTINOMERIS

We have a single species of *Actinomeris.*

WINGSTEM, A. ALTERNIFOLIA, is a tall plant, up to 6 feet or even more, with leaves mostly borne singly and "running down" (decurrent) on the stem; that is, the edges of a leaf are continuous with two narrow flanges or "wings" extending vertically below it. The flower-head has a few yellow rays of unequal size (this is the meaning of the name of the genus), extending out and down; and the disk-flowers point in all directions so that the general effect is that of a mop. The seedlike fruits, achenes, are flat, usually margined, with two (or three) spines (the pappus) at the end.

August and September: on roadsides and in waste places with other weeds, and in woodland, from New York to Iowa and southward to Florida and Louisiana. *Plate 145.* By some botanists this is placed in the following genus, *Verbesina;* but its appearance is sufficiently distinct to make identification easier if it is kept in a genus to itself.

CROWNBEARDS OR WINGSTEMS (VERBESINA)

The crownbeards have mostly yellow flowers (white in one of our species) and flat, margined achenes with commonly two spines (pappus) like those of *Actinomeris* (but the bracts and rays do not direct themselves downward; the disk is convex or conical but not globe-like). In most of our species the edges of the leaves "run down" on the stem, again as in *Actinomeris;* in all but one species they are borne singly.

TICKSEED or FROSTWEED, V. VIRGINICA, is distinguished by its white rays. The plant may reach a height of 6 feet or more. The leaves are ovate and may be toothed.

August to October: in dry woods from Pennsylvania to Kansas and southward to Florida and Texas.

V. HELIANTHOIDES has only a few heads, rarely more than ten. Stem and leaves are hairy. It rarely exceeds 3 feet in height.

June to October: in dry woods and prairies from

PLATE 145

Bidens beckii *Voss*

Helianthus annuus *Johnson*

Bidens comosa *Murray*

Bidens frondosa *Rickett*

Verbesina occidentalis *Uttal*

Actinomeris alternifolia *Rickett*

Verbesina helianthoides *Johnson*

Ohio to Iowa and Kansas and southward to Georgia and Texas. *Plate 145.*

V. OCCIDENTALIS is alone among our species in having paired leaves; they are ovate and toothed. It grows from 3 to 7 feet tall. The achenes have no marginal "wing."

August to October: in woodland from Pennsylvania to Missouri and southward to Florida and Texas. *Plate 145.*

V. ENCELIOIDES is from 8 to 30 inches tall. The leaves are ovate, toothed, and sometimes (especially in the South) have a pair of lobes or larger teeth at the base. The rays are three-toothed at the end.

June to October: in open ground from Kansas to Arizona, Texas, and Mexico, and occasionally straying eastward to New England and Missouri.

THE SUNFLOWERS (HELIANTHUS)

Most persons know the tall plants with immense flower-heads which are grown from seed or spring up as volunteers and from which one obtains "sunflower seed" for feeding birds and for other purposes. Many tourists also know the smaller sunflowers that line the roads in much of the Midwest; they are the state flower of Kansas. It may come as a surprise to these and other persons that there are twenty or twenty-five species (depending on what one calls a species) growing wild in the northeastern United States, most of them native. They all have long yellow (or occasionally reddish) rays and yellow or brown-purple disks. The pappus at the base of the corolla of each flower is a pair of scales (in some species toothed) or a pair of bristles. The disk on which the disk-flowers stand is flat. Alongside each flower is a thin bract, constituting the "chaff." The bracts are green and leaflike, in many overlapping rows. The leaves may be borne singly or paired, stalked or stalkless; they are mostly ovate or lanceolate, and generally have a pair of veins branching from the midrib at the base of the blade. Many of them are rough-hairy plants. All are handsome and hardy, worthy of more attention from gardeners.

Identification of some species is difficult. We group them first by the color of the disk.

I. *Species with dark-colored (brown or purplish, not yellow) disk. The first two are annual, growing each year from seed.*

COMMON SUNFLOWER, H. ANNUUS, is the species mentioned above as the most familiar one. It may be of any height from 1 to 10 feet or even taller. The leaves are mostly single, stalked, with broad, ovate blades. The whole plant is rough with bristly hairs. The bracts are long-pointed and edged with bristles. The "seed" is, of course, the seedlike fruit or achene; a single true seed is within each achene.

July to November: native in prairies from Minnesota and Missouri westward to the Pacific, but quite common in dry places throughout our area. *Plate 145.* The heads vary greatly in diameter, the disk often reaching an enormous size in cultivation and the head becoming too heavy for its stalk.

H. PETIOLARIS resembles *H. annuus* and reaches a similar height but is slimmer, with narrower and often toothless leaf-blades. The bracts are not conspicuously edged with bristles.

June to October: in prairies and sandy open places from Minnesota westward to Washington and southward to Louisiana and California. *Plate 146.*

H. ATRORUBENS bears its principal leaves in pairs near the base of the stem; the upper leaves are much smaller. The stem is hairy, from 2 to 7 feet tall. The larger leaves have broad flat stalks. The flowers are on long stems.

July to October: in dry woodlands from Virginia to Missouri and southward to Florida and Louisiana. *Plate 180.*

H. LAETIFLORUS has several varieties, some with dark disk and some with yellow disk. They grow from 6 inches to 8 feet tall, with leaves all the way up to the long flower-stalks. The leaves are mostly paired, with gray-green, lanceolate or ovate blades tapering to a very short stalk and with or without teeth. Stem and leaves are rough.

August to October: in open woodland and prairies from Quebec to Saskatchewan and southward to Georgia and Texas; native in the western half of this range, but appearing fairly frequently to the east. *Plate 146.* The varieties with dark disk are generally western and southwestern; the yellow disk is seen throughout the range.

H. ANGUSTIFOLIUS has, as the name indicates narrow leaves, often very narrow; they are mostly borne singly. The stem is from 2 to 6 feet tall. Both stem and leaves are rough.

August to October: in wet soil, bogs, etc. near the coast from Long Island to Florida and Texas and inland from New Jersey to Missouri and Kentucky.

PLATE 146

Helianthus decapetalus *Rickett*

Helianthus grosseserratus *Johnson*

Helianthus divaricatus *Rickett*

Helianthus petiolaris *Johnson*

Helianthus laetiflorus *Johnson*

H. SILPHIOIDES (to 10 feet) has broad, often almost round leaf-blades on slender stalks, mostly in pairs, extending up the whole height of the stem. Stem and leaves are rough.

August and September: in woodlands from Indiana to Minnesota and southward to Alabama and Louisiana.

Besides these, two other species with dark disks may be found in our range. *H. debilis* is southern and sometimes escapes from cultivation; it has weak (*debilis*) stems and triangular leaf-blades. *H. salicifolius* is a western species found in Missouri; it is smooth, with numerous narrow leaves.

II. *Species with yellow disk. These are all perennial, growing from underground stems (rhizomes, tubers). For convenience they are here grouped by the hairiness or roughness or smoothness of their stems.*

A. Species with yellow disk and smooth stems. (Compare B.) The first two of these have leaves also generally smooth, more or less three-ribbed, and mainly paired. The last three have leaves rough on the upper surface.

H. LAEVIGATUS (3–7 feet) has lanceolate blades on a short stalk or sometimes with no stalk. They may be plain-edged or toothed. The disk is from ⅖ to ½ inch across, the rays (from five to ten) ½ inch long or somewhat longer.

August to October: in woods in the mountains from Pennsylvania and West Virginia southward to North Carolina.

H. DECAPETALUS (2–5 feet) has mostly paired, lanceolate or ovate leaves on a fairly distinct stalk, generally sharply toothed, and sometimes slightly rough on the upper surface. The disk is about ½ inch wide, the rays (from eight to fifteen) an inch or more long.

August to October: in woods and along streams from Maine to Minnesota and southward to North Carolina, Kentucky, and Missouri. *Plate 146*. The name means "with ten petals," referring to the rays.

H. GROSSESERRATUS (3–15 feet) has lanceolate leaf-blades on short stalks, with coarse teeth (*grosse-serratus*) or almost none; they are mostly singly borne. The disk is from ⅗ to 1 inch wide, the rays (from ten to twenty) from an inch to nearly 2 inches long.

July to October: in damp prairies from Ohio to Saskatchewan and southward to Arkansas and Texas, and escaped from cultivation farther to the east. *Plate 146*. *H. kellermani*, found in Ohio and Wisconsin, is said to be a hybrid between *H. grosseserratus* and *H. salicifolius*.

H. MICROCEPHALUS (3–6 feet) has paired leaves with lanceolate blades on short stalks, tapering gradually to the tip, more or less toothed. The disk is only about ¼ inch broad, the rays (from five to eight) from ⅖ to ⅗ inch long (the name means "with small heads").

August and September: in woods from Pennsylvania to Missouri and southward to Florida, Mississippi, and Louisiana.

H. DIVARICATUS (2–5 feet) has lanceolate leaves mostly without stalks and with broad base. The disk is about ½ inch wide, the rays (from eight to fifteen) from ⅗ to over 1 inch long.

July to October: in dry woodland from Maine to Saskatchewan and southward to Florida, Louisiana, and Nebraska. *Plate 146*.

B. Species with yellow disk and stems rough or downy or hairy – not smooth.

1. Of these, three species have leaves without stalks or with very short stalks. Here we may place the forms of *H. laetiflorus* that have a yellow disk. See page 454.

H. HIRSUTUS has paired, lanceolate leaves. Except for the hairs on the stem it resembles *H. divaricatus*. The disk is about ⅘ inch wide, the rays (from ten to fifteen) from ⅗ to 1⅖ inches long.

July to October: in dry woodland from Pennsylvania to Minnesota and Nebraska and southward to Florida and Texas. *Plate 147*.

H. MOLLIS (2–3 feet) is distinguished by the thick white coating of hairs on the stem (*mollis* means "soft"). The leaves are mostly paired and rather broad, without stalks, often indented at the base, rather rough on the upper surface and hairy on the lower. The disk is about an inch wide, the rays (from fifteen to thirty) from ⅗ to 1⅖ inches long.

July to October: in dry open places from Ohio to Michigan and Iowa and southward to Georgia and Texas, and occasionally found in the Atlantic states.

H. GIGANTEUS earns its name by growing up to 10 feet tall. The leaves are mostly borne singly; they are rather narrow, toothed or plain, with short stalks or none. The numerous heads are small for the size of the plant, the disk an inch wide or less, the rays (from ten to twenty) from ⅗ to 1 inch long.

July to October: in damp thickets, roadside ditches, etc. from Maine to Alberta and southward to Florida and Colorado. *Plate 147*. *H. tomentosus*, with larger heads, is found in southeastern Virginia.

2. The remaining species with yellow disk and stems not smooth have distinctly stalked leaves.

JERUSALEM-ARTICHOKE, H. TUBEROSUS (3–10 feet), has nothing to do with Palestine. The English name is a corruption of the Italian *girasole*, mean-

PLATE 147

Helianthus tuberosus *Johnson*

Helianthus hirsutus *Johnson*

Helianthus maximiliani *Johnson*

Helianthus strumosus *Johnson*

Helianthus giganteus *Johnson*

ing "turning to the sun" – as all the sunflowers are reputed to do. The botanical name refers to the large edible tubers, for which it was cultivated by Indians before the advent of the white man. It is a tall rough plant with broad, ovate leaf-blades on distinct stalks, mostly borne singly. The disk is often an inch across, and the rays (from ten to twenty) up to 1⅗ inches long.

August to October: in waste land and damp places throughout our range and westward to Saskatchewan, southward to Georgia and Arkansas. *Plate 147*.

H. STRUMOSUS (3–6 feet) is named for the swollen bases of its stiff, pointed hairs, which give the plant a very harsh feeling. The leaves are mostly paired, with lanceolate or ovate blades on short stalks. The disk is less than an inch across, the rays (from eight to fifteen) from ⅗ to 1⅗ inches long.

July to September: in woodland from Maine to North Dakota and southward to Florida and Texas. *Plate 147*.

H. TRACHELIFOLIUS has rather smooth leaves, with coarsely toothed blades on stalks. Otherwise it is like *H. strumosus* and by some is considered a form of that species.

H. DORONICOIDES (2–5 feet) has lanceolate, rough, paired leaves with very short stalks and few or no teeth. The disk is about ½ inch wide, the rays (from thirteen to eighteen) about an inch long.

July to October: in dry woodland from New Jersey to Minnesota and southward to Pennsylvania, Ohio, Missouri, and Texas.

H. MAXIMILIANI (2–10 feet) has rough, lanceolate, stalked leaf-blades, mostly borne singly. The disk is up to 1 inch wide, the rays (from ten to twenty-five) from ⅗ to 1⅗ inches long.

June to October: in prairies from Minnesota to Saskatchewan and southward to Missouri and Texas; and occasionally in the Atlantic states. *Plate 147*.

H. OCCIDENTALIS (18 inches–5 feet) has a stem hairy near the base. The leaves are in several pairs near the base of the stem, their ovate blades mostly without teeth and tapering to the stalk. The disk is about ½ inch wide, the rays (from ten to twenty) from ½ to 1 inch long.

July to October: in dry woods from Ohio to Minnesota and southward to Florida and Texas. *Plate 148*.

CHRYSOGONUM

There is a single species of *Chrysogonum*, apparently with no English name.

C. VIRGINIANUM is noticeable among the composites with yellow rays for a small number of rays: about five. The plants are from 2 to 20 inches tall. The leaves are paired, with small, ovate, scalloped blades on rather long stalks. The flower-heads are on long stalks. The disk is less than ½ inch wide, the rays up to ⅗ inch long.

April to July: in woods from Pennsylvania and Ohio southward to Florida and Louisiana. *Plate 148*.

THE GUMWEEDS (GRINDELIA)

The gumweeds are western plants, one species having strayed into the northeastern states and a second just touching our western boundary. They get their name from the sticky resin that they exude, in these two species from the bracts; it is clearly visible before the flower-heads open. The leaves are narrow, toothed, without stalks, borne singly.

GUMWEED or TARWEED, G. SQUARROSA, is from 1 to 5 feet tall. The leaves are marked with translucent dots; their teeth are blunt. The disk is about ½ inch wide and the numerous rays ½ inch long or shorter. The pappus is from two to eight spines. The bracts are sharply bent outward and downward (they are "squarrose").

July to September: in prairies from Minnesota to British Columbia and southward to Nevada and Texas, and common in waste land from New York to Pennsylvania and westward. *Plate 148*.

G. LANCEOLATA is less resinous and the leaves have few or no dots. The teeth of the leaves end in sharp bristles. The flower-head resembles that of *G. squarrosa* but the bracts merely spread apart loosely, they are not bent down.

August to October: in dry prairies and barren places from Missouri and Tennessee to Kansas and southward to Alabama and Texas.

PLATE 148

Rudbeckia serotina *Johnson*

Rudbeckia hirta *Rickett*

Helianthus occidentalis *Johnson*

Chrysogonum virginianum *Williamson*

Grindelia squarrosa *Johnson*

Chrysogonum virginianum *Johnson*

DYSSODIA

We have one species of this southern genus.

FETID-MARIGOLD, D. PAPPOSA, is a small, ill-smelling plant from 2 to 20 inches tall. The leaves are paired, each pinnately cleft or almost divided into narrow, sometimes threadlike lobes. The flower-heads are numerous, with a disk about ⅖ inch wide and a few very inconspicuous yellow rays scarcely longer than the bracts around them. The pappus is a circle of scales which end in long bristles.

July to October: on dry open soil from southern Ontario to Montana and southward to Louisiana, Mexico, and Arizona. The smell comes from the large translucent glands with which the herbage is dotted.

CONEFLOWERS (RUDBECKIA)

The genus *Rudbeckia* furnishes several of our showiest and most abundant weeds of late summer. They have yellow or orange rays and brown or yellow disk. The English name describes the shape of the disk, which forms a hemisphere or cone. There is practically no pappus.

The leaves are various and the species may be grouped by their division or lobing.

I. *Species with leaves not lobed or divided.*

BLACK-EYED- (or BROWN-EYED-) SUSAN, R. HIRTA, is familiar nearly throughout our range. It is from 1 to 3 feet tall or taller, a rough-hairy plant. The leaves are more or less ovate, those at the base of the stem tapering down into stalks, and coarsely toothed.

June to October: in open woodland and fields and on roadsides from Massachusetts to Illinois and southward to Georgia and Alabama. *Plate 148.* See the note under the following species.

BLACK-EYED-SUSAN or NIGGER-HEADS, R. SEROTINA, resembles *R. hirta*, the two species being often confused. Its leaves are not much more than an inch wide and rarely toothed. (There are also several more technical differences.) The rays are commonly well over an inch long.

June to October: in fields and on roadsides, throughout our range but native farther to the west. *Plate 148.* Of this species Fernald wrote, "a most aggressive weedy species chiefly originating west of our range and . . . rapidly spreading eastward." Like most plants that invade new territory, it is very variable. Some botanists seem disposed to deny its existence, lumping all the plants under the name *R. hirta*.

R. FULGIDA has a rather hairy stem from 1 to 4 feet tall, bearing leaves with rough, elliptic blades on long stalks; several veins commonly run lengthwise in the blade. The species differs from *R. serotina* chiefly in the presence of a minute pappus, a toothed ridge at the base of the corolla and crowning the achene. The scales among the disk-flowers (the chaff) also differ, being blunt at the end instead of tipped with a bristle.

July to October: in various situations, dry and moist, from New Jersey to Indiana and southward to North Carolina and Kentucky. Associated with this species are at least six others which can be distinguished only by minute technical characteristics. These range from New York and Michigan to Missouri and southward to Florida, Alabama, and Texas. Some have broad, toothed leaf-blades with pinnate veining; others have very narrow, toothless leaves.

R. BICOLOR is a roughly hairy plant from 1 to 3 feet tall, with short leaves which have no stalks or teeth. The rays may have brown or purplish bases.

June to September: in dry soil from Indiana to Missouri and southward to Alabama and Texas.

R. maxima is a southwestern species found in Missouri. It is distinguished chiefly by the high, cylindric "disk," and by being smooth and whitish. *R. heliopsidis* is southern, extending northward to Virginia. The leaves have very long stalks. *R. grandiflora* is another westerner found in Missouri. The disk forms a tall cone. The leaves have rough, three- or five-ribbed blades on long stalks. *R. amplexicaulis* is smooth, with leaves that "clasp" the stem and rays toothed and often brown at the base; it is southern and western, extending to Missouri. By some botanists it is placed in a separate genus as *Dracopis amplexicaulis*.

II. *Species with some or all leaves divided, cleft, or lobed.*

R. TRILOBA is usually easily recognized by the presence of some leaves cleft palmately into three (or sometimes five or more) lobes. Most of the leaves are ovate, those at the base with heart-shaped blades on long stalks; careful examination will reveal some of the cleft leaves. The rays are comparatively rather short, generally less than an inch, and orange.

June to October: in open woods and fields from New York to Minnesota and southward to Florida, Louisiana, and Oklahoma. *Plate 149.*

SWEET CONEFLOWER, R. SUBTOMENTOSA, grows from 2 to 5 feet tall, with a downy stem. Many of the leaves (sometimes all) have blades deeply cleft into

PLATE 149

Ratibida columnifera *Rickett*

Rudbeckia triloba *Johnson*

Rudbeckia laciniata 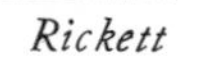 *Rickett*

Helenium nudiflorum *Gottscho*

Rudbeckia laciniata *Johnson*

Ratibida pinnata *Johnson*

three; they are downy also. The heads have the odor of anise. The rays are long, generally more than an inch.

July to September: in prairies and low fields from Indiana to Iowa and southward to Louisiana and Texas.

CONEFLOWER, R. LACINIATA, is generally a tall, coarse plant, from 2 to 10 feet. The stem and leaves are smooth. The leaves are irregularly and jaggedly divided, cleft, lobed, and toothed – "incised" is the botanical term. The yellow rays turn downward, exposing the conical disk which is also yellow (or greenish).

July to September: in wet places, roadside ditches, etc. from Quebec to Montana and southward to Florida, Louisiana, Texas, and Arizona. *Plate 149.*

PRAIRIE CONEFLOWERS (RATIBIDA)

The genus *Ratibida* is distinguished from other coneflowers largely by the cylindric receptacle on which the central flowers are borne – we must, to be consistent, still call this a disk and its flowers disk-flowers though it has no resemblance to our usual idea of a disk. The rays are yellow and tend to hang down around the "disk." We have two species; the others are southwestern.

R. PINNATA stands from 2 to 5 feet tall. The leaves are pinnately divided into from three to seven narrow, lanceolate segments. The "disk" is a brown cylinder about half as thick as it is high. There is no pappus.

June to September: in prairies and dry woods and on roadsides from southern Ontario to Minnesota and South Dakota and southward to Georgia, Arkansas, and Oklahoma. *Plate 149.* The heads when bruised are said to have the odor of anise.

R. COLUMNIFERA has its name from its "disk" which may be nearly 2 inches long and four times as tall as it is thick. The rays are only about an inch long; they may be yellow or purple. The plants are from 1 to 4 feet tall, with leaves pinnately divided into from five to nine narrow segments.

June to September: in prairies, etc. from Minnesota to Alberta and southward to Illinois, Arkansas, and Texas; and occasionally straying farther to the east. *Plate 149.*

SNEEZEWEEDS (HELENIUM)

The sneezeweeds are characterized by broad, fan-shaped rays with three or more lobes at the end; they are usually yellow, and generally directed downward, the dome of disk-flowers being conspicuous. The leaves are narrow, borne singly, and generally "decurrent," that is, their edges are continued down the stem as narrow projecting flanges or "wings." The pappus is several small scales, often tipped with a bristle. These are common in fields and prairies and on roadsides, mostly flowering in late summer and early autumn (from June or July to September or October).

H. NUDIFLORUM (8–40 inches) has a brown or purplish disk. The rays are sometimes purple-tinged at the base. In moist soil and waste land from New England to Michigan and southward to Florida and Texas. *Plate 149.*

H. AUTUMNALE (1–5 feet) is distinguished from *H. nudiflorum* by its yellow disk. In moist ground, swamps, waste places, etc. practically throughout the United States and southern Canada. *Plate 150.*

H. AMARUM (6 inches–2 feet) differs markedly from the two preceding species in its leaves; they are narrow, almost threadlike, very numerous and crowded. On waste ground and roadsides from New England to Michigan and southward to Florida, Texas, and Mexico; most abundant southward but rapidly spreading northward. *Plate 150.*

Several other species are found just within our southern borders: *H. brevifolium*, with few leaves and long-stalked heads with brown disks; *H. virginicum*, like *H. autumnale*, but with leaves mostly at the base of the stem; *H. curtisii*, a bog plant like *H. brevifolium*.

HELIOPSIS

We have one species of *Heliopsis*.

OX-EYE, H. HELIANTHOIDES, has a smooth stem, from 1 to 5 feet tall, bearing pairs of stalked leaves with ovate, toothed blades. In general aspect it resembles a sunflower (which is what *helianthoides*

PLATE 150

Helenium autumnale *Johnson*

Heliopsis helianthoides *Johnson*

Hymenoxys acaulis *D. Richards*

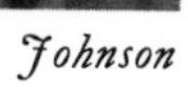

Silphium perfoliatum *Johnson*

Helenium amarum *Johnson*

means), but is distinguished by the conical shape of the "disk,"–the mass of disk-flowers. There is generally no pappus.

June to October: in open woods and prairies from Quebec to British Columbia and southward to Georgia and New Mexico. *Plate 150.*

HYMENOXYS

One species of *Hymenoxys* strays into our range from the West.

H. ACAULIS is a low plant (scarcely a foot tall) with a tuft of narrow leaves at the ground level, and a single flower-head on each leafless erect stem. *Acaulis* means "without a stem," since the only *visible* stem is the flower-stalk. There is a short, branching stem just at the ground level, which forms the leaves and flower-stalks. The rays are yellow, three-toothed at the end, up to $\frac{4}{5}$ inch long. The pappus consists of small scales.

May to July: in prairies and other dry, open places from Ontario and Ohio to Idaho and Arizona. *Plate 150.*

ROSINWEEDS (SILPHIUM)

The genus *Silphium* includes a number of species with very different types of leaves, but readily recognizable as all related by their flower-heads. The peculiarity of these is that only the rays are fertile, forming several rows of broad flat achenes ("seeds") around the disk (which yields no fruit). The bracts are loose and leaflike. The achenes are flat, "winged" (i.e. with a thin marginal flange), in some species crowned with the two teeth of the pappus; in other species there is no pappus. These are tall, coarse plants with much resin ("rosin") and large yellow flower-heads.

I. *Species with leaves borne on the flowering stems.*

A. In one species the leaves, which are paired, meet around the stem, forming a cup.

CUP-PLANT, INDIAN-CUP, or CARPENTER-WEED, S. PERFOLIATUM, has a four-angled stem from 3 to 8 feet tall. The leaf-blades are rough, ovate, coarsely toothed. The lower leaves are stalked. The numerous rays are generally an inch long or longer.

July to September: in woodland and waste land and on moist prairies, roadsides, and river-banks from Ontario to South Dakota and southward to Georgia and Oklahoma; occasionally straying farther eastward. *Plate 150.* A related species of the southeastern mountains, *S. connatum*, differs in having a hairy stem.

B. In the remaining species of group I the leaves are variously disposed but not *united* in pairs.

S. INTEGRIFOLIUM (2–6 feet) has very rough, paired leaves without stalks and without teeth (a plain-edged leaf is, in medieval Latin, *integer*, from which the modern botanist has derived "entire").

July to September: in prairies and woodland and on roadsides from Ohio to Minnesota and Nebraska and southward to Alabama and Oklahoma. *Plate 151.*

S. BRACHIATUM, a very smooth-stemmed southern species with paired, rough, ovate, toothed leaf-blades on stalks, is found in Kentucky.

S. ASTERISCUS resembles *S. integrifolium* but the leaves are sharply toothed and mostly borne singly.

June to September: in dry woodland and prairies from Indiana and Missouri southward to Florida and Texas. *Plate 151.* In Missouri the southern *S. asperrimum* is found; a very rough plant with paired leaves. Several other southern and western species are found in our range.

S. TRIFOLIATUM (2–10 feet) generally has leaves in circles of three, four, or five (the name means "three-leaved"); but they may also be single or paired. The blades tend to be lanceolate and sharply toothed, on short stalks.

June to September: in woodland from Pennsylvania to Indiana and southward to Georgia and Alabama. The leaves are generally rough, but some southern plants have smooth leaves. *S. atropurpureum*, differing in having hairy leaf-stalks and in certain characteristics of the achenes, occurs in Virginia and West Virginia.

COMPASS-PLANT, S. LACINIATUM (3–12 feet), has large pinnately cleft leaves scattered up the often very tall stem; their lobes are narrow and pointed and generally toothed or even themselves lobed. The whole plant is bristly and rough.

July to September: in prairies and on open roadsides from Ohio to Minnesota and North Dakota and southward to Alabama and Texas. *Plate 151.* The leaves tend to stand in a vertical plane with their edges north and south; hence the English name. The Latin name describes the "laciniate" or cut leaves.

PLATE 151

Silphium integrifolium *Johnson*

Silphium laciniatum *Rickett*

Silphium laciniatum *Johnson*

Silphium terebinthinaceum *Johnson*

Silphium asteriscus *Johnson*

II. *Other species have all their leaves at or near the base of the stem.*

PRAIRIE-DOCK, S. TEREBINTHINACEUM (2–10 feet), has a hairy or bristly stem bearing several flower-heads at the summit. The leaves have ovate or heart-shaped blades, coarsely toothed, thick and rough, on slender stalks.

July to October: in prairies and fields and on open roadsides from southern Ontario to Minnesota and southward to Georgia, Louisiana, and Missouri. *Plate 151*. This is an imposing plant on the open prairie, with its 8- or 10-foot stalks crowned with large, sunflower-like heads.

S. compositum is similar to *S. terebinthinaceum*, with variously lobed or cleft leaf-blades. It is a southern species, extending northward to Virginia.

LEAFCUPS (POLYMNIA)

The leafcups are tall, sticky-hairy plants with broad, lobed or cleft leaves; they are rank-smelling. The flower-heads are small for such large plants; the rays are pale yellow or almost white, in one species often minute or lacking. The disk is yellow. There is no pappus.

P. CANADENSIS is from 2 to 5 feet tall. The leaves may be a foot long, the blade ovate in general outline, and variously lobed or cleft pinnately as well as toothed; stalked. The rays are whitish, minute or lacking or sometimes up to $\frac{2}{5}$ inch long.

June to October: in damp woods from Vermont and Ontario to Minnesota and southward to Georgia and Oklahoma. *Plate 152*.

BEARSFOOT, P. UVEDALIA, is from 3 to 10 feet tall. The leaves are sometimes more than a foot long, the blades ovate and irregularly angular and more or less palmately cleft or lobed; those of the lower leaves taper into a broad stalk. The rays are yellow, up to $\frac{4}{5}$ inch long.

July to September: in woods and meadows from New York to Missouri and southward to Florida and Texas. *Plate 152*.

MADIA

One species of this western genus has become a roadside and barnyard weed from Quebec to Indiana and Delaware, flowering all summer and into autumn.

TARWEED, M. SATIVA, is from less than 1 to 5 feet tall, sticky-hairy and rank-smelling. The leaves are narrow, borne singly. The flower-heads are generally clustered at and near the tips of the branches. The rays are honey-colored and very short.

THE WILD ASTERS (ASTER)

The remarks made in the introduction to the goldenrods (*Solidago*) apply equally to the asters; this is no genus for the amateur except he be unusually ambitious and painstaking. However, it is possible (a) to place all our species in a few groups by easily seen characteristics; and (b) to characterize at least some species so that they may be quickly and easily recognized. Certain groups, however, are almost impossible even for the botanist. The number of species in our range is nearly as great as that of the goldenrods. The species perhaps do not interbreed quite so freely and hence are a little more sharply separated. But the separation often depends on what seem trivial characteristics – as the exact shape, texture, hairiness of the bracts around the flower-head – and one must be equipped with a good hand magnifier, an accurate rule, and plenty of patience. The drawings of bracts will help.

The exact wording of the group-headings below must be carefully attended to, or error will result. (It probably will anyway.)

The genus as a whole is fairly easily recognized. The bracts overlap in several circles. The rays are purple, lavender, pink, blue, or white. The disk is in general rather small, with yellow flowers which in many species turn reddish. The pappus consists of fine bristles.

Some of our species, and hybrids between them, are cultivated in England as Michaelmas daisies. They are also the ancestors of the "hardy asters" (as distinct from China-aster, *Callistephus*) of our own gardens.

I. *Species whose lower leaves, at least (and perhaps the upper leaves also), have heart-shaped blades on distinct stalks.*

PLATE 152

Aster azureus *Johnson*

Aster macrophyllus *Johnson*

Polymnia canadensis *D. Richards*

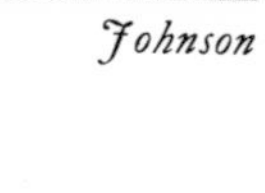

Polymnia uvedalia *D. Richards*

Aster sagittifolius *Johnson*

Aster shortii *Johnson*

We may separate these into species with colored rays (blue, purple, pink, lavender) and those with white rays. However, this separation is not absolute: *the rays may vary from colored to white in one species. It is better than no separation at all – but not much.*

A. Species of group I with colored rays.

LARGE-LEAVED ASTER, A. MACROPHYLLUS (1–5 feet), has thick, coarsely toothed leaf-blades tapering to a sharp point. The flowering branches are glandular. The bracts are downy, with broad, blunt tips. The rays are violet or pale blue.

July to October: in woodland from Quebec to Minnesota and southward to Georgia and Illinois. *Plate 152.*

A. AZUREUS (1–4 feet) has thick, rough leaf-blades without teeth, the lower ones perhaps scalloped. The branches of the inflorescence bear many small, sharp-pointed bracts. The bracts of the heads are smooth with broad green tips. The rays are generally deep blue or violet.

July to October: in prairies and open woods from New York and southern Ontario to Minnesota and southward to Georgia and Texas. *Plate 152.*

A. SHORTII (1–4 feet) has a nearly smooth stem bearing rather narrow leaves, mostly heart-shaped. The inflorescence is leafy. The bracts of the heads are narrow, with green tips, and very finely downy. The rays are usually pale violet.

August to October: in open woodland from Pennsylvania and Virginia to Minnesota and Iowa and southward to Georgia and Alabama. *Plate 152.*

HEART-LEAVED ASTER, A. CORDIFOLIUS (1–6 feet), has a nearly smooth stem with thin, ovate, sharply toothed leaf-blades on slender stalks. The bracts are smooth and rather narrow, with blunt tips. The rays are pale blue or violet or white, quite short (less than ½ inch). The disk tends to turn red quite early.

August to October: in open woodland and on slopes from Quebec to Minnesota and southward to Georgia, Alabama, and Missouri. *Plate 153.*

A. SAGITTIFOLIUS (1–6 feet) somewhat resembles *A. cordifolius* but the leaf-blades are narrower and the leaf-stalks are broad and flat; the blades are generally rough. The bracts are very narrow, pointed, and smooth. The rays vary from pale blue to pink or white, and are up to ½ inch long.

August to October: in dry woodland from Vermont to North Dakota and southward to Florida, Alabama, and Missouri. *Plate 152.*

A. DRUMMONDII (1–4 feet) is sometimes considered a variety of *A. sagittifolius*, from which it differs chiefly in the white down that covers the lower surface of the leaves and the stem. It is found from Ohio to Minnesota and southward to Tennessee and Texas.

A. LOWRIEANUS resembles *A. cordifolius* but has very smooth and soft leaves which have a whitish bloom; marginal teeth are generally lacking.

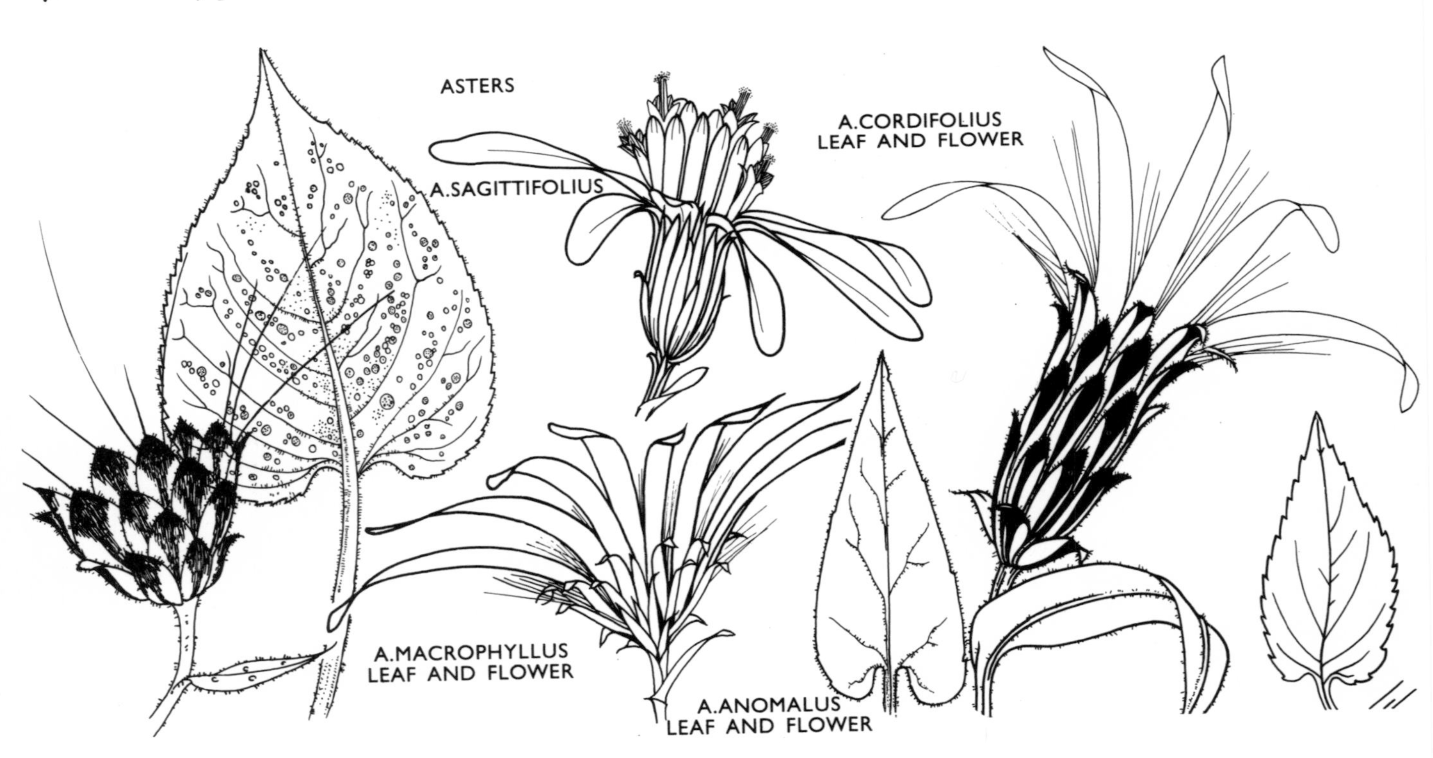

PLATE 153

Aster divaricatus *Gottscho*

Aster novae-angliae *Rickett*

Aster novae-angliae *Rickett*

Aster schreberi *Uttal*

Aster cordifolius *Rickett*

August to October: in open woodland from Connecticut to southwestern Ontario and southward to Georgia and Tennessee. *Plate 154*.

A. UNDULATUS (1–4 feet) is covered with pale hairs. The stalks of the leaves are expanded where they join the stem into two lobes which partly surround or "clasp" the stem. The edges of the leaves usually lack teeth. The bracts are downy. The rays are pale blue or violet.

August to November: in dry open woodland from Nova Scotia to Minnesota and southward to Florida, Alabama, and Arkansas. *Plate 157*.

A. CILIOLATUS (1–4 feet) has rather narrow, sharply toothed leaf-blades on wide, flat stalks; the upper ones are not indented at the base. The stem and leaves may be hairy, downy, or smooth; the young leaves are edged with fine hairs (*cilia:* "eyelashes"). The bracts are smooth and narrow.

July to October: in open woodland from Maine to northwestern Canada and southward to northern New England and New York, Michigan, South Dakota, Wyoming, and British Columbia.

A. ANOMALUS (1–5 feet) is readily distinguished from all other species with stalked, heart-shaped leaf-blades by the bracts around the flower-heads: their tips are sharply bent out and down. Stem and leaves are rough. The blades are rather narrow and generally lack teeth.

August to October in dry woods and on limestone bluffs from Illinois to Kansas, Arkansas, and Oklahoma.

B. Species of group I with white rays.

A. DIVARICATUS (1–3 feet) has thin, smooth, very sharply toothed blades on slender stalks. The stem is frequently zigzag. The bracts are rather broad and blunt. The rays are few (usually about ten) and ½ inch long; the general effect is rather untidy.

July to October: in dry woods and on shady roadsides from Maine to Ohio and southward to Georgia and Alabama. *Plate 153*. A very common and variable species.

A. SCHREBERI (1–3 feet) is much like *A. macrophyllus* but may be distinguished by its smooth bracts. It is generally rather smooth, with thin leaves. The rays are white.

July to October: in damp woods from New Hampshire to Wisconsin and southward to Virginia, Pennsylvania, and Kentucky. *Plate 153*.

A. FURCATUS (1–3 feet) is harsh and stiff, with rather narrow leaf-blades scalloped rather than toothed. The bracts are rather broad with very short green tips. The rays are fairly numerous.

August to October: in dry woods from Indiana to Wisconsin and Missouri.

II. *Species that have no heart-shaped leaf-blades on stalks. These we may group by the way in which the leaves are attached to the stem.*

A. Species whose leaves "clasp" the stem: i.e. the blade or the stalk has two lobes which extend partly around the stem at the point of attachment. The rays in this group are generally colored.

NEW ENGLAND ASTER, A. NOVAE-ANGLIAE (1–8 feet), easily takes the prize for the most beautiful wild aster. The numerous rays vary from a deep purple (the most common shade) to pink, pale lavender, or even white. The species is easily recognized, no matter what the color of its rays, by the lanceolate, clasping leaves which lack stalks and teeth and by the abundant stalked glands on the stalks and bracts of the flower-heads (the rest of the plant being hairy and perhaps glandular also).

July to October: in old fields and meadows and on roadsides from Quebec to Alberta and southward to Maryland, the mountains of North Carolina, Kentucky, Arkansas, and Colorado. *Plate 153*. This species has been used in the production of the "hardy asters" of our gardens, and is well worth cultivating in its own right.

NEW-YORK ASTER, A. NOVI-BELGII (1–4 feet), often rivals the New England aster in color, but differs in many ways. The stem is finely downy or nearly smooth. The narrow, generally lanceolate leaves are sharply toothed and smooth. The bracts of the flower-heads have loose tips which tend to turn outwards or even downwards.

July to October: in moist soil, including salt marshes, mostly near the coast from Newfoundland to Georgia.

A. PUNICEUS (1–8 feet) is one of the most variable of species, very common and very handsome. The stem and leaves are bristly, hairy, or almost smooth. The leaves vary greatly in shape. The bracts are narrow and sharp-pointed.

August to November: in moist meadows, roadside ditches, etc. from Newfoundland to Manitoba and southward to Georgia, Alabama, and Iowa. *Plate 154*. I have walked through a swampy field full of this species and others and seen scarcely two plants of *A. puniceus* which appeared completely alike. Generally it may be recognized by the combination of hairy or bristly stem and "clasping" leaves.

PLATE 154

Aster patens *Elbert*

Aster lowrieanus *Rhein*

Aster puniceus *Rickett*

Aster prenanthoides *Rhein*

Aster novi-belgii *Elbert*

Aster laevis *Johnson*

A. PRENANTHOIDES (1–3 feet) is recognizable by the odd shape of its leaves. These have an ovate or lanceolate blade which tapers sharply to a broad flat stalk, and this in turn has a broader base with two lobes which "clasp" the stem. The leaves are rough. The bracts are very narrow and translucent. The rays are pale.

August to October: in woodland and meadows from Massachusetts to Minnesota and southward to Virginia, Kentucky, and Iowa. *Plate 154.*

A. PATENS (1–3 feet) is known by its rather small, blunt leaves, lacking teeth and stalks, with lobes at the base which almost encircle the stem. The bracts are often glandular and downy, with spreading green tips. The rays vary from violet to pink.

August to October: in dry woodland from Maine to Minnesota and southward to Florida and Texas. *Plate 154.*

A. LAEVIS (1–3 feet) is named for its smooth surfaces; it is often whitish with a bloom. The leaves have mostly lanceolate or elliptic blades, the lower ones with stalks, the upper stalkless and "clasping" the stem. The bracts are narrow and whitish with green tips. The rays are violet.

August to October: in dry open places from Maine to British Columbia and southward to Georgia, Louisiana, New Mexico, and Oregon. *Plate 154.*

B. Species with no heart-shaped leaf-blades on stalks and no leaves that noticeably "clasp" the stem. This is the most difficult and confusing group of asters. Again we here separate them into those with colored rays and those with white rays; and again the separation is not sharp, and when one has difficulty in identifying an aster, one must look in both groups. (In fact, this grouping is more for the convenience of the author than for that of the reader.)

1. Species of group B with colored rays.

a. In order to facilitate identification of the species in this group, we here select, first, those species whose leaves are less than ten times as long as they are wide (length divided by width is less than 10). These all grow from 4 to 40 inches tall. (Compare b.)

Atlantic species.

SHOWY ASTER, A. SPECTABILIS, is a plant of dry woodland from Massachusetts to Maryland and North Carolina, flowering from August to October. It spreads by slender runners. The lower leaves are stalked, with lanceolate or ovate blades. The rays are violet and quite long (up to $\frac{4}{5}$ inch). *A. herveyi*, found from Massachusetts to Long Island, is similar, with the lower leaf-blades more ovate, the upper more toothed. Both species may be glandular.

A. RADULA has a smooth stem bearing numerous veiny, toothed, rough leaves without stalks. It is found in wet woods and bogs from Newfoundland to Virginia and inland to Pennsylvania and West Virginia, flowering from July to September.

A. NEMORALIS is a plant of peat and bogs from Newfoundland to Ontario and Michigan and southward to Delaware. It grows from threadlike runners. It is very leafy, with short, narrow leaves.

A. CONCOLOR has a stem silky but rather thinly so, or sometimes merely with hairs lying flat. The leaves also are more or less silky, and without stalks. It grows in dry sandy soil, often under pines, from Massachusetts to Florida and Mississippi; also inland in Kentucky and Tennessee. *Plate 156.*

Midwestern species.

A. OBLONGIFOLIUS is a stiff, bushy plant, with hairy stems and rough leaves (which almost "clasp" the stem with their blunt bases). The bracts are narrow, glandular, with tips that curve outward. The species is found from Pennsylvania to Wisconsin and North Dakota and southward to North Carolina and Texas, flowering in September and October. *Plate 155.* Worthy of garden culture.

A. SERICEUS, a plant of dry prairies and open woods from Michigan to Manitoba and southward to Tennessee and Texas, is easily recognized by its silky leaves. It flowers from August to October.

PRAIRIE ASTER, A. TURBINELLUS, is a smooth plant with lanceolate leaves. The flowering branches bear many small bracts below the actual flower-heads. The violet rays range up to $\frac{4}{5}$ inch in length. This grows in prairies and open woodland from Illinois to Nebraska and Louisiana, flowering in September and October.

A. MODESTUS is a northwestern species found in Michigan and Minnesota. It resembles *A. novae-angliae*, having purple rays and being glandular and hairy. The leaves are slightly clasping. It flowers in August and September.

A southern species.

A. GRACILIS is a small, stiff plant (rarely more than 2 feet tall) with only few and short rays (from eight to fourteen, $\frac{1}{5}$–$\frac{2}{5}$ inch long). The stem grows from a sort

PLATE 155

Aster dumosus *Rickett*

Aster linariifolius *Johnson*

Aster junciformis *Elbert*

Aster acuminatus *Uttal*

Aster oblongifolius *Johnson*

Aster gracilis *V. Richard*

of corm, an enlarged base. The leaf-blades are elliptic, stalked, rough, without teeth. It inhabits pineland, bogs, and barrens on the coastal plain from New Jersey to South Carolina, flowering from July to September. *Plate 155.*

b. Species of group B with colored rays and leaves ten or more times as long as wide (length divided by width is 10 or more). They are mostly from 6 to 36 inches tall.

Species of wide distribution.

BRISTLE-LEAVED ASTER, A. LINARIIFOLIUS, is rarely more than 2 feet tall. The stem is more or less downy, with numerous, stiff, rough, needle-like leaves. The stem often bears a single head. This flowers from July to October from Maine to Wisconsin and southward to Florida and Texas. *Plate 155.* One of the most beautiful asters.

A. JUNCIFORMIS has leaves smooth except for rough edges. The rays are numerous and pale. It grows from Quebec to British Columbia and southward to New Jersey and California, flowering from August to October. *Plate 155.*

A. DUMOSUS somewhat resembles the preceding species. The stem is smooth, often branched, the leaves from very narrow to lanceolate or elliptic, and rough. The flowerheads are on stems which bear many small bracts. The rays are short and pale. This is found in sandy or other soil from Massachusetts to Florida and from Ontario and Ohio to Wisconsin, Kentucky, Illinois, and Arkansas, flowering from August to October. *Plate 155.*

Atlantic species.

A. SUBULATUS is found in coastal marshes from New Brunswick and New Hampshire to Florida and Alabama, flowering from August to October. It is smooth and fleshy throughout. *Plate 156.*

A. TENUIFOLIUS, like the preceding species, is a plant of coastal marshes, rarely more than 2 feet tall, smooth, with few and fleshy leaves. The upper leaves are much shorter than the lower, seeming like bracts. The rays are numerous and short. The range is from Massachusetts to Florida and Mississippi, the blooming period from August to October. *Plate 156.*

Western and southern species.

A. PALUDOSUS, only up to 2 feet tall, has thick, firm leaves. The pappus is tawny or reddish. This enters our range only in Missouri, extending to Florida and Texas.

A. BRACHYACTIS is a smooth western plant which occurs in Minnesota and western New York, and is reported from Missouri. The leaves are rough and bear marginal hairs towards the base. Rays are often lacking.

A. BLAKEI is similar to *A. acuminatus* except for the lilac rays. The leaves are less obviously toothed and more numerous. It grows in damp woods from Newfoundland to New Jersey. *Plate 156.*

2. Of species of *Aster* whose leaves do not clasp the stem there are a number regularly with white rays. This is perhaps the most baffling group in this difficult genus. Several of the species are weedy and scarcely merit a place in this book; but since they are indeed asters they must be here included.

As in the preceding group we can make a distinction between species with very narrow principal leaves and those with broader principal leaves.

a. White-flowered asters with principal leaves (not clasping) less than ten times as long as broad (length divided by breadth is less than 10).

A. ACUMINATUS (8–40 inches) is fairly easily recognized by its leaves. These are generally elliptic or lanceolate, without stalks but tapering to a narrow base, and sharply toothed. The heads are quite large, the rays up to $\frac{3}{5}$ inch long.

July and October: in dry woods from Newfoundland to Ontario and southward to Georgia and Tennessee. *Plate 155.*

A. UMBELLATUS is – surprisingly – quite easy to recognize. It is a big, bushy plant from 1 to 8 feet tall, with rather narrow, lanceolate or elliptic leaves borne right up to the inflorescence. The flower-heads are numerous, on many rather erect branches (*not*, however, in an umbel), and all forming a rather flat or convex array. The rays are quite wide, $\frac{1}{4}$ inch long or longer.

July to September: in thickets and on the borders of woods, usually in moist soil, from Newfoundland to Minnesota and southward to Georgia, Illinois, and Nebraska. *Plate 156.*

A. SIMPLEX (2–5 feet) is a common and variable species. It has narrowly lanceolate or elliptic leaves on the main stem and branches, and many much smaller leaves on the flowering branches. The flower-heads are numerous, scattered all over the plant. The rays may be $\frac{1}{2}$ inch long, and are sometimes lavender. To be sure of the species, one must dissect out a few disk-flowers and note the shape of the corolla: it begins as a slender tube and expands suddenly to a wider tube which ends in five erect teeth; in *A. simplex the teeth are shorter than the wide tube below them.*

PLATE 156

Aster umbellatus *Johnson*

Aster subulatus *Gottscho*

Aster blakei *Scribner*

Aster pilosus *Allen*

Aster tenuifolius *Uttal*

Aster concolor *Uttal*

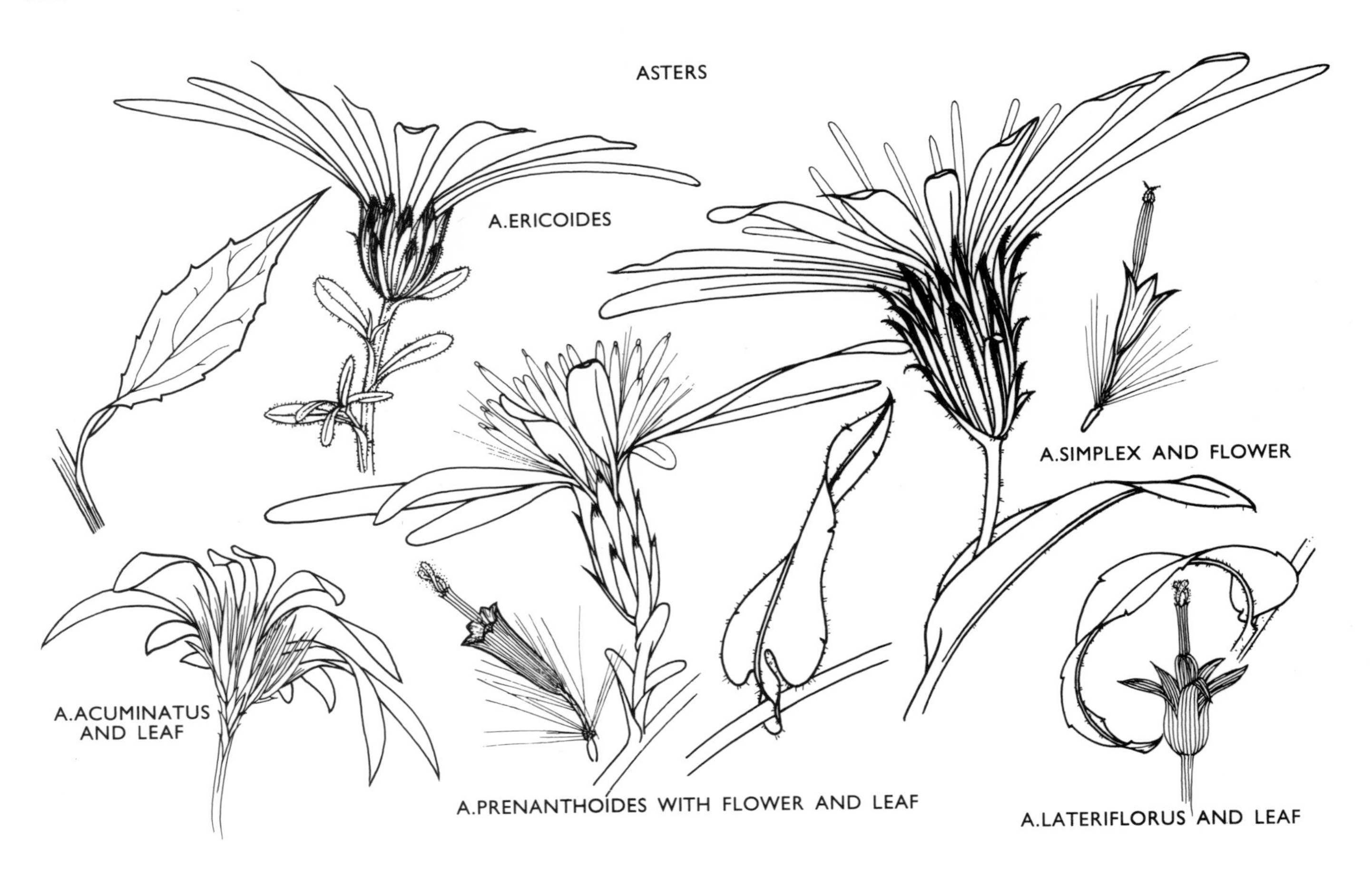

August to October: in meadows and thickets and on roadsides from Newfoundland to Saskatchewan and southward to North Carolina, Kentucky, and Texas. A variety found from Quebec to Virginia and westward through the central states has leaves whose length is more than ten times their width. This has often gone under the name *A. paniculatus*. *Plate 157*.

A. LATERIFLORUS (1–4 feet) somewhat resembles the preceding species. The type of leaves and their disposition are similar. The flower-heads are usually on long branches, sometimes all along one side of the branch (which is the meaning of the Latin). The rays are generally shorter than those of *A. simplex*, only about $\frac{1}{4}$ inch long, or less. To clinch its identity, one must dissect out some disk-flowers and note that the teeth are *as long as or longer than the wide part of the corolla-tube* immediately below them; they generally curve outward.

August to October: mostly in dry places, fields, open woodland, roadsides, etc. from Quebec to Minnesota and southward to Florida, Missouri, and Texas. *Plate 157*.

A. INFIRMUS (1–4 feet) has veiny, elliptic or even ovate leaves. The rather large flower-heads are in a widely branched cluster. The rays are wide and up to $\frac{2}{5}$ inch long.

July to September: in woodland from Massachusetts to Ohio and southward to Georgia and Alabama.

b. Species with white rays and leaves (not clasping) whose length is more than ten times their width. (See also *A. simplex* under a.)

A. PTARMICOIDES (4–28 inches) has several stems rising in a cluster among the fiber-like remains of old leaves. The leaves are stiff, erect, very narrow, and often rough. The flower-heads form a wide and rather flat cluster. The disk is white as well as the rays.

June to September: in prairies and rocky, dry places from Quebec to Saskatchewan and southward to Georgia, Ohio, Arkansas, and Colorado. *Plate 157*. Specimens have been reported with yellow rays, suggesting a goldenrod rather than an aster! The genera *Solidago* and *Aster* are separated mainly by the color of their rays and are certainly closely related.

WHITE-HEATH ASTER, A. PILOSUS (8 inches–5 feet), is misnamed, for several forms of the species have smooth stem and leaves ("pilose" means "with long hairs"). The stem bears numerous short, narrow leaves, often almost awl-shaped, those on the

PLATE 157

Aster vimineus *Rhein*

Aster undulatus *Rhein*

Aster simplex (variety) *Johnson*

Aster ptarmicoides *Johnson*

Aster lateriflorus *Uttal*

Aster ericoides *Johnson*

flowering branches so small as to be called bracts rather than leaves. The flower-heads are generally numerous with narrow, sharp-pointed bracts.

August to October: in dry thickets and old fields and on roadsides from Maine to Ontario and Minnesota and southward to Georgia, Mississippi, Arkansas, and Kansas. *Plates 156, 158*. Dr. Fernald says "a highly variable but discouragingly intergradient series of variations and forms"; some of them have been confused with *A. ericoides*.

A. PARVICEPS (1–3 feet) has a generally hairy stem and narrow leaves. The flower-heads are smaller than those of *A. pilosus*, only about $\frac{1}{6}$ inch high, with narrow, green-tipped bracts and ten or twelve rays.

August to October: in dry woodland and prairies from Illinois to Iowa and Missouri.

A. DEPAUPERATUS (4–16 inches) is like *A. parviceps* but smaller, smooth, with very short leaves.

July to October: in barrens from Pennsylvania and West Virginia to Delaware and Maryland. On serpentine rocks.

WHITE-HEATH or WREATH ASTER, A. ERICOIDES (1–4 feet), is usually covered with minute gray hairs. The leaves are narrow, stiff, and crowded (whence the likeness to *Erica*, heath). The flower-heads are numerous and small (only about $\frac{1}{8}$ inch high). The bracts are distinctive as seen through a good hand magnifier; they have broad green tips with a minute spine on the end.

July to October: in dry open places and thickets from Maine to British Columbia and southward to Georgia, Mississippi, Oklahoma, and Arizona. *Plate 157*. The heathlike plants are often so covered with the small white flower-heads as to be quite attractive.

A. VIMINEUS (1–5 feet) is smooth or nearly so, with rather long and narrow leaves which may have a few teeth on the edges. The stem is often purplish, and may arch instead of standing erect. The flower-heads often stand along one side of the long flowering branches. They are tiny ($\frac{1}{8}$ inch high), with narrow, green-tipped bracts. The corolla-lobes of the disk-flowers are nearly as long as the wide part below (compare *A. simplex*).

August to October: in open places, mostly damp, from Maine to Michigan and southward chiefly near the coast to Florida and Texas. *Plate 157*.

A. TRADESCANTI (6 inches–2 feet) resembles *A. simplex* but is smaller. The erect stem rises from slim creeping stems, it is generally smooth. The leaves are narrowly lanceolate or elliptic, sometimes toothed. The flower-heads are few and scattered, about $\frac{1}{5}$ inch tall, with narrow bracts.

July to September: on rocky banks of streams and lakes from Newfoundland to Michigan and southward to Massachusetts and New York.

A. PRAEALTUS (1–6 feet) is much like *A. simplex* in general aspect, with leafy branches and fairly large flower-heads. Its distinctive feature is the pattern made by the conspicuous veins on the under surface of the leaves; they form a network enclosing small areas no longer than broad. The corolla-lobes of the disk-flowers are even smaller than in *A. simplex*.

August to October: in moist thickets, meadows, and prairies from Massachusetts to Manitoba and southward to Georgia, Kentucky, Oklahoma, Arizona, and Mexico.

Besides all these a number of other species occur in our range: *A. ontarionis*, like *A. lateriflorus*, from Quebec to Minnesota and southward to North Carolina, Mississippi, Oklahoma, and Texas; *A. elliottii*, from Virginia to Florida; *A. falcatus*, from Wisconsin to Missouri and westward; *A. concinnus*, like *A. laevis*, from Pennsylvania to Alabama; *A. hesperius*, like *A. simplex*, from Wisconsin to Missouri and westward; and others that just reach our borders.

THE WHITE-TOPPED ASTERS (SERICOCARPUS)

Here are two species regarded as true asters by at least one authority but long treated as a separate genus and certainly easily distinguished in the field. They have only about five rays, which are rather broad and short, and white. The disk-flowers are cream-colored. The bracts also are whitish, with green tips. The fruits (achenes) are covered with silky hairs, which give the genus its Latin name. The pappus is composed of white bristles. (All these characteristics are found in some species of *Aster*, but not the whole combination.) The flower-heads are in flat clusters.

S. ASTEROIDES (6 inches–2 feet) has leaves of various shapes, often toothed on the edges, sometimes stalked, and veiny.

June to September: in dry woods from southern Maine to Michigan and southward to Florida, Alabama, and Mississippi. *Plates 158, 176*.

S. LINIFOLIUS (8 inches–2 feet) has narrow leaves without teeth (the name means "flax-leaved").

June to September: in dry woods and clearings from Massachusetts and New Hampshire to Ohio and southward to Georgia and Louisiana. *Plate 176*.

THE FLEABANES (ERIGERON)

There is really no satisfactory way of distinguishing *Erigeron* from *Aster*, when all species are considered. However, our eastern species of *Erigeron* may be recognized as (1) mostly blooming in spring and early summer, and (2) having bracts around the flower-head in a single circle and of approximately the same length rather than in several overlapping circles of different lengths as in *Aster*. There are many more species of *Erigeron* in the West – where there are fewer asters. In the East we have some species with pretty flowers and some common weeds. The English name tells its own story, though I have no modern evidence of its efficacy in repelling fleas. The botanical name is from two Greek words meaning "early" and "old man," apparently originally applied to some plant with white hair that blooms early.

The species are quite readily separated into two groups by their general aspect; but it is not easy to state their "general aspect" in precise terms.

I. *Species with the disk of the flower-heads at least $\frac{2}{5}$ inch and generally more than $\frac{1}{2}$ inch across; the rays generally tinted with pink or lavender but sometimes white.*

ROBIN'S-PLANTAIN, E. PULCHELLUS, is our handsomest species (*pulchellus* means "beautiful"). It has a hairy stem (6 inches to 2 feet tall) which forms runners at the base; these develop tufts of basal leaves from which new erect stems rise. The basal leaves are widest in their outer half, and toothed. The leaves on the stem are few, ovate or lanceolate, toothed or plain. At the summit of the stem there is a single flower-head, or a small group of heads on longish stalks. The rays are numerous.

April to July: in meadows and open woodland from Maine to Minnesota and southward to Florida and Texas. *Plate 158*.

DAISY FLEABANE, E. PHILADELPHICUS, grows a bit taller than *E. pulchellus*, with a hairy stem which forms offsets at the base. The basal leaves are rather narrow and toothed or scalloped. The leaves on the stem have basal lobes or "ears" which extend on either side of the stem. There may sometimes be only one flower-head but more commonly there are many. The rays are very numerous.

April to August: in thickets, fields, and woodland, on shores and roadsides from Newfoundland to British Columbia and southward to Florida and Texas. *Plate 158*.

E. glabellus may be mentioned here. It is a western species which is found in our northwestern borders. It has flower-heads much like those of *E. philadelphicus* but on long stalks; the leaves on the stem are much smaller.

II. *Species with the disk of the flower-heads mostly less than $\frac{2}{5}$ inch across; the rays commonly white but sometimes tinged with pink or blue.*

E. HYSSOPIFOLIUS forms dense tufts of narrow leaves and almost threadlike stems about a foot tall. The heads generally have about thirty rays.

June to August: in rocky and gravelly places from Newfoundland to Mackenzie and southward to New York and Michigan.

DAISY FLEABANE, E. STRIGOSUS, is perhaps the commonest species of the genus in the northeast. It grows from 1 to 4 feet tall with a slender stem branched at the top into an inflorescence (often a single flower-head terminates the main stem and three or four branches grow out just below, each ending in a flower-head, and so on). The disk of the flower-head is usually less than $\frac{2}{5}$ inch but more than $\frac{1}{5}$ inch across, and there are more than twenty disk-flowers. The leaves on the stem are narrow, more or less lanceolate, and mostly without teeth.

May to September: in fields and on roadsides across southern Canada and southward practically throughout the United States. *Plate 158*.

DAISY FLEABANE, E. ANNUUS, is a plant from 1 to 5 feet tall, somewhat like *E. strigosus* but coarser. The stem bears numerous leaves which are commonly ovate and sharply toothed.

June to October: in fields and waste land across southern Canada and practically throughout the United States. *Plate 158*. A form exists which lacks rays.

MULE-TAIL or HORSE-WEED, E. CANADENSIS, is an unattractive and in some places very abundant weed. The stem is from 1 to 6 feet tall or even taller, more or less hairy and bearing numerous narrow, bristly leaves. At the summit is an inflorescence of many small flower-heads, the disk generally less than $\frac{1}{5}$ inch across and containing less than twenty flowers. The rays are numerous and very small.

June to November: in waste land, fields, etc. throughout southern Canada and the United States. In certain parts of the Midwest fields may be entirely covered with this weed. It has also reversed the usual pattern of weed-travel and migrated eastward to the Old World; it flourished on the ruins created in London by German bombs during the Second World War. By some authors it is placed in another genus, *Conyza*.

Besides the above species, several southerners reach our borders: *E. vernus* and *E. quercifolius* in Virginia and *E. tenuis* in southwestern Missouri. The northern *E. angulosus*, a "circumpolar" species with very narrow pink or white rays, is found in the northern parts of Maine, Michigan, and Minnesota.

BOLTONIA

These plants have the general appearance of asters, with rather small flower-heads. They are distinguished by their pappus, which is mainly of two or four spines instead of numerous fine bristles (lacking in some southern species); and by the shape of the "disk" which is a dome or a cone instead of a flat surface. The rays are white, lilac, or pink; the disk-flowers yellow. The leaves are smooth, rather thick, narrow, and generally without teeth.

B. ASTEROIDES has its flower-heads generally in a broad flat cluster. The leaves vary from very narrow to lanceolate.

July to October: in sandy, gravelly, and wet soil from New York to Manitoba and North Dakota and southward to Florida and Texas. Western plants with broad bracts may be a distinct species, *B. latisquama*. In Missouri and Illinois plants are found whose rather broad leaves have edges running down as ridges on the stem. These also may be a separate species, *B. decurrens*.

B. DIFFUSA has flower-heads widely scattered on stalks which bear many very small leaves or bracts. The leaves are very narrow.

July to October: in woods and fields from Kentucky to Missouri and southward to Florida and Texas. Two other southern species, *B. caroliniana* and *B. ravenelii* extend to Virginia; their pappus is minute or lacking.

GALLANT SOLDIERS (GALINSOGA)

The plants that bear this name scarcely deserve it, for they are rather mean little weeds with inconspicuous flowers. They owe their English name to the Latin – slightly mispronounced! The botanical name is that of a Spanish botanist. The common species have paired leaves (occasionally in threes) with ovate, toothed blades on short stalks. They grow up to about 2 feet tall. The flower-heads are mostly a conical yellow "disk" surrounded by four or five bracts and the same number of small, white, three-toothed rays. The pappus is of small scales split into several minute spines; or it may be almost lacking.

G. CILIATA is the commonest species. It is roughly hairy. The sharp tips of the pappus are easily seen with a lens. *Plate 159.*

G. PARVIFLORA has hairs generally lying flat on the surface of the stem. The ray-flowers have no pappus; the pappus of the disk-flowers forms fine, short bristles. *Plate 159.*

Both species are weeds around dwellings and in cultivated soil throughout our range, flowering through summer and autumn. *G. ciliata* is cooked and eaten as "greens" by some Asian peoples.

PURPLE CONEFLOWERS (ECHINACEA)

The purple coneflowers are handsome plants, often cultivated in the border. The rays are generally a dull or very pale purple, sometimes almost white, and in one midwestern species orange-yellow; they hang down from the flower-head. In the midst rises the "disk," a dome rather than a cone. The disk-flowers are mixed with bracts (chaff) which have sharp tips longer than the flowers, so that the "disk" resembles a hedgehog; the Greek word for hedgehog is *echinos*. The leaves are near the base of the stem (which bears a single flower-head). They generally have three or five main veins, and are mostly borne singly.

These are characteristically plants of the Midwest and South.

E. PURPUREA is from 1 to 5 feet tall. The leaves have lanceolate or ovate blades on longish stalks.

June to October: in woodland and prairies from Virginia and Pennsylvania to Iowa and southward to Georgia and Louisiana. *Plate 159.*

E. PALLIDA grows up to 40 inches tall. The leaves are narrow, the blade tapering gradually to the stalk. The plant is generally roughly hairy.

May to August: in dry open places from Illinois to Minnesota and Montana and southward to Georgia and Texas. *Plate 159.* The rays are often 3 inches long. The more western plants, with rays less than $1\frac{1}{2}$ inches long, are by some considered a distinct species, *E. angustifolia*.

E. paradoxa, which has orange-yellow rays, is found in Missouri and Arkansas.

PLATE 158

Erigeron annuus *Johnson*

Erigeron pulchellus *V. Richard*

Sericocarpus asteroides *Uttal*

Erigeron philadelphicus *Johnson*

Erigeron annuus *Williamson*

Erigeron strigosus *Scribner*

Aster pilosus *Johnson*

PARTHENIUM

The flower-heads of *Parthenium* are small, the white or whitish rays scarcely extending above the bracts. Most of the species grow south of our range. *P. argentatum* is guayule, of Mexico, which attained a certain importance during the Second World War as a source of rubber. We have one fairly widespread species, and two others, one from the West and one from tropical America, occasionally enter our range. The name is derived from the Greek word for "maiden"; why it is applied to this genus I do not know.

WILD QUININE, P. INTEGRIFOLIUM, is a plant from 2 to 4 feet tall, bearing leaves singly. The lowest leaves have long stalks, the upper ones no stalks; all have ovate leaves, rough to the touch, with rather short, blunt teeth (they are not *integri-*, which means without teeth).

June to September: in dry woodland and prairies from southern New York to Minnesota and southward to Georgia and Texas. *Plate 159.*

P. HISPIDUM has a hairy stem less than 3 feet tall. The leaves are much like those of *P. integrifolium* but hairy and with sharper teeth.

May to August: in woods and prairies and on rocky ledges from Missouri and Kansas to Arkansas and Texas; also along railroads in Michigan. *Plate 159.* A variety with leaves that "clasp" the stem (with lobes extending on either side of the stem) occurs in Virginia and North Carolina.

ECLIPTA

One species of *Eclipta* grows almost throughout our range.

YERBA-DE-TAGO, E. ALBA is a small weedy plant found around the world in its warmer parts. It is much-branched, sometimes reaching a height of 3 feet. The leaves, which are paired, are narrow, pointed, and finely toothed. The flower-heads are small, in little clusters. The white rays scarcely project beyond the bracts.

August to October: in waste land, moist soil, etc. from southern New York to Nebraska and southward to Florida, Texas, and the tropics.

CHRYSANTHEMUM

In our range one species of *Chrysanthemum* is widespread and familiar, and several others have escaped from cultivation in scattered places. All have come from their original homes across the Atlantic.

OX-EYE DAISY or MARGUERITE, C. LEUCANTHEMUM, forms handsome colonies in fields and along roadsides. It stands from 1 to 3 feet tall, or taller, the flower-heads single at the tips of the branches or at the summit of the unbranched stem. The lower leaves are ovate with the broader part outward; the upper leaves are much narrower, perhaps with parallel sides; all are unevenly toothed or lobed with narrow round-ended projections, the upper ones often "clasping" the stem with their lowest lobes. The rays are white, the disk yellow.

May to October: in fields and waste places and on roadsides practically throughout the country. *Plate 160.* Dr Fernald called it a "pernicious but beautiful weed." Farmers have called it whiteweed. The name marguerite (or Margaret) comes from the Greek word for "pearl." The "ox-eye" part of the name must refer to its size rather than its color. "Daisy" is said to be derived from "day's-eye," as applied to the common English daisy (*Bellis*) which opens in the morning. *Chrysanthemum* means "golden flower," having been first applied to the yellow-rayed corn-marigold, *C. segetum*. *Leucanthemum* means "white flower." So our plant is the "white-flowered golden flower"! Corn-marigold, feverfew (*C. parthenium; Plate 160*), and costmary (*C. balsamita*) are occasionally found growing wild.

ACHILLEA

The flower-heads of *Achillea* are numerous and small, clustered in a flat inflorescence. Each has only a few short, white or pink rays. There is no pappus.

YARROW or MILFOIL, A. LANULOSA, is a common weed. It stands generally from 1 to 3 feet tall. The leaves, borne singly, are narrow, without stalks,

PLATE 159

Parthenium hispidum *Johnson*

Echinacea pallida *Johnson*

Galinsoga ciliata *Johnson*

Galinsoga parviflora *D. Richards*

Erigeron canadensis *Johnson*

Parthenium integrifolium *Johnson*

Echinacea purpurea *Rickett*

and cut into innumerable fine divisions. The plant has a characteristic odor. Both white- and pink-flowered forms occur.

June to November: in fields and waste land practically throughout the country. *Plate 160.* This is a native American species which resembles an European species so closely that the two can be distinguished only by microscopic techniques that reveal the chromosomes. The two have been generally confused and both have passed as one, bearing the name of the European species, *A. millefolium.* This does grow in North America but is apparently less common than our native species. It has been called nosebleed, from a belief that it can cause this affliction. It is certainly irritating to delicate membranes, but this particular effect seems to lack confirmation. It has had some medicinal uses in the control of visceral hemorrhages.

SNEEZEWEED, A. PTARMICA, sometimes escapes from cultivation. It has narrow leaves, toothed or plain but not finely cleft. The rays are longer than the bracts, making a handsome show. Cultivated (and escaped) varieties often have "double" flower-heads; i.e. the disk-flowers are partly or wholly replaced by rays. It is found from Labrador to Ontario and southward to New England and Michigan. Evidently the irritating effect of *A. millefolium* is shared by this species.

The northern *A. borealis*, with finely cut leaves like those of *A. lanulosa* but showier ray-flowers, grows on our northern borders and high mountains.

CHAMOMILE AND DOG-FENNEL (ANTHEMIS)

The genus *Anthemis* contains a number of Old-World species, some of which have become common weeds in North America. They have leaves pinnately cleft into many threadlike parts, and a solitary flower-head at the tip of each stem. The foliage of some species is strong-scented, rather unpleasantly so. The rays are white or yellow. The bracts of the flower-head have papery, translucent margins. There is no pappus or practically none. The disk-flowers are mingled with the bracts known as chaff.

STINKING MAYWEED or STINKING CHAMOMILE, or DOG-FENNEL, A. COTULA, has stems about 2 feet tall. The foliage has a strong unpleasant odor, as the first names above suggest. The rays are white.

May to October: in waste places and barnyards and on roadsides practically throughout the country. *Plate 160.* The leaves not only smell bad; they raised blisters on the hands and bare skin of men who got in the harvest without our modern machinery. It was not named for blooming in May – if it had bloomed only then it would not have been so hated – but the "may" in the name is a corruption of the Anglo-Saxon name for the weed.

CHAMOMILE or CORN CHAMOMILE, A. ARVENSIS, resembles the mayweed in leaves and flowers, but is not ill-scented. It occurs less commonly, throughout our range. *Plate 160.*

YELLOW CHAMOMILE, A. TINCTORIA, is distinguished by its yellow rays. It is found occasionally and sparingly throughout our range. The botanical name indicates that it has been used for dyeing.

MATRICARIA

Our species of *Matricaria* resemble those of *Anthemis* in having foliage divided into many very narrow parts, and solitary flowers at the tips of the stems; also in the papery, translucent margins of the bracts. They differ in such technical characteristics as the absence of chaff and the margins of the achenes. The rays, when present, are white.

WILD-CHAMOMILE, M. MARITIMA, is a spreading, branching herb up to 2 feet tall. The disk is from $\frac{1}{3}$ to $\frac{3}{5}$ inch across. There are from twelve to twenty-five rays.

June to October: in fields and waste land from Newfoundland to Ontario and southward to Pennsylvania, Kentucky, and Kansas; and on the Pacific Coast. *Plate 173.* From Europe. *M. chamomile* is similar but has the same pineapple fragrance as the following species.

PINEAPPLE-WEED, M. MATRICARIOIDES, lacks rays. The foliage when crushed emits the fragrance of pineapple.

May to October: in old fields and waste land from Newfoundland to Manitoba and southward to Delaware, Ohio, and Missouri; also in the West, whence it originally came. *Plate 160.*

PLATE 160

Chrysanthemum parthenium *Gottscho*

Chrysanthemum leucanthemum *Johnson*

Matricaria matricarioides *D. Richards*

Anthemis arvensis *Murray*

Achillea lanulosa *Gottscho*

Mikania scandens *Uttal*

Anthemis cotula *Johnson*

EVERLASTING, PUSSY-TOES, AND LADIES'-TOBACCO (ANTENNARIA)

The genus *Antennaria* is not difficult to recognize, with its heads of small tubular flowers surrounded by bracts of several lengths which are dry and papery and white or colored at least at the tips. The pappus is composed of bristles, and the plants of most species are woolly or silky with generally white hairs. Most of the names are easily explained, some having being applied to all the species. Everlasting, because the dry bracts remain unchanged after the plant is picked (for a winter bouquet). Pussy-toes obviously from the little round flower-heads with projecting pappus. But I can find no record of the use of this plant in pipes or cigarettes, by ladies or others. The botanical name refers to the projecting stamens of certain flower-heads, fancifully likened to the antennae of insects.

To distinguish *Antennaria* from its close relatives *Gnaphalium* and *Anaphalis* perhaps requires a little care. The principal leaves of *Antennaria* are at the base, those higher on the stem being much smaller. The pistils and stamens of *Antennaria* are in separate flowers, and the pistillate and staminate flowers are on separate plants. This is not readily apparent, since the staminate flowers often have a style; but careful examination will reveal that this style is not functional, lacking the two branches at the tip which catch the pollen in a truly pistillate flower.

It is, however, difficult to know what to do about the species. Most plants of *Antennaria* seem to have attained the peculiar ability to form seeds without pollen (indeed in some species staminate flowers are unknown). This has the effect of forming many races differing in very minor characteristics but of course not losing their identity through interbreeding. Whether these should all be called species, or varieties of a few species, is a matter of opinion. One authority, with an unexcelled knowledge of the plants in the field, lists thirty-two species in the northeastern states; to another student the same plants constitute only six species! In this book it is possible to describe only the kinds that differ most obviously and to keep in mind that many others are intermediate between them.

I. *Species with basal leaves from $\frac{1}{2}$ to 2 inches broad, and with several main veins.*

PUSSY-TOES or LADIES'-TOBACCO, A. PLANTAGINIFOLIA, has rather large leaves at the base with broad blades on long stalks. The blades have from three to seven conspicuous veins, and in this, and their shape, resemble leaves of plantain (hence the tongue-twisting botanical name). From this basal rosette rises the flowering stem, up to a foot tall, bearing several narrow leaves and a tight cluster of flower-heads. Runners grow out from the base bearing leaves and forming new rosettes at their tips.

April to June: in dry woodland and fields from Maine to Minnesota and southward to Georgia and Missouri. *Plate 165.* The involucres – i.e. the bracts taken together – typically measure not more than $\frac{1}{3}$ inch in height. The basal leaves are nearly smooth when mature. *A. fallax* has somewhat taller involucres (from $\frac{1}{3}$ to $\frac{2}{5}$ inch), and the basal leaves keep their hair longer. *A. parlinii* (*Plate 161*) has basal leaves smooth from the beginning, and often purple stems.

A. SOLITARIA has basal leaves much like those of *A. plantaginifolia* but more woolly. The runners are long and slender, with very small scale-like leaves. The flowering stem may be more than a foot tall, with very narrow leaves and a single ("solitary") flower-head.

April and May: in woods from Maryland and western Pennsylvania to Indiana and southward to Georgia, Alabama, and Louisiana. *Plate 161.*

II. *Species with basal leaves not more than $\frac{4}{5}$ inch broad, usually about $\frac{1}{2}$ inch, and with one main vein.*

A. NEGLECTA spreads by long, slender runners. The leaves of the basal cluster have blades widest near the tip and tapering to a stalk-like portion. The flowering stem rises to nearly a foot in height (that of staminate plants not so high), bearing very narrow leaves and several crowded flower-heads. The bracts have rather showy white tips.

April to July: in dry fields and prairies and open woodland from Nova Scotia to Minnesota and southward to Virginia, Indiana, and Kansas. *Plate 161.* A form is known with only one flower-head. *A. petaloidea* is very similar.

A. NEODIOICA resembles *A. neglecta* in the shape of its basal leaves, but forms offshoots rather than runners, so that many clusters of leaves exist side by side in a dense mat. The bracts are often purplish with paler tips.

May to July: in dry fields and woodlands and rocky places from Newfoundland to Ontario and southward to Virginia, Indiana, Wisconsin, and Minnesota. *Plate 161.* This species is very variable. In many places, particularly northward and westward, no staminate flowers have been found; but smaller plants of Virginia, West Virginia, and Pennsylvania, and perhaps farther north, with abundant staminate plants, are allied to this species; they have been named *A. virginica*. *A. canadensis* is distinguished by a slender papery appendage at the tip of each stem-leaf, in this resembling *A. neglecta*.

PLATE 161

Antennaria neodioica *Rickett*

Antennaria solitaria *Johnson*

Anaphalis margaritacea *Ryker*

Antennaria neglecta *Murray*

Gnaphalium purpureum *Johnson*

Antennaria parlinii *Elbert*

Gnaphalium obtusifolium *Johnson*

CUDWEEDS AND CATFOOT (GNAPHALIUM)

The cudweeds and their relatives have two kinds of flowers in the same head. The central ones have both stamens and pistil; the outer ones lack stamens. The bracts around the head are dry and white or colored, and these plants, like *Antennaria*, are sometimes called everlasting. The plants are generally covered with white or dingy wool. The leaves are distributed fairly evenly over the stem, not clustered at the base as in *Antennaria* (but the lower leaves may be larger and closer together, and some species form a rosette the first year).

CATFOOT, G. OBTUSIFOLIUM, stands from 4 inches to 3 feet tall, with flower-heads clustered at the tips of branches; the clusters are flattish. The leaves are without stalks, narrow and sharp-pointed.

August to November: in dry fields and borders of woods and on roadsides from Quebec to Manitoba and southward to Florida and Texas. *Plate 161.*

G. VISCOSUM is similar to the preceding species. It is distinguished by the presence of glandular hairs on the stem and by ridges that run down the stem from the edges of the leaves.

July to October: in fields and the borders of woods from Quebec to British Columbia and southward to New England, West Virginia, Tennessee, Wisconsin, South Dakota, New Mexico, Arizona, and Mexico.

LOW CATFOOT, G. ULIGINOSUM, is commonly widely branched and does not exceed a foot in height. The leaves are numerous, and the uppermost ones extend above the flower-heads. The heads are closely clustered and enveloped in wool.

July to October: in damp places from Newfoundland to British Columbia and southward to Virginia, Indiana, Kansas, Utah, and Oregon. Often a nuisance in gardens.

CUDWEED, G. PURPUREUM, grows from 2 to 20 inches tall. The leaves are rather blunt and mostly wider towards the tip. The flower-heads are clustered in the axils of the upper, smaller leaves or bracts and so form a tall narrow inflorescence. The bracts are brown or purplish.

April to October: in dry soil practically throughout the United States and southern Canada, and in Mexico and South America. *Plate 161.* A cudweed was something given to a cow to chew when her own cud was lost. In England the name is applied to the somewhat similar cotton-rose (*Filago germanica*) used in the same way. This is less common in America. (See page 506).

Several far-northern species invade the high mountains of New England: *G. supinum*, only 4 inches tall and leaves not more than an inch long; *G. sylvaticum*, like *G. purpureum* but with a tuft of narrow leaves at the base, and all leaves sharp-pointed. *G. saxicola*, like a small catfoot, is known only from certain rock ledges in Wisconsin. The southern *G. calviceps* is found in Virginia.

ANAPHALIS

We have one species of this mainly Asian genus.

PEARLY EVERLASTING, A. MARGARITACEA, is an attractive plant, up to 3 feet tall, with long narrow leaves projecting from the stem, all covered with white wool. The flower-heads are in a cluster on short stalks at the tip of the stem, which has usually no other branches. The bracts are numerous, blunt, and pearly-white. The pistils and stamens usually occupy separate flowers on separate plants, as in *Antennaria;* but pistillate flower-heads may have a few flowers with stamens, in the center.

July to September: in dry open places, especially in sand and gravel, from Newfoundland to Alaska and southward to New York, the mountains of North Carolina, Wisconsin, South Dakota, New Mexico, and California; occasionally found farther south; also in Asia, and in Europe. *Plate 161.*

THE THISTLES (CIRSIUM)

The thistles – some of them at least – are only too familiar as weeds in cultivated fields, pastures, and gardens, and in waste places. Some may question their right to inclusion in this book; but a glance at the photographs will reveal many beauties not at once apparent to the cultivator. They form a large genus, some of the species not easy to distinguish. The outstanding characteristic is, of course, the presence of prickles – on stem, leaves, and in most species on the bracts around the flower-heads. They are distinguished from other prickly composites by two features: the hairs or bristles with which the flowers are mixed; and the pappus, which is composed of hairs which themselves bear hairs along their sides (they are feathery or

PLATE 162

Cirsium pitcheri *Voss*

Cirsium vulgare *V. Richard*

Filago arvensis *Voss*

Cirsium hillii *Dobbs*

Cirsium arvense *Johnson*

Cirsium flodmanii *Johnson*

"plumose"). The flowers are very narrow and all alike (there are no rays), usually blue-purple, but in some species white or yellow.

I. *Two of our species have prickly flanges or "wings" running down the stem from the edges of each leaf.* (*See also* Carduus *and* Onopordon.)

BULL THISTLE, C. VULGARE, is biennial, the first year forming a flat rosette of leaves, the second a flowering stem up to 5 feet tall or even taller. The leaves are coarsely toothed or pinnately lobed; they are usually woolly or webby on the under surface. The flower-heads are 2 or 3 inches tall; the bracts are all tipped with spines.

June to October: in fields, pastures, and waste places practically throughout the country; introduced from the Old World. *Plate 162.* This is called boar thistle in parts of England; why the association with farm animals? It is one of the species possibly used as the origin of the emblem of Scottish royalty, the "Scotch thistle"; another is *Onopordon acanthium.*

C. PALUSTRE is biennial (see *C. vulgare*). The leaves are pinnately cleft, somewhat downy and often webbed with hairs. The flower-heads are about an inch tall, and tightly clustered. The outer bracts are tipped with a very minute prickle; the inner ones have long soft tips.

June to September: in damp woods and clearings from Newfoundland to Michigan and southward to New York.

II. *The remaining species of our range have no prickly "wings" on the stem; the stems are consequently in general not prickly.*

A. Several of these species have white wool on the stems. (See also *C. arvense* under B.)

C. PITCHERI grows about 3 feet tall. The leaves are stalked, the blades pinnately cleft into narrow lobes which may end in small, weak prickles. The flowers are pale yellow or cream-colored. The outer bracts have prickles; the inner have soft tips.

May to September: in sand, around the Great Lakes, in Ontario, Michigan, Illinois, Indiana, and Wisconsin. *Plate 162.*

C. FLODMANII is a slender plant usually about 2 feet tall. The leaves are commonly very deeply pinnately cleft into lobes each ending in several spine-tipped teeth. The flowers are purple. The outer bracts bear spines; the inner have soft tips.

June to September: in fields, meadows, and prairies from Vermont to Saskatchewan and southward to Iowa, Colorado, and Arizona. This is a native of the West which has wandered northeastward into our range. *Plate 162.*

C. UNDULATUM resembles *C. flodmanii* but is larger, reaching 4 feet in height. The leaves are not so deeply cleft. The flower-heads are taller (usually more than 2 inches as against less than 2), and the spines on the outer bracts mostly longer ($\frac{1}{8}$ to $\frac{1}{3}$ inch as against $\frac{3}{16}$ or less).

June to October: in prairies and other open places from Michigan to British Columbia and southward to Indiana, Kansas, Texas, and Arizona.

B. In the remaining thistles without prickly-winged stems the stems are not covered with white wool (except in a form of *C. arvense*). The first four of these have small flower-heads, not usually more than $1\frac{1}{2}$ inches tall. The rest have heads generally 2 inches tall or taller.

CANADA THISTLE, C. ARVENSE, is one of the worst pests in grainfields and other cultivated land. Its flowering shoots, from 1 to 5 feet tall or taller, grow from deep underground stems; they bear many smallish, curly, cleft, prickly leaves and numerous flower-heads at the tips of slender branches. (They are white-woolly in one form). The heads are narrow, with prickle-tipped outer bracts. The flowers are lilac or pinkish (or occasionally white).

June to October: an European weed now naturalized practically throughout the country. *Plate 162.* It is misnamed, for it did not come from Canada; the English call it creeping thistle. Some plants bear flowers with stamens only; the pistils, with no stamens, are found on other plants; and some have "perfect" (bisexual) flowers.

C. CAROLINIANUM is a southern thistle generally from 2 to 5 feet tall, with rather few leaves which may be unlobed and untoothed but prickly-edged; they may be white with wool on the under surface. The flower-heads are borne on long leafless stalks. The outer bracts are prickle-tipped, the inner ones with soft tips. The flowers are red-purple.

May and June: in dry open woodland from southern Ohio to southern Missouri and southward to Florida and Texas.

C. virginianum, with forty or more leaves on the stem which reaches 4 feet or more in height, and small leaves on the stalks of the flower heads, is similar to *G. carolinianum*. It grows on the coastal plain from New Jersey to Florida. *C. nuttallii* reaches northward to southeastern Virginia. It may be 10 feet tall or even taller. It has no white wool on the mature leaves. The flowers are pink or white.

PASTURE THISTLE, C. PUMILUM, has a hairy stem from 8 inches to 3 feet tall, rising from a rosette of green, pinnately cleft leaves. The leaves on the stem

PLATE 163

Carduus nutans *Rhein*

Cirsium horridulum *Johnson*

Cirsium muticum *Johnson*

Cirsium discolor *Rickett*

Cirsium muticum *Scribner*

Carduus acanthoides *Johnson*

Cirsium pumilum *Rickett*

are very prickly. The heads are short-stalked, sometimes with leaves (or leaflike bracts) immediately below; they are large – sometimes 4 inches tall. The flowers are purple or white. Most of the bracts are prickle-tipped.

June to September: in old fields and pastures from southern Maine to Ohio and southward to Maryland and North Carolina. *Plate 163.*

C. HILLII resembles *C. pumilum*. The prickles on the outer bracts are only $\frac{1}{8}$ inch long or less. The bracts are marked by a dark sticky ridge.

June to August: in sandy soil from Ontario to Manitoba and southward to Pennsylvania and South Dakota. *Plate 162.*

C. DISCOLOR has numerous leaves deeply pinnately cleft and bearing a felt of white wool on the under surface. The uppermost leaves on each stem are close around the flower-head. The outer bracts have a weak prickle bent downward; the inner ones have a long colorless soft tip. The flowers are purple.

July to October: in prairies and woodland from Quebec to Manitoba and southward to Georgia, Tennessee, and Kansas. *Plate 163.*

C. ALTISSIMUM closely resembles *C. discolor*, but is taller (12 feet tall or taller as against 10 feet), with leaves less deeply cleft or lobed (sometimes unlobed).

July to October: in fields, river-bottoms, and woods from Massachusetts to North Dakota and southward to Florida and Texas.

SWAMP THISTLE, C. MUTICUM, has a rosette of long-stalked leaves from which arises a hollow flowering stem up to 10 feet tall, usually branched. The leaves on the stem are rather thin, deeply pinnately cleft. The bracts are covered with a web of hairs; the outer ones have minute spines at the tip, the inner ones papery tips. The flowers are purple.

July to September: in moist woods and swamps from Newfoundland to Saskatchewan and southward to Maryland, the mountains of North Carolina and Tennessee, and Louisiana. *Plate 163.*

YELLOW THISTLE, C. HORRIDULUM, grows from 1 to 5 feet tall, with very prickly, pinnately cleft leaves. It is at once identified not only by the pale yellowish flowers (pale purple in one form) but by the circle of prickly leaflike bracts which closely invest the flower-head. The bracts themselves are soft-tipped.

May to August: in sandy and peaty places, wet meadows, etc. on the coastal plain from southern Maine to Florida and Texas. *Plate 163. Horridus* in Latin means "bristly" or "thorny." *Horridulus* is the diminutive.

C. repandum is a southern thistle which extends to southeastern Virginia. It grows up to 2 feet tall. The leaves have wavy edges, and webbed hairs. The bracts of the large heads bear a broad sticky band and strong prickles. The flowers are purple.

CARDUUS

The thistles of genus *Carduus* differ from *Cirsium* only in having pappus-hairs that have no branch hairs – they are not plumose. The two genera have been united by some authors. The stem is beset with prickly "wings" – projecting flanges that continue the edges of the leaves. This is a large Old-World genus of which three species have become naturalized in this country.

NODDING or MUSK THISTLE, C. NUTANS, is distinguished by having a single drooping or "nodding" (*nutans*) flower-head. The plant stands from 1 to 3 feet tall. The bracts are broad and the outer ones are bent backward. The heads are large, up to 2 inches across, with purple flowers.

June to October: in fields and waste land from Quebec to Iowa and southward to Maryland and Missouri. *Plate 163.*

C. acanthoides (*Plate 163*), with erect flower-heads borne singly or in clusters, and *C. crispus*, welted thistle, with narrower flower-heads, occur less commonly through our range.

ONOPORDON

One species is sparingly established in the eastern United States.

SCOTCH or COTTON THISTLE, O. ACANTHIUM, differs from *Cirsium* in having unbranched pappus-hairs and from *Carduus* in having no bristles ("chaff") among the flowers. It grows from 3 to 10 feet tall. The stem is prickly-winged like that of *Cirsium vulgare* but more so. The flower-heads are large, from 1 to 2 inches across, with spine-tipped bracts and purple or crimson flowers.

July to October: in waste places throughout our range but uncommon. For the name Scotch thistle see also under *Cirsium vulgare*.

PLATE 164

Eupatorium serotinum *Johnson*

Eupatorium dubium *V. Richard*

Eupatorium rugosum *Johnson*

Eclipta alba *Elbert*

Eupatorium fistulosum *Rickett*

Eupatorium purpureum *Rickett*

Eupatorium maculatum *Johnson*

BURDOCKS (ARCTIUM)

Burdocks seem unlikely candidates for the status of wild flowers. They are weeds, often a nuisance, and sometimes poisonous to cattle. However, close examination reveals small flowers of considerable beauty and interest. These are largely enclosed in the bracts, and the fruits that they form – achenes – are likewise concealed. The bracts form the "bur" with its hooked prickles. There has been much confusion over the names and characteristics of the species.

COMMON BURDOCK, A. MINUS, is a bushy-branched plant up to 5 feet tall. The leaves have ovate blades on distinct stalks. The flower-heads have very short stalks or none, and are an inch across or less. The purple or pink flowers project from the bristly bracts.

July to October: an Old-World plant now established in waste places practically throughout this country. *Plate 167.* The young shoots have caused poisoning of cattle. Plants with larger heads (up to $1\frac{3}{5}$ inches across) and bracts longer than the flowers are called *A. nemorosum.* Species with heads on stalks are *A. tomentosum* (*Plate 167*) and *A. lappa;* the former with webbed hairs on the bracts. These are less common but widespread.

MIKANIA

There are many species of *Mikania* in the warmer parts of the New World. One has spread to North America and to the Old World.

CLIMBING HEMPWEED, M. SCANDENS, is the only vine among our composites. It twines and clambers among other plants, with a stem often 15 feet long or even more. The leaves are paired, stalked, the blades ovate, tapering to the sharp tip, indented at the base, sometimes toothed or wavy-edged. The flower-heads are in compact clusters; each contains only four flowers (there are no rays) surrounded by four bracts.

July to October: in thickets and swamps and moist hedgerows from Maine to Florida and Texas and inland in Ontario, Michigan, and Missouri; also in the tropics of Old and New Worlds. *Plate 160.*

JOE-PYE-WEEDS, BONESETS, AND THEIR RELATIVES (EUPATORIUM)

The genus *Eupatorium* includes many tall and handsome plants of late summer and autumn, flowering with the goldenrods and asters – as well as smaller woodland species. Some are easily recognized, others more difficult to identify. Some are confusing because of crossing. The leaves are mostly in pairs or circles, but some species bear leaves singly. The heads contain only tubular flowers of the "disk" type, no rays. Many heads are clustered to form a generally flat or round-topped inflorescence.

Different authorities list twenty-three or twenty-five species for our range (there are several hundred others elsewhere). Twenty are described below, with brief mention of the others. We here separate most of them into two groups by the arrangement of their leaves. One species is treated separately (under III), having divided leaves.

I. *Species whose leaves are mainly in circles of three or more (some of the upper leaves may be paired or single). The flowers are generally pinkish or pale purple (white in one species). The first four are generally known as Joe-Pye-weed. Joe Pye is said to have been an Indian who used one of these plants in curing fever. There actually was a Joseph Pye in the eighteenth century, perhaps of the same family as the Joe Pye of the legend. The plants, however, have no curative properties known to modern medicine.*

To distinguish these four species one must examine flower-heads carefully as well as the leaves. They have been often confused. All flower from July to September.

(*Compare both II and III.*)

SWEET JOE-PYE-WEED, E. PURPUREUM, may grow up to 7 feet tall. The leaves are mostly in threes and fours, short-stalked with lanceolate or ovate sharply toothed blades; they have one main vein. Each flower-head contains usually from three to seven flowers. The plant when bruised smells of vanilla.

In woodland (dry or wet) from New Hampshire to Minnesota and Nebraska and southward to Florida, Tennessee, and Oklahoma. *Plate 164.*

E. FISTULOSUM may reach 7 feet in height. The stem is hollow ("fistulose"). The leaves are in circles of from four to seven, with rather narrow elliptic or lanceolate, blunt-toothed, often rough blades on short stalks; they have a single main vein. Each flower-head contains from five to eight flowers.

PLATE 165

Eupatorium hyssopifolium *Gottscho*

Eupatorium perfoliatum *Johnson*

Eupatorium rotundifolium *Scribner*

Antennaria plantaginifolia (male) *Murray*

Antennaria plantaginifolia (female) *Johnson*

Eupatorium coelestinum *Johnson*

In damp, often wooded, places from Maine to Iowa and southward to Florida and Texas. *Plate 164.*

E. MACULATUM is from 2 to 7 feet tall. The leaves are mostly in fours and fives, with lanceolate or narrowly ovate blades on very short stalks, the teeth either sharp or blunt; there is one main vein. Each flower-head contains from eight to twenty or more flowers. The inflorescence is flatter than those of the other Joe-Pye-weeds.

In damp thickets and meadows from Newfoundland to British Columbia and southward to Maryland, the mountains of North Carolina, Ohio, Illinois, Nebraska, New Mexico, and Washington. *Plate 164.*

E. DUBIUM is the smallest of these species, rarely exceeding 5 feet in height. The leaves are in threes and fours, the coarsely toothed, ovate blades on short stalks; there are often three main veins. Each flower-head contains from five to twelve flowers.

In swamps and other moist places on the coastal plain from Nova Scotia to South Carolina. *Plate 164.*

E. HYSSOPIFOLIUM, from 1 to 4 feet tall, has very narrow leaves mostly in threes and fours. In most of the axils there are tufts of leaves (branches that do not grow longer). The flowers are white, about five in each head.

August to October: in dry open places from Massachusetts to Florida and thence to Texas; inland to Ohio and Tennessee. *Plate 165.*

II. *Species whose leaves are not in circles: they are mostly paired. The flowers are generally white (blue, violet, or pinkish-purple in two species). These may be classed in two groups according as their leaves have stalks or not.*

A. Species with paired leaves, at least the lower ones stalked.

WHITE SNAKEROOT, E. RUGOSUM, is a fairly common woodland plant, from 1 to 5 feet tall. The leaf-blades are ovate, with large teeth, the larger ones on stalks an inch or more long. The flowers are white. The bracts are sharp-pointed and nearly all of the same length.

July to October: in woodland from Quebec to Saskatchewan and southward to Virginia, the mountains of Georgia, and Texas. *Plate 164.* Many plants have been called snakeroot from their supposed powers of curing snakebite. This species is poisonous. Cattle that graze it become subject to the disease known as "trembles." The poison is soluble in milk and is thus transmitted to those who drink it; the resulting symptoms are called "milk sickness."

E. AROMATICUM closely resembles *E. rugosum* and has been confused with it. The leaf-blades are thicker and less sharply toothed, and have stalks less than an inch long – the upper ones have no stalks. The bracts of the flower-heads are of several lengths and tend to be broader towards the tip.

August to October: in dry woodland mostly on the coastal plain from Massachusetts to Florida and Louisiana; inland to Ohio and Tennessee. In spite of the name, the plant is not especially aromatic.

LATE BONESET, E. SEROTINUM, is from 3 to 7 feet tall. The leaf-blades are mostly lanceolate and sharply toothed, often with three or even five main veins; the stalks of the lower ones are an inch or more long. The flowers are white or pale lilac.

August to October: in moist woods and old fields from Massachusetts to Wisconsin and perhaps Minnesota and southward to Florida, Texas, and Mexico. *Plate 164. Serotinum* means "late (flowering)."

MISTFLOWER or BLUE BONESET, E. COELESTINUM, closely resembles the garden flower of tropical origin called ageratum and is sometimes so named. It grows from 1 to 3 feet tall, or taller. The leaf-blades are more or less triangular, rather wrinkled, with blunt teeth; the stalks are short. The flowers are light, bright blue or violet (forms are known with reddish or white flowers).

July to October: along streams and in moist woods and meadows from New Jersey to Kansas and southward to Florida and Texas; also in the islands of the Caribbean. *Plate 165.*

E. INCARNATUM is a plant with a straggling stem from 1 to 5 feet long. The leaf-blades are triangular, with blunt teeth; the stalks from $\frac{3}{5}$ to $1\frac{3}{5}$ inches long. The flowers are pink or lilac.

August to October: in swamps and wet woods from Virginia to Missouri and southward to Florida and Arizona and Mexico. The foliage has the fragrance of vanilla when dried.

B. Species with paired leaves that lack stalks or have very short stalks. The flowers are white.

THOROUGHWORT or BONESET, E. PERFOLIATUM, is easily distinguished by its leaves, which are joined around the stem so that the stem seems to grow through a leaf with two ends (*per*, "through"; *folia*, "leaves"). The leaves are long, narrow, tapering, much wrinkled, toothed, rather light green. The whole plant is hairy. The flowers are white.

July to October: in moist places from Quebec to Manitoba and southward to Florida and Texas. *Plate 165.* Forms are known with purplish flowers and with leaves not joined; it is possible that they are hybrids

PLATE 166

Vernonia missurica *Gerard*

Vernonia noveboracensis *V. Richard*

Elephantopus carolinianus *Core*

Liatris scariosa *Roche*

Vernonia baldwinii *Rickett*

Liatris spicata *Uttal*

with other species. The name thoroughwort is of course the same as "throughwort" and refers to the "perfoliate," "through-the-leaves," condition. The plant was formerly much used medicinally, the extract having tonic, cathartic, and emetic effects. Whether or not it was ever efficacious in the setting of broken bones I do not know. Several other species are similarly named.

E. ALBUM grows up to 3 feet tall, or taller. The leaves are narrow, of various shapes – elliptic or oblong or wider near the base or near the tip – and edged with large blunt teeth. There are about five white flowers in each head.

July to October: in dry woodland, often in sandy places near the coast from Long Island to Florida and Louisiana; inland from Maryland to South Carolina and Arkansas.

Another species of the coastal plain is *E. leucolepis*, a slender plant with thick, narrow leaves (often without teeth) covered with very fine down.

E. ALTISSIMUM is not the "tallest" (*altissimum*) of these plants, since it shares its 7-foot maximum height with several others. The leaves are numerous, narrow, with three or five main veins, tapering to the narrow base and generally sharply toothed towards the tip. The whole plant is usually hoary with a minute down. The flowers are white, about five to a head.

August to October: in woodland and prairies from Pennsylvania to Minnesota and Nebraska and southward to North Carolina and Texas.

E. ROTUNDIFOLIUM is from 1 to 5 feet tall, with leaves rather like those of *E. rugosum* except for the lack of stalks. The plant is generally hairy throughout. There are from five to seven white flowers in each flower-head.

July to October: in open woodland and plains from Long Island to Oklahoma and southward to Florida and Texas; perhaps also farther northeastward. *Plate 165*. Plants with sharply toothed and pointed leaves are by some considered a separate species, *E. pubescens*.

E. SESSILIFOLIUM is from 2 to 5 feet tall, with narrow, toothed leaves which are broadest at the round base. The leaves and stem are smooth or nearly so. The flowers are white; there are five or six in each head.

July to September: in woodland from New Hampshire to Minnesota and southward to Georgia and Arkansas.

E. PILOSUM is from 2 to 5 feet tall, with oblong or ovate leaves edged with a few large blunt teeth and sometimes cleft at the base. The plant is roughish. The flowers are white, about five to a head.

August and September: in bogs and wet, acid soils, from New England to Florida and thence to Louisiana, and inland to Kentucky.

III. E. CAPILLIFOLIUM has a rather rough stem up to 10 feet tall. It is easily distinguished from all our other species of this genus by its leaves, which are pinnately divided into threadlike segments. The inflorescence also is very distinct, being tall and pyramidal rather than flat or domed. The flower-heads are very numerous and small with from three to six greenish flowers in each.

September to November: in fields and borders of woods from Massachusetts to Florida and Texas, inland to Arkansas. A weedy species, unlike all the others and suggesting rather a species of *Artemisia*.

Several other species are of more limited occurrence in our range. *E. recurvans*, *E. cuneifolium*, *E. linearifolium*, *E. cordigerum*, *E. semiserratum* are southerners that extend up into Virginia. *E. saltuense* is known only from southeastern Virginia. It grows to about 4 feet in height, with narrow, stalkless, toothed leaves, mostly in pairs. It resembles *E. altissimum* but the leaves have only one strong vein, and are smooth or almost so. *E. luciaebrauniae* inhabits certain moist sandstone rocks in Kentucky. It resembles *E. rugosum* but does not exceed 4 feet in height, and has very delicate leaves. *E. resinosum*, with narrow paired leaves, may be identified by having usually more than ten white flowers to a head. It grows in wet places in pine barrens in New Jersey and Delaware.

THE IRONWEEDS (VERNONIA)

The ironweeds are mostly tall, handsome plants with leafy stems crowned by broad clusters of usually purple flower-heads. The leaves are without stalks or nearly so. The flowers are all tubular; there are no rays. The flower-heads are closely invested by their bracts, whose tips only may be loose and spreading; these are helpful in identification. The principal technical characteristic of the genus is in the pappus, which is double: an inner circle of long bristles surrounded by an outer circle of short bristles or scales. The flowers may be white or pink, and the pappus varies from purple to bronze, tawny, or white – even in the same species. Besides the species described below, *Vernonia* has hundreds of species in South America and in Africa and Asia.

Our species interbreed in nature, so that many plants are found of intermediate character. Fortunately for persons who live in the northeastern states, only one species is at all common there. Difficulties with the genus are mostly with midwestern plants.

PLATE 167

Liatris squarrosa *D. Richards*

Arctium minus *Gottscho*

Liatris cylindracea *Johnson*

Arctium tomentosum *Johnson*

Liatris punctata *Johnson*

Liatris squarrosa *V. Richard*

A curious difficulty exists in photographing these plants: the flowers are bright pink in the color transparencies, no matter what make of film is used or who is using it. The blue element of the natural purple color disappears in the process. The cause is unknown. I have seen a plant of *V. noveboracensis* in the evening with bright blue flowers; next morning they were the usual purple. Perhaps the blue is produced by refraction of light from the surface, while the actual pigment of the corolla is red.

All the species described below flower in late summer and autumn – from July to October.

I. *Species whose bracts end in loose, hairlike tips. These three species are easily distinguished by the ranges in which they occur.*

NEW YORK IRONWEED, V. NOVEBORACENSIS, has a smooth or downy stem often 6 or 7 feet in height with numerous lanceolate leaves, their edges toothed or plain. There are from thirty to fifty flowers in each head. The bracts narrow rather abruptly into the hairlike tip.

In low fields and marshy places near the coast from Massachusetts to Georgia and Mississippi, inland to West Virginia and Ohio. *Plate 166.*

V. GLAUCA has a smooth stem from 2 to 5 feet tall, with ovate, sharply toothed leaves. The distinctive characteristic is the color of the pappus, which is straw- or cream-colored; but this is not very satisfactory, since the preceding species may have pappus of a similar shade.

In woods, mostly in the mountains, from New Jersey and Pennsylvania to Georgia and Alabama.

V. CRINITA is reported to grow from 3 to 10 feet tall, but is usually nearer the lower of these heights, at least in our range. Its leaves are lanceolate, rather narrow, their edges finely toothed or plain. The flower-heads are large, with from fifty-five to ninety flowers. The bracts taper gradually into their hairlike tips.

In woods and along streams in Missouri, Kansas, Arkansas, and Oklahoma. This species is reported to hybridize with all the other ironweeds in its range!

II. *Species whose bracts are blunt or pointed but not with hairlike ends. Most of these are midwestern.*

V. BALDWINII is easy to recognize by the bracts of the flower-heads: they have narrow sharp tips bent outward and downward. They are commonly beset with webbed hairs. The plants are from 2 to 5 feet tall, with mostly ovate, sharply toothed leaves. The pappus is typically purple but may be tawny.

In fields and woodland from Illinois to Oklahoma and Arkansas. *Plate 166.*

V. MISSURICA is from 3 to 5 feet tall, with lanceolate leaves which may be coarsely toothed or plain on the edges. The heads are rather large, with up to fifty-five flowers. The bracts are closely placed around the head, often purple, with short, sharp points. The pappus is tawny or purple. *Plate 166.*

V. ALTISSIMA is commonly a very tall (*altissima*) plant, from 3 to 10 feet. The leaves are lanceolate, long and rather narrow, finely toothed. The flower-heads are numerous and small, with not more than thirty flowers and less than $\frac{1}{2}$ inch tall. The bracts are blunt or short-pointed and pressed closely together.

In moist woodland from western New York to Michigan and Nebraska and southward to Georgia and Louisiana. *Plate 169.*

V. FASCICULATA grows from 2 to 6 feet tall. The leaves are narrow, toothed, and marked with small pits on the under surface (these are most easily seen on dried leaves). The heads are small, with not more than thirty flowers. The bracts are blunt or bluntly pointed. The pappus is purple.

In moist prairies and woodland from Ohio to Minnesota and Manitoba and southward to Missouri and Texas.

THE BLAZING-STARS (LIATRIS)

Several quite different plants are called blazing-stars. Those in the genus *Liatris* are slender unbranched plants, generally 3 or 4 feet tall, with mostly narrow leaves in a close spiral, and a spike of rose-purple flower-heads. The leaves are generally "punctate" – i.e. marked with resinous dots. The genus is easily recognized, but the species have been – and may be – badly confused. The reason for this is their propensity for crossing in nature, with the formation of plants with intermediate combinations of characteristics. Moreover, the only valid means of identifying the species involves the hand magnifier: one must examine the pappus and even count the flowers in several flower-heads. The bracts of some species have colored margins and add to the handsome display; these also are useful in identification. The ranges of some species may help to identify them.

This is a native North American genus. Some species are cultivated, and more deserve to be. One English name serves for all. They all flower in late summer and autumn, from July to October. They are perennial, the flowering stem growing from a corm or rhizome underground.

PLATE 168

Liatris ligulistylis *Johnson*

Liatris intermedia *Johnson*

Liatris pycnostachya *Gottscho*

Liatris graminifolia *Elbert*

Liatris pycnostachya *Johnson*

Liatris aspera *Johnson*

It is easy to separate our species into two groups by the character of the pappus.

I. *Species whose pappus consists of bristles that themselves bear smaller bristles or hairs along their sides – they are feathery or "plumose."*

A. Of these, some species have more than ten flowers in each flower-head.

L. SQUARROSA (1–3 feet) has narrow, stiff leaves, the lowermost much longer than those above. There may be as many as sixty flowers in a head. The bracts are lanceolate or ovate, tapering to sharp, stiff points; the ends of the outer ones are bent outward and downward (they are "squarrose"); the inner ones are erect but rather loose.

In dry open woodland and fields from Delaware to South Dakota and southward to Florida and Texas. *Plate 167*. The plants vary in hairiness – the southwestern plants tending to be hairier – and in the degree to which the bracts are bent.

L. CYLINDRACEA (8 inches–2 feet) is similar to *L. squarrosa* in general form. The flowers are as numerous – up to sixty in a head. The bracts are quite different, being hard and shining, the inner ones broad at the end with a very short sharp point; they lie close one upon the other.

In dry open places from New York and southern Ontario to Minnesota and southward to Ohio and Arkansas. *Plate 167*.

B. Other species with plumose pappus have fewer than ten flowers in each head.

L. PUNCTATA (6 inches–3 feet) usually forms several stems from one corm or rhizome. The leaves are very numerous, very narrow, and stiff. The flower-heads are rather small, without stalks or nearly so, and form a dense spike; each has only from four to eight flowers. The bracts are oblong, with a triangular point, and fringed with hairs.

In dry prairies from Michigan to Alberta and southward to Arkansas and Mexico. *Plate 167*. Several similar midwestern and western plants have been given names: *L. densispicata* in Minnesota, *L. angustifolia* from Missouri to Nebraska, Arkansas, and Texas. *L. mucronata*, with bracts tipped with short, sharp points, is known from Missouri to Texas.

II. *Species whose pappus-bristles have no branch-hairs – they are not "plumose." (A good lens will reveal very small bristles along the sides.)*

A. Of these a number generally have fifteen or more flowers to a head. These form perhaps the most baffling group.

L. BOREALIS (1 to nearly 4 feet) has many leaves (up to eighty-five) which are somewhat broader than those of most blazing-stars, the lower often an inch wide or wider. There are from thirty to sixty or even more flowers to a head. The bracts are often reddish with a narrow, red, petal-like margin; they are roundish, and lie close around the flower-head.

In dry open woods and thickets from Maine to Michigan and southward to Pennsylvania, West Virginia, and Arkansas.

L. SCARIOSA (1–5 feet) has a smooth or finely downy stem bearing up to twenty leaves which are rather broad for the genus, the lowest often with blades up to 2 inches wide on long stalks. The heads are mostly distinctly stalked, with from twenty-five to sixty flowers in each. The bracts are green and leaflike, ovate or oblong, the outer ones loose or bent down, the inner with broad round ends, often with a narrow colored margin.

In dry woodland in the Appalachian mountains from Pennsylvania to Georgia. *Plate 166*.

L. SCABRA (1–3 feet) has a rough ("scabrous") stem. The leaves (up to forty) somewhat resemble those of *L. scariosa* but are narrower. The outer bracts have pointed tips bent outward, the inner are round at the ends; all have a very fine ashy down. There are from twenty to forty flowers to a head.

In dry open woodland from Ohio to Illinois and southward to Alabama, Louisiana, and Oklahoma.

L. ASPERA (1–4 feet) has a downy stem bearing narrow, rough (*asper*) leaves. The bracts are roundish with broad, colored, irregularly toothed edges. There are from twenty to forty flowers in a head.

In dry soil from Ontario to North Dakota and southward to Florida and Texas. *Plate 168*. The smoother southeastern plants have been sometimes known as *L. intermedia*.

L. LIGULISTYLIS (8 inches–4 feet) may have up to one hundred leaves. The bracts have thin, colored, toothed margins. There are up to seventy or even more flowers in a head.

Usually in damp soil from Michigan to Alberta and southward to Ohio, Missouri, and New Mexico. *Plate 168*. The plants from Ohio to Missouri are thought to be hybrids with some other species.

L. EARLEI (1–4 feet) may be known by its downy stem and crowded leaves; the basal leaves are often broad. The heads are comparatively few-flowered (fifteen to twenty-five). The bracts are green and downy, the outer ones lanceolate and sharp-pointed, the inner oblong with broad tips tapering to a short sharp point and usually several irregular teeth.

In dry soil from Indiana southward to Florida and Texas.

Pluchea purpurascens *Johnson*

Marshallia grandiflora *Elbert*

Pluchea foetida *Uttal*

Vernonia altissima *Horne*

Erechtites hieracifolia *Elbert*

Cacalia atriplicifolia *Uttal*

B. The remaining species whose pappus is not plumose have less than twenty flowers to a head.

L. PYCNOSTACHYA (2–5 feet) has very crowded narrow leaves. The flower-heads also are crowded (the meaning of the botanical name). The bracts are oblong with sharp tips which spread outward or are bent downward; they are ridged on the back. There are from five to twelve flowers to a head.

In moist prairies and woodland from Indiana to Minnesota and South Dakota and southward to Kentucky and Texas; and escaped from cultivation in Long Island. *Plate 168.*

L. GRAMINIFOLIA (8 inches–4 feet) has narrow-bladed, stalked basal leaves, their flat stalks edged with hairs (but these leaves may be lacking at flowering time). The leaves on the stem are grasslike (*gramini-* means "grass"). Each head has from five to fifteen flowers. The bracts are oblong, rather narrow, blunt, edged with hairs, ridged or striped on the back.

In woodland chiefly near the coast from New Jersey to Georgia. *Plate 168.*

L. microcephala, with heads only $\frac{1}{8}$ inch thick or even thinner, is a southern species found in Kentucky. Several other southern species extend into Virginia and West Virginia: *L. helleri*, *L. turgida*, *L. regimontis*.

L. SPICATA (1–6 feet) is the species oftenest seen in gardens. It has very many, narrow leaves (the lowest sometimes nearly an inch wide). The flower-heads are in a dense spike, sometimes mingled with long bracts. Each head has from three to eighteen flowers. The bracts are oblong and blunt, green or purplish, with a thin red margin, ridged on the back, closely pressed together.

In meadows, swamps, etc. from southern New York to southwestern Ontario and Wisconsin and southward to Florida and Louisiana. *Plate 166.*

MARSH-FLEABANES (PLUCHEA)

Our marsh-fleabanes have tubular pink or cream-colored flowers (no rays) in numerous small heads, The pappus is formed of numerous bristles. The outer flowers in each head lack stamens. The leaves are borne singly. Our species flowers in late summer, from July to September.

MARSH-FLEABANE, P. PURPURASCENS, is from 1 to 5 feet tall. The leaves are without stalks or with very short stalks and tend to be elliptic or lanceolate. The bracts are usually downy and glandular. The flower-heads form a rather broad and flat inflorescence.

In salt marshes near the coast from Maine to Florida and Mexico; reported also from western New York, Michigan, Kansas and California. *Plate 169.* One variety, which occurs inland and on the coast from Maryland southward, has hairy leaves.

MARSH- or STINKING-FLEABANE, P. FOETIDA, is distinguished by its leaves which are quite stalkless and commonly "clasping" with lobes extending on either side of the stem. They are oblong or lanceolate and finely toothed. The inflorescence is broad but scarcely flat. The bracts are somewhat sticky. The flowers are cream-colored.

In wet meadows, swamps, woods, etc. near the coast from New Jersey to Florida and Texas, and northward to southeastern Missouri. *Plate 169.* The plant has a disagreeable smell. Does it kill or repel fleas?

STINKWEED or CAMPHOR-WEED, P. CAMPHORATA, resembles *P. purpurascens*, but the leaves are more obviously stalked, with lanceolate, toothed blades. The pink flower-heads form a roundish cluster. The bracts are smooth.

In marshes (fresh or brackish) and other wet places from Delaware to Kansas and southward to Florida and Texas. The plant has a strong musky or skunk-like odor, especially when bruised.

KUHNIA

One species of this genus occurs in our range.

FALSE BONESET, K. EUPATORIOIDES, does, as both its names indicate, have something the aspect of an *Eupatorium*. It is a plant from 1 to 5 feet tall with paired, lanceolate leaves toothed or plain on the edges. The flowers are all tubular, cream-colored, up to thirty in a head. The pappus is of plumose bristles; i.e. the bristles bear smaller hairs along their sides and so have a feathery appearance. The bracts are narrow, blunt.

July to October: in prairies and dry open places from New Jersey to Montana and southward to Florida and Arizona. *Plate 173.* The plants vary in leaf shape and hairiness, in size of flower-heads, and in sharpness of bracts. The northwestern plants are more hairy, and have been treated by some botanists as distinct species. The genus name commemorates Dr. Adam Kuhn of Philadelphia, who carried the plants alive to Carl Linnaeus. The great Swedish botanist gave the species its name.

PLATE 170

Lapsana communis *Fischer*

Cichorium intybus *Johnson*

Tanacetum vulgare *Elbert*

Tanacetum huronense *Voss*

Tragopogon major *Williamson*

Sclerolepis uniflora *Rhein*

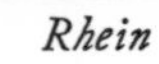

Voss

Tragopogon porrifolius (hybrid)

INDIAN-PLANTAINS (CACALIA)

Indian-plantains are smooth plants of rather weedy aspect with large heads clustered at the summit of the stem. The bracts of a head are all of the same length and in a single circle (but in one species there are some smaller bracts below). The pappus is composed of numerous bristles. The species are mostly distinguished by the shape of their leaves.

INDIAN-PLANTAIN, C. TUBEROSA, from 2 to 6 feet tall, has long-stalked leaves at and near the base, the blades nearly elliptic, mostly without teeth, and with five or seven ribs.

June to August: in damp prairies and bogs from southern Ontario to Minnesota and Nebraska and southward to Alabama and Texas. From the shape and veining of the leaves, this seems to have the best right to the English name.

PALE INDIAN-PLANTAIN, C. ATRIPLICIFOLIA, from 2 to 7 feet tall, has a smooth stem whitened with a bloom. The leaves are as broad as they are long, pale on the under surface, jaggedly toothed or irregularly lobed, with the main veins palmately disposed.

June to September: in dry open woods from New York to Minnesota and Nebraska and southward to Florida, Alabama, and Oklahoma. *Plate 169.*

GREAT INDIAN-PLANTAIN, C. MUHLENBERGII, is our largest species, from 3 to 10 feet tall. Stem and leaves are green, not whitish. The leaves are as broad as or broader than they are long, sometimes kidney-shaped, irregularly toothed, with main veins palmately arranged. The stem is obviously angular or grooved.

June to September: in woodland from New York to Minnesota and southward to Georgia, Alabama and Missouri.

C. SUAVEOLENS is very smooth, from 2 to 5 feet tall. The leaves are shaped like some old-fashioned spears, with a triangular blade which has two sharp triangular lobes at the base; the edge is finely toothed.

July to October: in moist woods from Rhode Island to Iowa and southward to Florida, Tennessee, and Missouri; also in Massachusetts. *Suaveolens* means "sweet-smelling."

ERECHTITES

We have one or perhaps two species of *Erechtites.*

FIREWEED or PILEWORT, E. HIERACIFOLIA, is a rather weedy plant from a few inches to 10 feet tall, with narrow, lanceolate, toothed leaves borne singly on short stalks or none. The tubular whitish flowers are in many small heads, each surrounded by a ring of bracts of equal length with a few smaller ones at the base. The outer flowers lack stamens. The pappus is composed of many white bristles.

July to October: in damp thickets, woods, and waste places from Newfoundland to Minnesota and Nebraska and southward to Florida and Texas. *Plate 169.* A fleshier type with green, brown-edged flowers, *E. megalocarpa,* grows on seashores in sand and in marshes from Massachusetts to Long Island.

FILAGO

Two or three European species of this genus have been found in North America.

COTTON-ROSE or CUDWEED, F. GERMANICA, somewhat resembles a *Gnaphalium,* having an erect stem about a foot tall bearing narrow leaves, all white with wool, and a terminal cluster of small woolly flower-heads. The flowers are all tubular and very narrow, those in the center of a head bearing stamens, the rest having pistils and no stamens. The flower-heads differ from those of *Gnaphalium* chiefly in the possession of "chaff," dry bracts mixed with the flowers. From just below the terminal cluster of heads there usually appear two branches, each terminated in the same way; and these may in turn bear branches.

May to October: occasional in dry fields and waste places from southern New York to Ohio and southward to Georgia. The herbalist Gerard called it "herbe impious" because of the flowering branches which rise above the elder clusters; "pious" (*pius*) in Latin meant not only reverent towards the gods but similarly dutiful towards one's parents and elders generally. For the name cudweed see *Gnaphalium.*

Plate 162 shows the much less common *F. arvensis,* found in Michigan. It has shorter branches, fewer flower-heads, and more wool on the bracts.

PLATE 171

Centaurea vochinensis *Rickett*

Centaurea nigra *Gottscho*

Centaurea vochinensis *Rickett*

Centaurea jacea *Johnson*

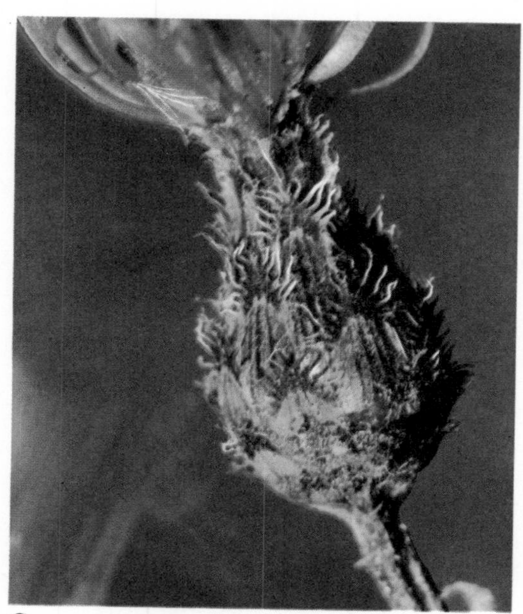

Centaurea maculosa *Rickett*

Centaurea maculosa *Johnson*

BARBARA'S-BUTTONS (MARSHALLIA)

The genus *Marshallia* includes several species of small plants with undivided leaves at or near the base of the stem, which bears a single head of pink, purplish, cream-colored, or white tubular flowers (there are no rays). The leaves have no teeth; they are ribbed lengthwise. The bracts are all nearly equal in length. The pappus is composed of several small scales.

The English name seems to be applied indiscriminately to any of the species, not to any one in particular. I have searched in vain for any clue to the identity of "Barbara"; the older American botanists do not mention her. Perhaps Saint Barbara? Or perhaps it should be Barbary-buttons, a name applied to a Mediterranean species of *Medicago* (related to alfalfa).

M. GRANDIFLORA is the only species at all widespread in our range. It grows from 8 inches to 3 feet tall. The leaves at the base are three-ribbed, about an inch wide and broadest towards the tip; those on the stem are smaller, narrower, and have one main vein. The flowers are pinkish-purple.

June to August: in woodland and along streams from Pennsylvania to Kentucky and North Carolina. *Plate 169.*

The southern *M. trinervia*, with all leaves about equal and three-ribbed, extends to Virginia; the southwestern *M. caespitosa*, not 2 feet tall, with narrow leaves and lavender, cream-colored, or white flowers, occurs in Missouri.

SCLEROLEPIS

There is only one species of *Sclerolepis*.

S. UNIFLORA is a creeping plant with mostly unbranched stems bearing narrow leaves usually less than an inch long in circles of from four to six. The flowering stems are erect, each terminated by a single small head of pale pink flowers. There are no rays. The pappus is composed of five scales.

July to November: in shallow water from New Hampshire to Florida and Alabama. *Plate 170.*

ELEPHANTOPUS

The genus *Elephantopus* – elephant's-foot – is mainly tropical. It is represented in temperate North America by species that in no way suggest the feet of a pachyderm. The peculiarity of the genus is in the arrangement of the flowers. There are from two to five in small heads which are themselves clustered among leafy bracts into a mass which may be mistaken for a single head. The individual flowers are all tubular and blue or purple, not quite radial in symmetry. The pappus is composed of a few small scales whose tips extend into bristles.

E. CAROLINIANUS is the only species widespread in our range. It grows from 8 inches to 3 feet tall, usually much branched; with leaves borne singly. The leaves are in general elliptic or ovate, scalloped on the edges, sometimes with short stalks.

August to October: in dry open woods from New Jersey to Kansas and southward to Florida and Texas. *Plate 166.*

Devil's-grandmother or tobacco-weed, *E. tomentosus*, with few or no leaves on the stem, those at and near the base velvety, is found from Maryland and Kentucky southward. *E. nudatus*, also with all leaves at or near the base but none velvety, grows along the coast from Delaware southward.

STAR-THISTLES AND KNAPWEEDS (CENTAUREA)

Most species of *Centaurea* are of European origin, but several have become common weeds in parts of North America. There is one native species that finds its way from the Southwest into our range. The leaves are either narrow and undivided or pinnately divided into narrow segments, all without teeth on the margins or nearly so and all borne singly. The flowers are generally blue or purple, sometimes pink or white, all tubular but the outer ones larger and not quite radially symmetric, their larger corolla-lobes sometimes simulating rays. The bracts are peculiar and must be examined with a lens for sure identification: each has a terminal appendage, in most species partly or wholly brown and bearing a characteristic fringe. The pappus may be bristles or scales or be lacking.

I. *In one common weed the bracts of the flower-heads end in stiff spines which spread in all directions.*

PLATE 172

Prenanthes alba *Johnson*

Prenanthes racemosa *Johnson*

Prenanthes nana *Scribner*

Prenanthes altissima *Rickett*

Prenanthes serpentaria *Rhein*

Prenanthes autumnalis *Allen*

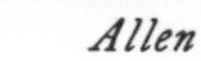

Prenanthes trifoliolata *Ryker*

Sonchus asper *Johnson*
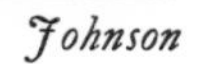

CALTROPS or STAR-THISTLE, C. CALCITRAPA, is from 8 to 20 inches tall, the lower leaves deeply pinnately cleft or divided, the lobes or segments more or less toothed. The few flowers in each head are tubular and purple.

June to October: in waste places from New York and Ontario southward, and probably farther to the west.

II. *The remaining species have bracts variously terminated but not by stiff spines.*

A. Of these, some may be distinguished by having leaves not divided or cleft. To identify them it is helpful to study the drawings of bracts.

CORNFLOWER, BLUEBOTTLE, or BACHELOR'S-BUTTONS, C. CYANUS, is from 1 to 4 feet tall, with long branches ending in single heads. The outer bracts have a toothed whitish margin, the inner ones often a jagged and colored tip. The flowers vary in color. The pappus is very short.

July to September: escaped from cultivation in fields and waste places. Cornflower because it grew in the "corn" – in England not maize but generally wheat; it is now rare there. Why bachelor's-buttons? They have been likened to certain ragged cloth buttons once worn by gentlemen. But the many "buttons" (flower-heads or flowers) known in England by this name are associated with the desire of a maid to win a susceptible bachelor, and with certain rites practiced to that end.

BROWN KNAPWEED, C. JACEA, from 1 to 3 feet tall, has stalked basal leaves with lanceolate or ovate, sometimes toothed blades. The bracts bear brown, rather jagged appendages. There is no pappus.

June to September: in fields and waste places from Quebec to Iowa and southward to Virginia, Ohio, and Illinois. *Plate 171.*

BLACK KNAPWEED or SPANISH-BUTTONS, C. NIGRA, is a common English weed now widespread in this country. It grows from 1 to nearly 3 feet tall. The lowest leaves are lanceolate and sometimes toothed or wavy-edged, the upper very narrow. The whole plant is rough. The outer bracts bear large, almost black, triangular appendages with long fringes. There is a minute pappus.

July to October: in fields and waste places almost throughout our range. *Plate 171.*

C. VOCHINENSIS resembles *C. nigra.* The bracts are shorter, with a smaller appendage fringed with from five to seven long hairs on each side; the inner ones have a colored, jagged tip. Pappus is lacking.

June to September: in fields and waste places from Maine to Ontario and southward to Virginia and Missouri. *Plate 171.*

C. repens and *C. nigrescens* are two other species that are somewhat common weeds. *C. repens* has bracts without a fringe on the appendage; *C. nigrescens* is like *C. nigra* with broader leaves which surround the flower-heads, and a shorter fringe on the bracts. *C. americana* is our one native species; it is southwestern and only in Missouri in our range. It has large flower-heads (often 2 or 3 inches tall). The bracts have a long, fringed appendage. The leaves are oblong. The stems are thickened just below each flower-head. There is a small pappus.

B. Our remaining species have at least their lower leaves pinnately divided or deeply cleft into narrow segments or lobes.

C. MACULOSA has a branching, wiry stem from 1 to 4 feet tall. The bracts are pale and ribbed, bearing a short, dark appendage fringed with from five to seven hairs on each side.

June to October: in fields and waste places throughout our range and westward. *Plate 171.*

C. scabiosa, with large flower-heads 2 inches or more across and the appendages $\frac{1}{6}$ to $\frac{1}{4}$ inch long, has been found in a few places in Quebec, Ontario, Ohio, and Iowa.

ADENOCAULON

Of this small genus we have one species; there are others in the West, South America, and Asia.

TRAIL-PLANT, A. BICOLOR, has stalked leaves with white wool on the under surface of the blades, which are triangular and toothed. The bracts are all equal in length; the flowers are whitish. Flowers appear in June and July, from the region of the Great Lakes to South Dakota; and in the West. The name of the genus means "gland-stem," from the numerous stalked glands in the inflorescence.

TANACETUM

This genus consists of strong-scented plants with leaves pinnately divided into segments cut into many fine lobes, and numerous small yellow flower-heads without rays in a dense, rather flat cluster. There is no evident pappus.

PLATE 173

Tragopogon major *Johnson*

Taraxacum officinale *Rickett*

Leontodon autumnalis *Smith*

Matricaria maritima *Johnson*

Kuhnia eupatorioides *Johnson*

Hypochoeris radicata *Johnson*

TANSY, T. VULGARE, is a well-known herb brought from Europe and escaped from cultivation in this country. The leaves are smooth. The very numerous heads are $\frac{2}{5}$ inch wide or narrower. The plant is up to 5 feet tall.

July to October: on roadsides and in fields and waste land, even in woods, through most of the United States. *Plate 170.*

Tansy is a famous herb. The bitter-tasting infusion, tansy tea, has been used to prevent miscarriage; the oil, which is poisonous, has procured abortion (sometimes with disastrous results). Tansy puddings and sausages have reputed medicinal virtues; the bruised leaves were laid on sprains and bruises. Some English names are bachelor's-buttons, bitter-buttons, golden-buttons, scented-daisies, stinking-Willie, and traveller's-rest.

T. HURONENSE rarely exceeds 3 feet in height. It is hairy or woolly throughout. The heads are less numerous, rarely more than fifteen, $\frac{2}{5}$ inch wide or wider.

June to August: on sandy and gravelly beaches and shores from Newfoundland to Hudson Bay and Alberta and southward to Maine and the Great Lakes. *Plate 170.*

CICHORIUM

One species of this European genus is now a common roadside weed in North America. Another is cultivated as endive.

CHICORY, C. INTYBUS, is one of the few composites of our third group (with no tubular flowers) that have blue flowers. The plants grow up to 5 feet tall, with small oblong or lanceolate, often toothed, leaves without stalks, sometimes "clasping" the stem; the basal leaves are generally pinnately cleft. The flower-heads are scattered along the stem, with practically no stalks. The flowers may be pink or white.

June to October: in fields and on roadsides practically throughout the country. *Plate 170.* The root furnishes a flavoring for coffee esteemed in France and other countries. Another name is blue-sailors. In England the plant is called also succory.

THE RATTLESNAKE-ROOTS (PRENANTHES)

Various species of *Prenanthes* were reputed to furnish effective remedies for snake-bite – probably with no good basis of fact. The very bitter roots have been used in making tonic (perhaps on the principle that anything that tastes so bad must be good for you). These are tall, leafy plants (the leaves borne singly) with milky juice and narrow flower-heads which usually hang in a loose or narrow cluster near the tip of the stem. The bracts are of two sizes, the longer forming a cylinder around the flowers, the others, much shorter, at the base of this cylinder. The pappus is composed of numerous bristles. The flowers are all of the raylike kind, and vary in color.

For identification we may first group the species by their leaves.

I. *Species that have at least some of the leaves on the stem stalked.*

A. Of these, one has fewer than eight flowers in each head.

P. ALTISSIMA is from 2 to 7 feet tall. The leaves have ovate or triangular blades, sometimes indented at the base; often deeply lobed or cleft palmately, or toothed. There are five long bracts in each head and five or six greenish-white flowers. The pappus is generally creamy-white but in one form (from Indiana to Missouri and Louisiana) it is reddish or brown.

July to October: in moist woods from Quebec to Manitoba and southward to Georgia, Tennessee, and Louisiana. *Plate 172.*

B. The other species with stalked leaves have eight or more flowers to a head.

WHITE-LETTUCE or RATTLESNAKE-ROOT, P. ALBA, is from 2 to 5 feet tall, with more or less triangular leaf-blades which may be variously wavy-toothed or palmately lobed or cleft. There are from eight to twelve whitish flowers in a head, surrounded by from six to eight bracts. The pappus is reddish-brown.

August and September: in moist woods from Quebec to Saskatchewan and southward to Georgia, Tennessee, Missouri, and South Dakota. *Plate 172.*

LION'S-FOOT or GALL-OF-THE-EARTH, P. SERPENTARIA, may be up to 4 feet tall. The leaves are ovate and blunt, often pinnately lobed. The flowers are whitish or pink, about 10 to a head, surrounded by about 8 bracts. The pappus is cream-colored.

PLATE 174

Sonchus oleraceus *Johnson*

Hieracium aurantiacum *White*

Pyrrhopappus carolinianus *Rickett*

Hieracium pilosella *Uttal*

Hieracium floribundum *Elbert*

Hieracium vulgatum *Uttal*

V. Richard

Hieracium pratense and Hieracium venosum

August to October: in woodland from Massachusetts to Florida, Tennessee, and Mississippi. *Plate 172*. See the note under *P. trifoliolata*. The name lion's-foot recalls *Leonurus*, "lion's-tail"; certain kinds of lobed leaves apparently reminded early botanists of the king of beasts.

GALL-OF-THE-EARTH, P. TRIFOLIOLATA, grows from 6 inches to 5 feet tall. The leaves vary greatly in shape, commonly lobed, cleft, or divided palmately into three and mostly stalked. In number of bracts and flowers and color of pappus this species resembles *P. serpentaria*.

July to September: in woodland from Labrador and Newfoundland to Maine and Ohio and southward to Georgia and Tennessee. *Plate 172*. These two similar species may be distinguished to some extent by their ranges, one being mainly northern, the other southern. Smaller, northern plants with blackish bracts are sometimes considered distinct as *P. nana*. (*Plate 172*.) A dwarf species, *P. boottii*, only a foot tall or less, with flower-heads forming a raceme, is found in the mountains of New England and New York.

P. CREPIDINEA is a stout plant up to 9 feet tall with numerous flower-heads in a loose cluster. The leaves are large, with more or less triangular or ovate, toothed blades on flat stalks. The bracts are hairy. There are from twenty to thirty-five cream-colored flowers in each head. The pappus is brownish.

August to October: in moist woodland from western New York to Minnesota and southward to Tennessee and Missouri.

II. *Species with leaves on the stem not stalked.*

P. RACEMOSA, up to 5 feet tall, is very smooth and pale. The lowermost leaves are stalked, with more or less elliptic blades. The leaves on the stem mostly "clasp" the stem with the lobes at the base of the blade. The flowers are pink or white, up to twenty-five in a head. The pappus is yellowish.

August and September: in moist thickets and meadows from Quebec to Alberta and southward to New Jersey, Ohio, Missouri, and Colorado. *Plate 172*.

P. ASPERA is similar to *P. racemosa* but rough with minute hairs. The leaves on the stem do not "clasp" it. The inflorescence is a tall, narrow cluster of flower-heads.

August and September; in dry prairies from Ohio to Minnesota and Nebraska and southward to Tennessee, Louisiana, and Oklahoma.

P. AUTUMNALIS is a slender, smooth plant up to 4 feet tall or a little taller. The leaves are lanceolate, without stalks, the lower pinnately cleft, the upper toothed. The inflorescence is narrow. There are from eight to twelve pink flowers to a head, surrounded by about eight bracts. The pappus is a dirty white.

August to October: in sandy soil from New Jersey to Florida and Mississippi. *Plate 172*.

WILD LETTUCE (LACTUCA)

Cultivated lettuce is *Lactuca sativa*. Its wild relatives are unlovely weeds and merit only brief notice in this work. The botanical name is derived from the Greek word for "milk," referring to the milky juice. The English name is a corruption of the Latin. Our species are tall, leafy-stemmed plants, some reaching a height of 10 feet. The leaves are sharply toothed and often pinnately lobed. The flower-heads are small but numerous, the flowers yellow, white, or blue. The pappus is a mass of numerous soft white bristles, in most species at the summit of a long narrow beak which extends from the "seed" (achene). The species named below are widely distributed throughout the United States.

I. *Species with yellow flowers* (*they may turn blue as they wither*)*:* L. canadensis, L. serriola, *sometimes* L. ludoviciana, *with long-beaked achenes.*

II. *Species with blue flowers:* L. ludoviciana *sometimes, with long-beaked achenes;* L. floridana, *with very short-beaked achenes; and* L. pulchella, *with achene-beak intermediate in length.*

TRAGOPOGON

Three species of this Asian and African genus are found in North America. They are readily identified by their smooth stems bearing single large flower-heads and long grasslike leaves which sheathe the stem at their bases. The flowers are all of the "ray" type, none tubular; they are purple or yellow. The bracts are equal in length. The pappus is of feathery or "plumose" bristles – the bristles bearing hairs along their length; it stands at the summit of a slender beak and is very conspicuous – and beautiful – in fruit.

PLATE 175

Krigia dandelion *D. Richards*

Hieracium paniculatum *Rickett*

Krigia virginica *Johnson*

Picris hieracioides *Elbert*

Serinia oppositifolia *Johnson*

Krigia biflora *D. Richards*

SALSIFY or OYSTER-PLANT, T. PORRIFOLIUS, has purple flowers on a stem less than 4 feet tall. The stem is swollen and hollow just below the flower-head. The bracts are an inch long or longer.

April to August: escaped from cultivation and naturalized in fields and on roadsides throughout our range. The root furnishes a well-known vegetable. *Plate 170.*

GOAT'S-BEARD, T. PRATENSIS, has yellow flowers. The stem is not swollen beneath the flower-head. The bracts are mostly less than an inch long.

May to August: naturalized in fields and waste places throughout our range and beyond. The English name is a literal translation of *Tragopogon*. Because it closes at midday, it is known also (in England) as Jack-go-to-bed-at-noon.

T. MAJOR has yellow flowers like those of *T. pratensis* but a swollen stem like that of *T. porrifolius*. The bracts are 2 inches long or longer.

May to July: naturalized in fields and waste places through most of North America except the southeastern states. *Plates 170, 173.*

HYPOCHOERIS

One species of this Old-World genus is widely naturalized in North America.

CAT'S-EAR, H. RADICATA, resembles *Leontodon* in general aspect, being distinguished chiefly by the long slim beak which rises from the achene ("seed") and bears the pappus. In this respect it resembles a dandelion, but the pappus is "plumose" as in *Leontodon*. The bracts are of several lengths and overlap. There are also bracts ("chaff") mixed with the flowers.

May to September: on roadsides and in fields and waste places from Newfoundland to Ontario and southward to North Carolina and Illinois; also in the Pacific states. *Plate 173.* In some places in Linnaeus' works the printer substituted *ae* for *oe*, and the misspelling *Hypochaeris* has lasted to this day in some works. *Hypochoeris* means "little pig" or perhaps "under a pig" – pigs are said to like the root; *Hypochaeris* means nothing.

HAWKBITS (LEONTODON)

The hawkbits resemble small-flowered dandelions; and indeed their botanical name means "lion's tooth," which is also the meaning of "dandelion" (dent de lion). Just what part of such plants is comparable to a feline tooth seems not to have been explained; perhaps the jagged leaves at the base, characteristic of both genera. The principal bracts of *Leontodon* are more or less equal, with a few smaller outer ones at the base. The pappus is of "plumose" bristles – i.e. they bear smaller bristles along their sides and so have a feathery appearance; they are not borne up from the achene on a long beak as in the dandelions.

FALL-DANDELION, L. AUTUMNALIS, is up to a foot tall or a little more. The basal leaves are deeply cleft pinnately into narrow lobes. The pappus is tawny. The stem may branch at the tip and so bear more than one flower-head; it bears also small scales.

May to November: in fields and waste land and on roadsides from Newfoundland to Michigan and southward to Pennsylvania; from the Old World. *Plate 173.* In the Atlantic states we find also *L. leysseri*, only 8 inches tall, with basal leaves less deeply cut, and bristly. *L. hispidus*, with bristly stem and leaves, is less common in the same range.

PYRRHOPAPPUS

Most of the species of *Pyrrhopappus* inhabit our southern states and Mexico. One extends into the northeastern states.

FALSE DANDELION, P. CAROLINIANUS, has heads of yellow, ray-like flowers which resemble those of true dandelions (*Taraxacum*), but the pappus is a light reddish-brown instead of white, and the leaves extend up on the flowering stem instead of being all at the base. The longer bracts have a distinct appendage at the tip, like that of a dandelion. There are much smaller bracts at the base of these long bracts.

May to October: in fields and open woodland from Delaware to Kansas and southward to Florida and Texas. *Plate 174. Pus* in Greek means "fire"; *purrho* (pyrrho), consequently, "fire-colored."

PLATE 176

Sericocarpus asteroides *Scribner*

Valeriana officinalis *Murray*

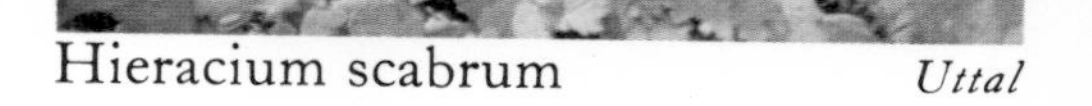

Hieracium scabrum *Uttal*

Sericocarpus linifolius *Uttal*

Aster sericeus *Johnson*

Physalis virginiana *Johnson*

Physalis heterophylla *Johnson*

LAPSANA

One species of this Old-World genus is naturalized in America.

NIPPLEWORT, L. COMMUNIS, is a slender branching plant from 6 to 60 inches tall. The leaves are borne singly; their blades are ovate, toothed; the lower have stalks. The small flower-heads form a loose inflorescence. The flowers are yellow, only about ten to a head, all of the "ray" type. There is no pappus.

June to September: in waste places through most of our range. *Plate 170.*

SOW-THISTLES (SONCHUS)

The sow-thistles are as unbeautiful as their name: disagreeable and troublesome weeds, more or less prickly, with small heads of yellow flowers which are all of the ray-like type, none tubular. Our species range from 1 to 6 feet tall or taller. Those named below are common throughout the United States.

S. asper has prickly-edged leaves, with large rounded lobes at the base which extend on either side of the stem (*Plate 172*). *S. oleraceus* has pinnately lobed or cleft, prickly-edged leaves, with pointed lobes at the base extending on either side of the stem (*Plate 174*). *S. arvensis* has pinnately lobed or cleft, prickly-edged leaves with small "ears" at the base.

DWARF-DANDELIONS (KRIGIA)

Dwarf-dandelions are distinguished from the genuine article by their pappus, which has a ring of scales surrounding the few long bristles; and by the bracts, which are all approximately equal in length. The heads of yellow flowers are smaller than those of dandelions. The leaves are chiefly at the base, the flower-heads borne singly on leafless stems. These are among the relatively few composites that bloom in the spring.

I. *Species with not more than ten pappus-bristles to a flower or achene.*

K. VIRGINICA has a dense cluster of pinnately lobed leaves from which arises one or several leafless stems up to a foot tall.

March to August: in dry soil from Maine to Iowa and southward to Florida and Texas. *Plate 175.*

K. OCCIDENTALIS is a dwarf, the flowering stems rarely exceeding 6 inches in height. The leaves are pinnately lobed. There are not more than eight bracts to a head.

March to May: in prairies and other open places from Missouri to Texas and Arkansas.

II. *Species with fifteen or more pappus-bristles to each flower or achene.*

K. BIFLORA grows from 8 to 30 inches tall. The smooth stem forks and bears from one to three small leaves, without stalks and mostly without teeth. The basal leaves are sometimes pinnately lobed or cleft but more often without lobes.

May to August: in fields, meadows, and open woods from Massachusetts to Manitoba and southward to Georgia, Tennessee, Missouri, and Arizona. *Plate 175.*

POTATO DANDELION, K. DANDELION, has all leaves basal and toothed or pinnately lobed like those of its namesake. The leafless flowering stem rises from 6 to 20 inches tall. The bracts are numerous.

April to June: in prairies and meadows from New Jersey to Kansas and southward to Florida and Texas. *Plate 175.*

SERINIA

The genus *Serinia* is represented with us by one species, differing from *Krigia* in its lack of pappus.

S. OPPOSITIFOLIA is a branched plant up to 18 inches tall with most leaves at the base but a few on the flowering stems. All leaves are narrow; they may be pinnately lobed, toothed, or with plain edges.

April to June: in moist fields and waste places from Virginia to Kansas and southward to Florida and Texas. *Plate 175.*

PLATE 177

Psoralea esculenta *Johnson*

Astragalus canadensis *Johnson*

Psoralea argophylla *Johnson*

Oxalis montana *Johnson*

Glycyrhiza lepidota *Johnson*

Anthyllis vulneraria *Johnson*

Psoralea onobrychis *Johnson*

THE HAWK'S-BEARDS (CREPIS)

These are plants with yellow ray-like flowers and a pappus of numerous white bristles. The leaves are mostly at the base. They resemble some species of *Hieracium;* the distinction between the two species is explained under *Hieracium*. Our species of *Crepis* are mostly weeds from Europe, but one native in the West has strayed into our range.

C. CAPILLARIS is a small, commonly branched plant, usually not more than 2 feet tall. The basal leaves have blades of various shapes on short stalks, mostly wider towards the tip. They may be toothed or pinnately lobed, cleft, or even divided with the segments sometimes themselves lobed, cleft, or divided. The leaves on the stem are without stalks and have pointed lobes at the base.

July to October: naturalized in fields and waste land throughout our range.

C. RUNCINATA is a western species with larger flower-heads. The basal leaves are commonly pinnately cleft with the lobes at the base pointing downwards (they are "runcinate").

June and July: in open soil from Minnesota to Washington and southward to Iowa, Arizona, California, and Mexico.

Several other European species are reported established in our range but are rather seldom found.

THE HAWKWEEDS (HIERACIUM)

The hawkweeds are a large group of mostly rather weedy plants, with generally small heads of mostly yellow flowers of the "ray" type (none tubular). The bracts around each head are in several overlapping circles (the botanist calls the arrangement "imbricated"); there are often much shorter bracts at the base of the head. The pappus is composed of a circle of bristles.

The genus in Europe includes an enormous number of forms which some would regard as species. In our range, fortunately, in spite of the presence of a number of immigrants, the number is not so formidable. Some species, however, are distinguished by rather minor characteristics. Some resemble species of *Crepis*, and together these two genera form a rather baffling group of plants. *Crepis* differs mainly in having a single circle of equal bracts around the flower-head, with a few much shorter ones at the base. *Hieracium* is from the Greek word for "hawk." An old belief held that the keen eyesight of hawks was due to their habit of swooping upon these plants.

I. *Species whose leaves are all or mostly at the base of the flowering stem; if there are leaves higher on the flowering stem, they are much smaller than the basal leaves. Most of these are Europeans, now naturalized in the United States.*

A. Of these, several species have flower-heads in a rather small or tight cluster at the tip of the stem (compare B). These plants generally spread rapidly by runners.

DEVIL'S-PAINTBRUSH, H. AURANTIACUM, is the familiar weed that colors so many eastern fields a bright red-orange in the late summer; a beautiful color but a pernicious weed. The plant is coarsely hairy. It stands from 8 to 30 inches tall.

June to October: naturalized in fields and meadows from Newfoundland to Minnesota and southward to Virginia, Ohio, and Iowa; also on the Pacific Coast; from Europe. *Plate 174.*

KING-DEVIL, H. PRATENSE, is the same as devil's-paintbrush in yellow: the same hairs, almost the same runners, from a few inches to 3 feet tall.

May to September: in fields and pastures from Quebec to Ontario and southward to Georgia and Tennessee; from Europe. *Plate 174.*

KING-DEVIL, H. FLORIBUNDUM, may be distinguished from *H. pratense* by its lack of hairs (except on the edges and midrib of the leaves); the foliage is gray-green. It grows from 1 to 3 feet tall. The bracts are blackish. *Plate 174.*

June to August: in fields from Newfoundland to Connecticut and westward to Ohio.

H. FLORENTINUM is less common than the preceding species, which it resembles. It is from 6 inches to 3 feet tall. The leaves are pale grayish-green, with a few hairs or none. Runners are generally lacking.

May to September: in fields and meadows from Newfoundland to Michigan and Iowa and southward to Virginia; from Europe.

MOUSE-EAR, H. PILOSELLA, is a small plant, rarely exceeding 10 inches in height. The slender stem often bears only one flower-head. The stem and leaves bear scattered long hairs. The runners are slender and leafy.

Cryptotaenia canadensis *Johnson*

Caltha palustris *Rickett*

Hyoscyamus niger *Johnson*

Halenia deflexa *Johnson*

Anemone caroliniana *Johnson*

Oenothera serrulata *Johnson*

May to September: naturalized from Europe in pastures and fields from Newfoundland to Minnesota and southward to North Carolina and Ohio; also in Oregon. *Plate 174.*

B. In other species with leaves mostly basal the heads are more loosely disposed in a widely branching cluster. These have no runners.

RATTLESNAKE-WEED or POOR-ROBIN'S-PLANTAIN, H. VENOSUM, is readily identified by the characteristic suggested by its botanical name: the veins are generally red or purple and stand out plainly in the green blade. They bear long hairs. The flowering stem is from 8 to 30 inches tall.

May to October: in open woods from Maine to Michigan and southward to Florida and Louisiana. *Plate 174.*

H. VULGATUM is from 6 inches to 3 feet tall. The basal leaves bear sparse long hairs; they are short-stalked, with generally long blunt teeth between the base and the middle of the blade. The heads are quite large, with from forty to eighty flowers in each. The bracts bear blackish hairs.

May to September: in fields and waste places from Newfoundland to Michigan and southward to New Jersey and Pennsylvania. *Plate 174.* A naturalized weed from Europe. *H. robinsonii* is similar but much smaller, generally about a foot tall. It is a northern species of rock ledges, extending southward to Maine and New Hampshire.

H. GRONOVII and H. LONGIPILUM are best considered together: they are both conspicuously hairy, and have tall narrow clusters of flower-heads. Their basal leaves are rather narrow, widest at the tip, tapering in to a stalk-like part, with no teeth on the edges. The hairs of *H. longipilum* are from $\frac{2}{5}$ to $\frac{4}{5}$ inch long; those of *H. gronovii* do not exceed $\frac{2}{5}$ inch. The flowers of *H. longipilum* run forty or more to a head; those of *H. gronovii* are fewer than forty.

July and August (and *H. gronovii* to October): in open woods; *H. gronovii* from Massachusetts to Kansas and southward to Florida and Texas; *H. longipilum* also in prairies from southwestern Ontario to Nebraska and southward to Indiana and Oklahoma.

II. *Species with leaves on the flowering stem not much smaller than those at the base (the basal leaves are usually lacking at flowering time).*

H. CANADENSE is from 1 to 5 feet tall, more or less hairy. The leaves are rather numerous and all much alike, lanceolate, without stalks, their edges variously toothed or plain. The flower-heads, each with forty or more flowers, are arranged in a widely branching cluster.

July to September: in woodland and fields and rocky and sandy places from Labrador to British Columbia and southward to New Jersey, Ohio, Iowa, Montana, and Oregon.

H. UMBELLATUM resembles *H. canadense.* The stem is not hairy. The leaves are narrow, tapering to the base, and rough. The cluster of flower-heads more or less resembles an umbel, with heads rising to about the same level.

July to September: in sandy open places from Ontario to Alaska and southward to Michigan, South Dakota, Montana, and Oregon. A "circumpolar" species, i.e. ranging around the pole in northern America, Europe, and Asia.

H. PANICULATUM is distinguished by the long and slender stalks of the flower-heads, which spread in all directions, even downward, and are rarely quite straight. The small flower-heads contain usually from ten to twenty flowers. The plant is nearly or quite smooth. The leaves are lanceolate or elliptic, paler on the lower side.

July to September: in woodland from Quebec to Michigan and southward to Georgia and Alabama. *Plate 175.*

H. SCABRUM has a stem from 1 to 5 feet tall, densely downy or hairy with short hairs, many of them tipped with glands. The leaves are rather thick, with few or no marginal teeth, sparsely hairy. The flower-heads are in a rather compact cone-shaped cluster, each with forty or more flowers.

June to September: in dry open woods and pastures from Quebec to Minnesota and southward to Maryland, the mountains of Georgia, Tennessee, and Missouri. *Plate 176.* This is said to cross with the preceding species. Several groups of plants of supposedly hybrid origin have been named as species.

DANDELIONS (TARAXACUM)

It may surprise the reader to learn that there are many kinds of dandelion. The genus contains many species – more than a thousand have been named. Fortunately only a few are known here. The differences among the species of the vast European, Asian, and South American hordes are slight, and some differences of opinion exist among authorities on our few species: one botanist has eight species in our

PLATE 179

Acorus calamus *Johnson*

Ornithogalum nutans *Gray*

Sparganium fluctuans *Johnson*

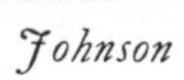

Elodea canadensis *Clewell*

Smilax ecirrhata *Johnson*

Smilax lasioneura *Johnson*

range; another treats the same plants as belonging to only three species. Since this book does not pretend to satisfy a desire for the ultimate details of classification, the simpler concept is here presented.

One reason for the variety of forms is the same as that already mentioned for *Antennaria:* the production of seed without fertilization. This results in the perpetuation of races differing in minor ways. When we consider the great number of races, the worldwide distribution, and the tough hold on life of these humble but often offensive plants, we are impelled to assign them a place, with the orchids, at the peak of the evolution of the plant world!

Dandelions are often gathered for "greens," particularly by Old-World peasants and their descendants in the New World. It is perhaps not always realized that they have medicinal properties; dandelion wine, dandelion beer have diuretic effects – one English name is piss-a-bed (and the same in other languages). The fruiting heads – the "seeds" crowned with the pappus – are used for telling the time or the future. Another English name is fairy-clocks.

As the yellow petals wither, the long bracts grow together, enclosing the fruits during their development; when the fruits ("seeds") are mature the bracts open and turn downward around the stem. There is a circle of small bracts at the base of the long ones. The achenes taper up into a long thin beak bearing the hairs of the pappus at the summit; the pappus is not plumose.

COMMON DANDELION, T. OFFICINALE, is the naturalized Old-World weed found everywhere in lawns. The pinnately lobed or cleft leaves taper into a flat stalk. The achenes ("seeds") are olive-brown.

March (or even earlier) to December: in lawns, grasslands, and open places generally throughout the country. *Plate 173.*

RED-SEEDED DANDELION, T. ERYTHROSPERMUM, is much less common than the preceding species. The bracts are generally tipped with a small hornlike appendage. The leaves are deeply cleft, with very little tendency towards a stalk. The achenes ("seeds") are bright red or purplish. The pappus is dirty-white.

April to July: naturalized in fields, lawns, etc. from Quebec to British Columbia and southward to Virginia, Kansas, and New Mexico.

T. CERATOPHORUM resembles *T. officinale* but has the hornlike appendages of the bracts like those of *T. erythrospermum.* The leaves are narrow and less deeply cut. The pappus is cream-colored.

June to August: from Labrador to British Columbia and southward in mountains to Massachusetts, New Hampshire, and California.

In addition to the genera described above with flowers (mostly yellow) of the "ray" type, a number of other weeds from the Old World are occasionally found in the northeastern states. *Arnoseris* and *Chondrilla* from Europe, *Ixeris* from Japan, must be sought in the manuals and floras. *Picris hieracioides* (*Plate 175*) has toothed leaves on the flowering stems, and a sparsely plumose pappus. It is a native of Europe naturalized in scattered places in most parts of our range, flowering from June to October. Ox-tongue, *P. echioides,* differs in having a circle of broad outer bracts, and in being prickly. It grows through much of the same range. Rush pink, *Lygodesmia juncea,* with pink or purplish flowers is found from Minnesota westward and southward. *Agoseris glauca* is another westerner found in Minnesota; it has yellow flowers in a head on a leafless stem and a cluster of very narrow, toothed leaves at the base. *A. cuspidata* is somewhat similar. The pappus consists of scales as well as bristles. The leaves at the base of the flowering stem have crisped edges (like lettuce but narrow). It grows from Illinois and Wisconsin westward.

PLATE 180

Senecio jacobaeus *Rickett*

Solidago odora *Johnson*

Nicandra physalodes *Johnson*

Cuscuta gronovii *Rickett*

Kickxia elatine *Johnson*

Helianthus atrorubens *Johnson*

Ipomoea purpurea *Johnson*

INDEX OF NAMES Common and Botanical

English names beginning with Black, Blue, Common, Early, False, Field, Red, Tall, White, Wild, Wood, Yellow are indexed under the second word, as Snakeroot, Black or Indigo, False – except when first and second words are joined by a hyphen.

Other names of two words not joined by hyphens are often indexed under both words, as Moth mullein and Mullein, Moth. The same is true of some names of words joined by hyphens, but in general these appear only under the initial letter of their first word: Black-eyed-Susan, Blue-eyed-grass, Wood-sorrel.

Botanical names of species are listed under the names of their genera: Acorus calamus, Agalinis maritima.

THE  NEW YORK BOTANICAL GARDEN

This second printing of the first edition of the first volume of five volumes of THE WILD FLOWERS OF THE UNITED STATES has been published in 1967

The illustrations and text have been printed by offset lithography by W. S. Cowell Ltd at their press in the Butter Market, Ipswich, England

The Typographical design is by Lewis F. White

The Typeface is English Monotype Caslon Old Face

The Paper has been specially made by John Dickinson, Croxley, Hertfordshire, England

C A N
Mt Olympus
WASHINGTON
Mt Rainier
Mt Adams
Mt Hood
Columbia R.
Coast Ranges
Cascade Ranges
OREGON
Siskiyou Mts
Mt Shasta
North Coast Ranges
Sacramento
Mt Lassen
L. Tahoe
Sierra Nevada
CALIFORNIA
San Joaquin
South Coast Ranges
Mt Whitney
Death Valley
Mojave Desert
Channel Islands
Colorado Desert
PACIFIC OCEAN
BAJA CALIFORNIA
IDAHO
Bitter Root Mountains
ROCKY MOUNTAINS
MONTANA
NORTH DAKOTA
SOUTH DAKOTA
Black Hills
Bad Lands
Big Horn Mts
WYOMING
NEBRASKA
Great Basin
Great Salt Lake
NEVADA
U N I T E D
UTAH
Wasatch Mts
Uinta Mts
Long's Peak
COLORADO
Arkansas R.
KANSA
Colorado R.
Grand Canyon
ARIZONA
NEW MEXICO
OKLA
T E X A
Edwards Plateau
Rio Grande
Rio Grande
M E X I C O